Indoor Wireless Communications

Indoor Wireless Communications

From Theory to Implementation

Alejandro Aragón-Zavala

Registered Offices
John Wiley & Sons, Inc., 111 River Street, Hoboken, NJ 07030, USA
John Wiley & Sons Ltd, The Atrium, Southern Gate, Chichester, West Sussex, PO19 8SQ, UK

Editorial Office
The Atrium, Southern Gate, Chichester, West Sussex, PO19 8SQ, UK

For details of our global editorial offices, customer services, and more information about Wiley products visit us at www.wiley.com.

Wiley also publishes its books in a variety of electronic formats and by print-on-demand. Some content that appears in standard print versions of this book may not be available in other formats.

Library of Congress Cataloging-in-Publication data is available for this book.

ISBN 9780470741160 (hardback)

Cover Design: Wiley
Cover Image: ©Wangwukong/Gettyimages

Set in 10/12 pt WarnockPro-Regular by Thomson Digital, Noida, India
Printed and bound in Malaysia by Vivar Printing Sdn Bhd

10 9 8 7 6 5 4 3 2 1

Lefis, Co, Oji and Fimbie, finally this is done. Thanks for keeping me up and running in this project. I could have not done it without your support, especially at those times of difficulties. I entirely dedicate this to you, Arafra.

Alejandro Aragón-Zavala

Contents

Preface

It all started in my early days of PhD work at the University of Surrey, UK. I remember that on my first year and while searching for literature related to in-building radio systems, to my surprise, there was not any book available on indoor radio networks at the time. I had quick chats with some of my professors and co-students to see if they have seen anything related at all; the answer was the same in all cases: there is nothing out there – only papers and articles, that was all.

Later on and while working at Cellular Design Services, I again needed to search for some reference material as we were in the middle of a very large project related to wireless infrastructure in airports. The only things I could find were chapters in some books, which had scattered information and was a jig-saw puzzle task to put all the pieces together. That is where I had the very first idea of having a single book that could cover all indoor radio aspects in a simple but useful way.

On my return to Mexico, I finally decided to start working on producing a book that could fulfil all the needs from an engineer or specialist willing to work on in-building radio design, covering many aspects of what I had learnt both in research and in industry.

This book includes materials and knowledge acquired after many years of experience and hard work in the field of indoor wireless communications. It goes from the very basics of building characteristics and 'things to look at' when designing a radio network in a certain type of building to more sophisticated propagation models, capacity calculations and a chapter on case studies that covers practical aspects of real indoor designs.

I believe this book can be very useful as a practical reference for scientists and engineers involved in the design, planning and operational aspects of new infrastructures for wireless communication systems inside buildings, a field that is growing very rapidly around the world. It could also be used as a text for graduate students or final year undergraduates willing to expand their knowledge in the field of in-building radio systems.

I am solely responsible not only for technical inaccuracies but also for typos. I would appreciate any feedback to aaragon@itesm.mx, where comments, corrections or recommendations are welcome. An errata file will be kept and sent to anyone interested, upon request.

First of all, I would like to thank Prof. Simon R. Saunders for all his help and support during the writing of this book and for taking the time to contribute to the Small Cells chapter. In particular, thanks Simon for your friendship all these years and for your valuable guidance.

I also like to thank Dr Vladan Jevremovic for contributing to the Case Studies chapter and for sharing those valuables calls where we discussed key aspects of what is going on in the indoor wireless industry.

Thanks to all the iBwave Solutions Inc. team for their help, in particular to Dr Ali Jemmali, Benoit Courchese, Dominique Gauthier and Peter Thalmeir. I have learnt much from you – the experts in in-building radio design!

I wish to thank the support received by Tecnológico de Monterrey, Campus Querétaro and Campus Monterrey authorities and especially thanks to Dr Héctor Morelos-Borja, Dr David Muñoz-Rodríguez and Dr Manuel I. Zertuche. Special thanks also to all my graduate students who helped me during the elaboration of this work and for those who have shared with me useful discussions after lectures.

A very special group of people who helped me with bits and pieces of this book throughout the years is that from Real Wireless Ltd. Thanks to Mark Keenan, Kostas Konstantinou and Julie Bradford; we have worked in many really interesting in-building projects from which you have shared the best of your knowledge and attitude with me.

This book is dedicated to all my students, both graduate and undergraduate, from Tecnológico de Monterrey, Campus Querétaro; you are truly my inspiration and my motor to keep myself updated.

Finally, I wish to thank the valuable support from John Wiley & Sons editorial team, especially to Mark Hammond, Tiina Wigley, Ashmita Thomas Rajaprathapan, Sarah Tilley, Susan Barclay, Victoria Taylor, Liz Wingett and Sandra Grayson, for all their patience, enthusiasm and support given to me throughout these years. Thanks for always keeping a nice smile and encouraging comments!

Santiago de Querétaro, Mexico   *Dr Alejandro Aragón-Zavala*
Autumn 2016

1

Introduction

1.1 Motivation

Currently around 70% of mobile usage is inside buildings and some analysts predict that in the next few years around 90% of mobile usage will take place indoors (Paolini, 2011), be it at home, in the office or in public buildings, with the majority of that being in an individual user's home and main office locations. Traditionally such services have been provided on an 'outside-in' basis from macrocells, which were originally deployed to provide voice coverage to vehicles travelling along major roads. This conventional architecture is increasingly limited in its ability to meet modern mobile users' needs for reliable indoor mobile service because:

- The increased volume of mobile data use puts a greater load on macrocell capacity to deliver a reliable service.
- User expectations of the minimum data rate that constitutes a viable service are continuing to increase and are dominated by the indoor locations in which they mostly consume services.
- Modern mobile devices have to support a wide range of frequency bands in a small-form factor, reducing their sensitivity and hence increasing the signal strength needed to achieve a given coverage.
- Improved thermal insulation properties of buildings lead to an increase in the use of denser and more conductive external construction materials, including metallized windows, increasing the losses that radio waves encounter in penetrating a building.
- Increased use of high-frequency bands at 2.1 GHz, 2.4 GHz, 2.6 GHz, 3.5 GHz and beyond, which suffer greater losses than frequencies below 1 GHz.
- Economic incentives for mobile operators to share macrocell networks, reducing the diversity of options for users to switch to operators who do have macrocell coverage close to the locations they care most about.

In consumer surveys, mobile users frequently cite poor in-building coverage as the number one network-related reason to turn to other operators. Increased mobile data usage has changed customer priorities and expectations. Smartphone users rate the importance of messaging and Internet quality 2.5 times higher than standard phone users. Even the average phone user rated network quality as the most important reason for choosing a mobile operator. Mobile Internet is much more demanding than a few years ago, requiring Wi-Fi coverage as a 'default' commodity. All these factors increase

Indoor Wireless Communications: From Theory to Implementation, First Edition. Alejandro Aragón-Zavala.
© 2017 John Wiley & Sons Ltd. Published 2017 by John Wiley & Sons Ltd.

the need of specialized indoor radio technologies to satisfy such user expectations and demands.

A recent global mobile consumer survey audit conducted by Deloitte (2012) examined the use of Wi-Fi by mobile users and found:

- In the UK, Wi-Fi is the main Internet connection for smartphones at nearly 60%.
- The desire for faster, more reliable connectivity is the principal driver of Wi-Fi usage over cellular mobile.
- Almost 90% of both smartphone and tablet users connect to the Internet using Wi-Fi from home.

Given this and the potential loss in consumer and societal benefits arising from poor in-building services, it is very relevant to consider the planning and design of in-building radio systems following a methodological and engineering approach, considering:

- The types of indoor wireless technologies and their characteristics
- Design requirements and standards for coverage and capacity
- Voice and data traffic considerations when deploying an indoor radio solution
- The physics of in-building radio propagation
- Channel modelling options that are available to estimate coverage in an indoor radio system
- Available RF and antenna equipment for indoor systems
- Measurement techniques and systems used to design and validate in-building networks
- Practical issues related to indoor radio design and deployment that could be useful for engineers, students and scientists.

1.2 Evolution of Macro to Heterogeneous Networks

The changing shape of mobile networks from the traditional macrocell approach to what is known as heterogeneous networks or 'HetNets' (a mix of macro and small-cell architectures) can both resolve issues for the operator but at the same time create potential policy problems for telecommunications regulators, notably:

- **Future demand for spectrum.** The indoor layer of the network starts to soak up the rapid increase in demand and the value of future mobile spectrum allocations may be reduced.
- **Interference management.** Current policy stipulates that coordination between multioperators should be resolved by each party. A market-driven change into the network architecture may impact the regulatory policy in this area.
- **Competition between MNOs.** Competition may start to be reduced by the growing use of multioperator collaborations on both the core network and on the Radio Access Network.
- **Incentives for investment.** There may come a time when operators reduce their level of investment in the macronetwork to focus on serving the indoor consumers with smaller more cost effective building solutions.
- **Consumer switching.** This is encouraged by operators marketing their new products and services.

Traditionally, when MNOs first deployed their mobile networks they were designed for:

- Wide area coverage
- Targeting mobile voice
- Roads/carphones.

However, in today's network topology the demand from users has dramatically shifted towards the consumption of data anywhere, anytime, which has led to the following trends:

- Usage is predominantly indoors (70–90% of the traffic by volume) (Analysis Mason Ltd, 2011).
- The indoor locations of relevance are predominantly just two per person (my home, my office), that is not just geographical coverage.
- Indoor coverage is often cited by operators as the number one network-related cause of churn.
- Smartphones have poorer sensitivity than traditional phones – and it gets worse as more bands are added.
- Even voice is increasingly on 3G and LTE: better link budget, but mainly 2.1 GHz today.
- User expectation of what constitutes the minimum acceptable data rate increases with time (as does expectation of typical rates).
- Building regulations increasingly specify thermal insulation requirements, which are increasingly being met by metallized glass, significantly increasing attenuation.

There are three broad indoor user environments to consider for in-building systems. These are in the home, in the office or in a public building/venue. There are distinct differences in terms of both achieving coverage and satisfying capacity in each of these environments that impact the scale of the market and the optimum technical solution when considering indoor solutions.

1.3 Challenges

This book aims at presenting methods and techniques of how in-building coverage can be enhanced relative to that provided by the existing (predominantly macrocell-based) networks at predominantly a technical level but including also some commercial level detail. For example, one particular challenge to be addressed for anyone interested in designing an in-building radio system may be *openness* of any systems and we ask the question: 'What do we mean by "allowing access to different operators"?' The following points attempt to provide some possible solutions:

- One solution can simultaneously handle users with different operators.
- Solutions are cheap, unobtrusive and simple enough to permit multiples of them in one location if needed.
- It could work for any operator and can be changed between operators at the user's choice (but only phones from one operator at a time).
- It is easy and cheap to switch operators even if one 'box' cannot handle multiple operators.

- Allow consumers to take action themselves rather than 'begging' an operator; for example, three currently limit femtocells to those who they consider have a valid coverage problem.
- How does this compare and contrast with being able to move your phone between providers, move contracts between providers or move gateways/home hubs between providers.

The increased use of Wi-Fi presents further challenges and issues to cellular operators and good knowledge of the technology is mandatory for deploying a whole radio network, as nowadays Wi-Fi needs to coexist with cellular and other wireless technologies. Following this trend, mobile operators are beginning to adopt Wi-Fi as a complementary service as an in-building solution, principally to help offloading capacity constrained parts of the network. However, there are still some technical issues from Wi-Fi that, once overcome, will provide a more integrated solution for mobile operators.

The overriding issue for operators is one of cost: while macrocells have technical challenges in addressing remaining in-building demand, they have the benefit of spreading the costs over large numbers of users, resulting in the former operator mantra 'outside-in always wins'. Provision of in-building systems for every building with a need has hitherto been excessively expensive, in both equipment and professional services.

New technologies do provide opportunities to significantly reduce the cost of provision, but significant work remains to encourage widespread roll-out of these technologies, including consumer understanding, commercial incentives and regulatory clarity. Some of these issues are to be reviewed throughout this book, illustrating some concepts with case studies to make understanding much easier.

Finally, it is well known that buildings have a very large variety of shapes, partitions, floors, materials, etc., which makes them very difficult to be categorized for design purposes. On top of this, propagation inside buildings is much more complex than in open spaces, thus making the planning of an indoor radio network more challenging. This book presents in a simple and complete way the physics of radio propagation inside buildings as well as methods to design and plan indoor networks, considering various technologies and a handful of mathematical models to use. Sufficient references are given at the end of each chapter for the interested reader to investigate further.

1.4 Structure of the Book

The book is organized into eleven chapters, including this chapter, the introduction to the world of in-building radio systems. An overview of various indoor wireless technologies is given to the reader in Chapter 2, for a better perspective on the current available technologies and their characteristics.

From a design perspective, the first step is to establish a clear set of requirements. Therefore, the reader is introduced in Chapter 3 to the specific requirements that any indoor wireless system may have. These issues are discussed in detail, to make sure the reader understands each and foresees the importance of taking them into account for a successful system deployment. This includes the different types of indoor environments present (corporate office, airport, theatre, shopping centre, etc.) for a radio designer to identify specific characteristics and additional requirements.

Once all of the main requirements for an indoor system have been established, a study of the propagation phenomena occurring in an in-building environment is needed. From the propagation mechanisms to the scales of mobile signal variation, all aspects to be considered in terms of radio propagation are covered in Chapter 4.

Channel models are explained in Chapter 5, for both narrowband and wideband fast fading as well as for median path loss, specifically for indoor systems. The models are not exclusive for any technology, but will cover all ranges of indoor wireless technologies, for the reader to have a useful reference.

The antenna requirements listed in Chapter 6 will be discussed for the case of in-building systems, referencing to some characteristics that indoor antennas must have. Antenna theory is included for completeness, especially to avoid the need to refer to specialized antenna books in addition to the information provided here.

Due to the importance that indoor radio measurements have in the in-building design process, Chapter 7 is dedicated to measurements, starting with basic concepts and highlighting important issues to be considered, from my experience over 10 years on indoor wireless measurements. A section on indoor model tuning is included at the end of the chapter, since this is a common practice in practical indoor wireless designs.

Capacity is presented in Chapter 8, which is one of the essential requirements for an indoor wireless system. Careful planning and dimensioning will allow system resources to be used efficiently, especially for those high-demanding applications involving indoor scenarios. Data transmission-related issues that affect some indoor wireless systems are also discussed here.

Chapter 9 is about RF and distribution systems and components used in in-building radio systems. The aim of the chapter is to provide the reader with a deep survey on existing equipment available for indoor wireless design, covering all technologies and levels. Transmission lines have been included here to illustrate what can be used to interconnect elements within the architecture of the distribution system, and not to show any connection with users. Examples of commercial equipment are shown.

Chapter 10 presents a brief overview on small cells and their importance inside buildings. Finally, Chapter 11 includes case studies for in-building designs, where a design approach is applied for each study, having different parameters and technologies.

References

Analysis Mason Ltd (2011) Wireless Network Traffic 2010–2015: Forecasts and Analysis, A Research Forecast report by Terry Norman, Analysis Mason Ltd.

Deloitte (2012) *The Data Capacity Crunch: Challenges and Strategies*, Mobile Broadband Specialist Interest Group, presentation by Deloitte to Cambridge Wireless event on 10 October 2012, http://www.cambridgewireless.co.uk/Presentation/Mobile.Broadband_David.Griffin.Intro_10.10.12.pdf.

Paolini, M. (2011) Mobile data move indoors, *Mobile Europe*, retrieved 14 September 2011. URL: http://www.senzafiliconsulting.com/Blog/tabid/64/articleType/ArticleView/articleId/59/M%20obile-data-move-indoors.aspx.

2

Indoor Wireless Technologies

This book is not exclusive of any indoor wireless technology; instead, an effort has been made to present an overview of design, planning and deployment issues that may be applicable to many wireless technologies. An understanding of the various standards and protocols is therefore a necessary asset for any in-building radio planner or designer, since nowadays the deployment of multioperator, multitechnology networks is becoming more frequent.

This chapter aims at providing a brief overview of wireless technologies that are frequently encountered inside buildings. Special emphasis is given to cellular and Wi-Fi, two of the most important wireless technologies deployed inside buildings for which coexistence and design issues need to be carefully considered.

2.1 Cellular

Perhaps one of the most revolutionizing developments in the communications industry in the twentieth century has been the invention of the cellular concept. Since the creation of fixed telephony and other wired communication systems such as the telegraph, the possibility of having mobility while holding a conversation seemed possible but maybe not so feasible, as the size of electronic equipment still made it prohibitive for a device to be portable within a reasonable cost. Nowadays, not only are mobile voice communications possible but also data communications are a reality with increasing data rates, utilizing the spectrum more efficiently. It could be said that cellular has been one of the wireless technologies that has mostly driven research and development efforts in the last century, as users are demanding higher speeds, with connectivity everywhere and at every time.

Interestingly enough, in-building cellular has also become a major breakthrough in the history of cellular telephony. In the early days, the main objective was to deliver sufficient signal strength in geographical regions denoted as *cells*, which were mainly outdoors. In-building coverage was achieved by flooding the building with power from surrounding base stations. As voice was the main driver and capacity was not a big issue, this seemed to be enough for a few years. However, as larger facilities such as airports, shopping centres, etc., demanded more capacity, the need to deploy dedicated in-building networks was inevitable.

There have been various generations of cellular telephony that have been developed throughout the years, including: GSM, AMPS, IS-95, WCDMA, IMT-2000, LTE,

Indoor Wireless Communications: From Theory to Implementation, First Edition. Alejandro Aragón-Zavala.
© 2017 John Wiley & Sons Ltd. Published 2017 by John Wiley & Sons Ltd.

HSPA, etc. All of them have their characteristics and particularities that need to be understood when deploying in-building radio networks. Since the aim of this book is not to provide a deep understanding of each of these standards, the key elements and characteristics of each standard will be highlighted, especially those that are relevant for indoor systems. For a nice and expanded overview of each of these technologies please refer to Tolstrup (2011).

2.1.1 The Cellular Concept

In the early days of mobile telephony around the 1950s, coverage and capacity were provided by a single transceiver, also known as BTS, capable of radiating sufficient power to illuminate a large area and thus provide service to users. Each BTS, generically known as a base station (BS), was designed to cover, as completely as possible, a designated area or cell. The main problem with this was the use of the available radio resources: even having a trunked capacity, the number of users was very restricted. That is when the concept of *frequency reuse* was introduced.

The frequency reuse strategy involved the division of the radio access network into overlapping *cells*, as depicted in Figure 2.1, so that a cluster of frequencies could be reused within a geographical area, minimizing the levels of interference and therefore providing service to more users. In Figure 2.1 a cluster is made of seven cells, each having a different channel (A–G) and can be reused. The idea of *handover* was also possible since now users within an area could have seamless mobility between cells.

As will be discussed in Chapter 8, *capacity* in a cellular network aims at providing service to users in a designated area. For in-building networks, a method commonly employed to increase capacity is known as *cell splitting*, where smaller cells are utilised to increase the frequency reuse pattern and take advantage of the propagation conditions in the building to limit coverage levels. *Sectorization* is also employed, especially in stadiums where high capacity is required, by the use of directional antennas to decrease the co-channel interference at certain directions and accommodate more users.

For most cellular networks, regardless of the specific standard, there are two main types of cell configurations. In an *omnidirectional* cell, the BS is depicted at the centre of the cell and omnidirectional antennas are used (Figure 2.2a). For *sector* cells, the BS is located on three of the six cell vertices, as shown in Figure 2.2b. Three- or six-cell sectors are often found in cellular networks, many of them in a similar configuration as the base station shown in Figure 2.3.

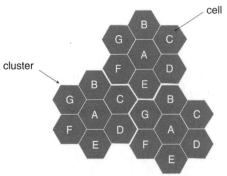

Figure 2.1 Frequency reuse.

(a)   (b)

Figure 2.2 Cell configurations: (a) omnidirectional; (b) sector.

Figure 2.3 Cellular base station.

As for many other radio communication systems, noise and interference are to be minimized in order to guarantee an optimal service performance to provide a desired quality-of-service. In particular, interference is of special relevance since the frequency reuse permits the assignment of co-channel frequencies to cells at a minimum distance so that interference levels can be kept within tolerable limits. This is particularly important for indoor networks, since sector configurations are commonly deployed and interference control is essential there. On the other hand, noise limits the capacity in some cellular systems, as in the case of the uplink in WCDMA, so it should be properly controlled.

A very brief overview of the key cellular technologies is presented in the following sections, highlighting relevant aspects that need to be taken into account when designing in-building radio networks.

2.1.2 GSM

Figure 2.4 shows the key elements of a standard GSM cellular network. The central hub of the network is the mobile switching centre (MSC), often simply called the switch. This provides connection between the cellular network and the public switched telephone network (PSTN) and also between cellular subscribers. Details of the subscribers for whom this network is the home network are held on a database called the home location

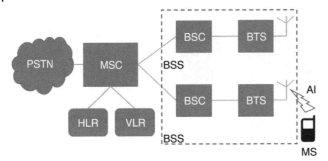

Figure 2.4 Elements of a standard cellular system, using GSM terminology.

register (HLR), while the details of subscribers who have entered the network from elsewhere are on the visitor location register (VLR). These details include authentication and billing details, plus the current location and status of the subscriber. The coverage area of the network is handled by a large number of base stations. The base station subsystem (BSS) is composed of a base station controller (BSC), which handles the logical functionality, plus one or several base transceiver stations (BTS) containing the actual RF and baseband parts of the BSS. The BTSs communicate over the air interface (AI) with the mobile stations (MS). The air interface includes all of the channel effects as well as the modulation, demodulation and channel allocation procedures within the MS and BTS. A single BSS may handle 50 calls and an MSC may handle some 100 BSSs.

GSM was originally licensed to operate in the 900 MHz band. More spectra were allocated later on, in 1800 MHz, for a standard known as DCS1800. Both were deployed in Europe, and many other countries in the world use the same frequency bands. In the Americas, the bands 800 MHz and 1900 MHz are used for GSM.

Although GSM was mainly designed to provide voice services at a speed of 9.6 kbps, some overhead was left in the data resources, so this overhead was used for transmitting short messages, leading to the SMS service. Later on, some limited data capabilities were included in the standard using packet data, introducing GPRS and EDGE. By using more sophisticated modulation schemes, data rates of up to 200–300 kbps could be achieved.

GSM uses separate frequency bands for the uplink (UL) and downlink (DL). This scheme is known as Frequency Division Duplex (FDD). The two bands are separated by 45 MHz on GSM900 and by 95 MHz on DCS1800. The spectrum allocated to GSM is divided into 200-kHz channels, and each of these channels is divided into eight time slots to be used as logical and traffic channels.

Handovers in GSM are 'hard', which means that the mobile monitors the Rx level of neighbouring cells and if the handover criteria is fulfilled (e.g. insufficient signal level, poor signal quality, etc.) the network commands the mobile to hand over to a new serving cell, using a different channel. Handovers are possible between cells or between sectors in a cell.

For GSM in-building systems, there are some considerations that are relevant to take into account when planning and designing the network, and could be summarized as follows:

- Coverage is achieved either by macrocell penetration or a dedicated indoor cell using distribution systems (antennas or leaky feeders) and adjusted so that dominance is achieved inside the building.

- Handover overlapping areas are desired to allow mobiles to have sufficient time to handoff between cells/sectors, especially in venues like airports where, for example, handover from the macrocell to the indoor cell needs to be achieved right at the entrance of the terminal building.
- Voice traffic is dimensioned using trunking theory, as discussed in Chapter 8, and more channels can be allocated depending on Grade of Service (GoS) requirements. For larger traffic demands in densely populated venues, such as airports, the system can be upgraded by adding more capacity to the base stations and by performing zoning or further cell splitting, thus enhancing the frequency reuse. Special care should be taken to maintain C/I levels within specs.
- Capacity is limited by the number of available resources in the BTS and by traffic demands, which are ruled by trunking theory.
- Co-channel and adjacent-channel interference need to be considered and taken into account, especially if sectors (zones) are deployed inside a building.

2.1.3 UMTS

The Universal Mobile Telecommunications System (UMTS) was specified and selected for 3G since the use of the spectrum is very efficient. UMTS has a high rejection to narrowband interference using Wideband Code Division Multiple Access (WCDMA), being thus very robust against frequency selective fading.

For UMTS cells, users share the same frequency, having a distinct spreading code. Thus, most of radio planning for UMTS is based on noise and power control. Unlike GSM where cells are assigned a different frequency, the intercell interference needs to be minimized.

There are two main types of UMTS radio systems: TDD and FDD. For TDD systems, the same frequency is used for the UL and the DL, whereas for FDD systems different frequencies are used for UL and DL. The latter is the most frequently used WCDMA radio system and requires a paired set of bands and an equal bandwidth separated 95 MHz duplex distance throughout the band.

The frequency band 1920–1980 MHz is assigned for the UL for WCDMA-FDD and 2110–2170 MHz for DL. This is used worldwide, although WCDMA-TDD is used in a few countries. Most operators are assigned two or three carriers in the 2.1 GHz band per license, but considerations for reusing the GSM900 spectrum are being made to utilize it for UMTS (Tolstrup, 2011).

In UMTS a spread spectrum signal is used, having a bandwidth of 5 MHz. A spreading code is employed to spread the original narrowband signal throughout the spectrum. Thus the signal becomes less sensitive to selective interference; for example, inter-modulation products from narrowband services.

The concept of frequency reuse in UMTS can be understood in a slightly different way to GSM. For UMTS the frequency reuse factor is 1 and different primary scrambling codes (highly orthogonal) are used per cell.

One of the key parameters for UMTS signal quality is the energy-bit-per-noise density ratio (E_b/N_0). It is the reference point for link budget calculations and defines the maximum data rate possible with a given noise – the higher the data rates, the stricter the E_b/N_0 requirements.

On the other hand, the quality of the pilot channel is measured as E_c/I_0, which is the energy per chip/interference density ratio measured on the pilot channel (CPICH).

When the user equipment (UE) detects two or more CPICH with similar levels, it will enter *soft handover*. Thus, the UE constantly monitors the E_c/I_0 of the serving cell and adjacent cells and compares the quality of the E_c/I_0 of the serving CPICH against the quality of other measured CPICHs and trigger levels or thresholds to add or remove cells from the neighbour list. The CPICH thus defines the cell size.

There are two types of handovers in UMTS. The *softer handover* occurs when a UE is within the service area of cells originating from the same NodeB at the same power level, using both RF links, using two separate codes in the DL. On the other hand, *soft handover* occurs when a mobile is in the service area of two cells originating from a different NodeB, and thus the mobile will use one RF link to both base stations; that is macrodiversity.

Pilot pollution occurs when a mobile receives CPICH signals at similar levels from other cells that are not in the neighbour list. This is the case when a distant macrocell can be 'seen' in a high-rise building at one of the upper levels. This will cause interference of the serving cell's CPICH, the so-called pilot pollution. This problem can be solved if an indoor cell is installed inside the building and has dominance (Tolstrup, 2011).

UMTS is very sensitive to noise control, since all traffic is in the same frequency and all signals from active UE need to reach the NodeB at the same level. If one UE reaches the UL of the NodeB at a much higher level, it will interfere with all the other UEs in service on the same cell. Therefore, noise and power control are very important for UMTS networks.

The load of a UMTS cell determines to a great extent the soft capacity, as more traffic will bring more noise to the cell. Therefore, the load needs to be limited to around 60% to 65% in indoor cells as they are more isolated from macros and can in principle be loaded relatively high (Tolstrup, 2011).

The main issues that should be considered when designing UMTS indoor networks can be summarized as follows:

- Interference management and control should be very strict, as all cells share the same frequency.
- Power limits DL capacity, so sufficient antennas should be used to provide dedicated indoor coverage.
- Noise limits UL capacity; therefore low noise devices should be employed in the front end of any distribution system.
- Soft handover areas should be minimized, as this requires doubling the use of overhead and network resources.
- In-building cell dominance should be guaranteed if a dedicated indoor solution is being deployed, as this minimizes the risk of pilot pollution.

2.1.4   HSPA

HSPA was developed to improve the speed of data rates in 3G cellular networks. It consists of an addition to the existing 3G network infrastructure and utilizes the power headroom not used by UMTS traffic channels. In the DL, data rates of 14.4 Mbps can be achieved and its best performance is achieved by deploying indoor DAS (Tolstrup, 2011). Data speeds are related to SNR and good radio links are obtained if antennas are closer to mobiles, as is the case of indoor DAS.

2.1.5 LTE

With the use of advanced and adaptive modulation schemes and MIMO, much higher data rates have been achieved. This is one of the key elements of long-term evolution (LTE), with downlink data speeds in the range of 100 Mbps and uplink speeds of up to 50 Mbps, which is based on the existing GSM/EDGE and UMTS/HSPA standards. Flexible bandwidths from 1.4 MHz to 20 MHz are also employed in LTE, ensuring compatibility with older networks and enough flexibility for network roll-out and deployment. In terms of spectrum efficiency LTE is three to four times better than HSPA (Tolstrup, 2011).

Indoor environments, as discussed in Chapter 4, are rich multipath environments, which for GSM networks, for example, are a challenge for power control and selective fading issues. However, since LTE utilizes 2×2 MIMO, a strong multipath is much more beneficial, since it employs the scattering of the local clutter (short reflections) to create multipath parallel links at the same time, frequency and space, thus theoretically doubling the throughput of the channel. LTE uses advanced and adaptive modulation techniques so that for good radio links (closer to the base station or eNodeB), higher-order modulation is employed (64-QAM), whereas as the link quality is degraded (farther away from the eNodeB), lower modulation schemes are used such as QPSK.

In terms of mobility, LTE is optimized to support a maximum data rate for pedestrian speed (0–15 km/h) and still provide high-performance data throughput for 15–350 km/h, being functional for 350–500 km/h to support high-speed trains (Tolstrup, 2011).

LTE can operate in a paired spectrum for duplex operation (FDD), and therefore can be deployed in existing GSM or UMTS systems. It can also be deployed in an unpaired spectrum (TDD). The different LTE frequencies and bands used in different countries will mean that only multiband phones will be able to use LTE in all countries where it is supported.

Some key features of the LTE air interface can then be summarized as follows:

- Low data transfer latencies (sub-5 ms latency for small IP packets in optimal conditions), lower latencies for handover and connection setup time than with previous radio access technologies.
- Peak download rates up to 299.6 Mbit/s and upload rates up to 75.4 Mbit/s depending on the user equipment category (with 4×4 MIMO using 20 MHz of spectrum). Five different terminal classes have been defined from a voice centric class up to a high end terminal that supports the peak data rates. All terminals will be able to process 20 MHz bandwidth.
- Improved support for mobility, exemplified by support for terminals moving at up to 350 km/h (220 mph) or 500 km/h (310 mph) depending on the frequency band.
- OFDMA for the downlink and SC-FDMA for the uplink to save power.

- Support for both FDD and TDD communication systems as well as half-duplex FDD with the same radio access technology.
- Support for all frequency bands currently used by IMT systems by ITU-R.
- Increased spectrum flexibility: 1.4 MHz, 3 MHz, 5 MHz, 10 MHz, 15 MHz and 20 MHz wide cells are standardized.
- Support for cell sizes from tens of metres radius (femto- and picocells) up to 100 km (62 miles) radius macrocells. In the lower frequency bands to be used in rural areas, 5 km (3.1 miles) is the optimal cell size, 30 km (19 miles) having reasonable performance, and up to 100 km cell sizes supported with acceptable performance. In city and urban areas, higher frequency bands (such as 2.6 GHz in the EU) are used to support high-speed mobile broadband. In this case, cell sizes may be 1 km (0.62 miles) or even less.
- Simplified architecture. The network side of E-UTRAN is composed only of eNodeBs.
- Support for interoperation and coexistence with legacy standards (e.g., GSM/EDGE, UMTS and CDMA2000). Users can start a call or transfer data in an area using an LTE standard and, should coverage be unavailable, continue the operation without any action on their part using GSM/GPRS or W-CDMA-based UMTS or even 3GPP2 networks such as cdmaOne or CDMA2000.
- Packet-switched radio interface.
- Support for MBSFN (Multicast-Broadcast Single Frequency Network). This feature can deliver services such as Mobile TV using the LTE infrastructure, and is a competitor forDVB-H-based TV broadcast.

The LTE standard covers a range of many different bands, each of which is designated by both a frequency and a band number. In North America, 700, 750, 800, 850, 1900, 1700/2100 (AWS), 2500 and 2600 MHz (Rogers Communications, Bell Canada) are used (bands 2, 4, 7, 12, 13, 17, 25, 26, 41); 2500 MHz in South America; 700, 800, 900, 1800, 2600 MHz in Europe (bands 3, 7, 20); 800, 1800 and 2600 MHz in Asia (bands 1, 3, 5, 7, 8, 11, 13, 40) and 1800 MHz and 2300 MHz in Australia and New Zealand (bands 3, 40).

There are various challenges that are envisaged when designing LTE indoor systems, which can be summarized as follows:

- Leveraging the maximum performance from MIMO systems, requiring in-building systems designed to provide a sufficiently multipath rich environment while still achieving a high SNR.
- Designing LTE systems that can in principle co-exist with WCDMA or HSPA networks, having the capability to accommodate more capacity as the network grows.
- Performing a very careful interference management, especially in areas where LTE femtocells are deployed.

2.2 Wi-Fi

2.2.1 History

The Wireless Local Area Networks (WLANs) have been around for a few years, evolving from simple low data rate technologies towards more sophisticated and powerful

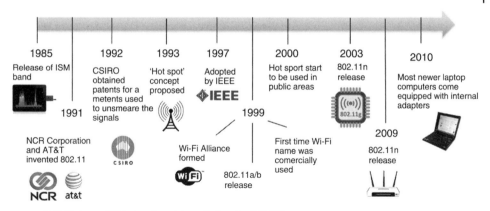

Figure 2.5 History of Wireless Local Area Networks (WLAN).

protocols. Let us have a brief timeline perspective in the evolution of this technology (Figure 2.5).

1985. The Federal Communications Commission (FCC) releases the ISM (Industrial, Scientific and Medical) band for the use of radio frequency (RF) energy for industrial, scientific and medical purposes other than communications. This is meant to be an unlicensed band. Over the years, the need to have more spectrum available for short-range, low-power communications systems has pushed regulatory bodies to allow its use for applications such as cordless phones, Bluetooth devices and WLAN systems.

1991. NCR Corporation and AT&T, in a joint venture, invented the precursor to 802.11 in Nieuwegein, The Netherlands, intended to be used in cashier systems. The very first wireless products were under the name WaveLAN with data rates of 1 Mbps and 2 Mbps.

1992. CSIRO (The Commonwealth Scientific and Industrial Research Organisation) obtained patents for a method used to 'unsmear' signals produced by WLAN networks.

1993. Public access WLANs were first proposed by Henrik Sjödin at the Net-World+Interop Conference in The Moscone Center in San Francisco in August 1993. Sjödin did not use the term 'hotspot' but referred to publicly accessible wireless LANs.

1997. The technology is adopted by the IEEE (Institute of Electrical and Electronic Engineers) and the working group 802.11 is created to create standards related to WLANs.

1999. The Wi-Fi Alliance was formed. The IEEE released standards 802.11a (5 GHz, 54 Mbps) and 802.11b (2.4 GHz, 11 Mbps). This year was the first time in which the term 'Wi-Fi' was used commercially.

2000. During the dot-com period in 2000, dozens of companies had the notion that Wi-Fi could become the payphone for broadband. The original notion was that users would pay for broadband access at hotspots. Since then, both paid and free hotspots continue to grow. Wireless networks that cover entire cities, such as municipal broadband, have mushroomed. Wi-Fi hotspots can be found in remote RV/Campground Parks across the US.

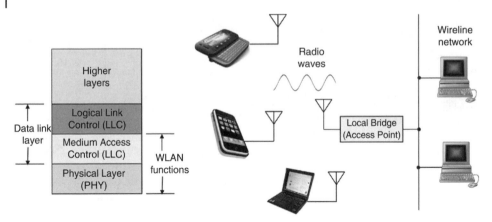

Figure 2.6 Generic Wi-Fi block diagram.

2003. The IEEE releases the 802.11g standard, which is compatible to the 802.11b but can work at higher data rates, up to 54 Mbps using the 2.4 GHz frequency band.

2009. The latest standard released by the IEEE related to WLAN is the 802.11n, having MIMO capabilities, with potential data rates of up to 600 Mbps using special combinations of modulation and coding schemes. It also operates at 2.4 GHz.

2010. Most laptop computers come equipped with internal WLAN adapters as well as integrated antennas, something that facilitates its use, especially in hotspots.

Most WLAN operate over unlicensed frequencies at near Ethernet speeds using carrier-sense protocols to share a radio wave. The majority of these devices are capable of transmitting information between computers within an open environment. Figure 2.6 (right) illustrates the concept of a WLAN interfacing with a wired network, whereas the logical architecture of a WLAN is also shown in Figure 2.6 (left).

In general, WLAN perform the following functions to enable the transfer of information from source to destination:

- The medium provides a bit pipe (a path for data flow) for the transmission of data.
- Medium access techniques facilitate the sharing of a common medium.
- Synchronization and error control mechanisms ensure that each link transfers the data intact.
- Routing mechanisms move the data from the originating source to the intended destination.
- Connectivity software interfaces an appliance to application software hosted on a server.

2.2.2 Medium Access Control (MAC) Sublayer

The MAC enables multiple appliances to share a common transmission medium via a carrier sense protocol similar to Ethernet. This protocol enables a group of wireless computers to share the same frequency and space. A WLAN MAC provides reliable delivery of data over somewhat error-prone wireless media. The protocol used for this purpose is Carrier Sense Multiple Access, where appliances can transmit only if the channel is 'idle' to avoid collisions.

2.2.3 Physical Layer

The physical layer provides for the transmission of bits through a communication channel by defining electrical, mechanical and procedural specifications. Modulation and multiple access methods are therefore defined as part of the physical layer.

2.2.4 Industry Bodies

The three main Wi-Fi industry bodies related to WLAN technology are: the Wi-Fi Alliance, the IEEE 802.11 working group, and the Wireless Broadband Alliance. A brief description of each follows.

2.2.4.1 Wi-Fi Alliance

The Wi-Fi Alliance (2015) is a trade association that promotes wireless LAN technology and certifies products if they conform to certain standards of interoperability. Not every IEEE 802.11-compliant device is submitted for certification to the Wi-Fi Alliance, sometimes because of costs associated with the certification process. The lack of the Wi-Fi logo does not necessarily imply a device is incompatible with Wi-Fi devices.

The Wi-Fi Alliance owns the Wi-Fi trademark. Manufacturers may use the trademark to brand certified products that belong to a class of wireless local area network (WLAN) devices based on the IEEE 802.11 standards. One of the benefits of certification of products towards the Wi-Fi Alliance is that interoperability amongst vendors is guaranteed.

2.2.4.2 IEEE 802.11

The IEEE 802.11 (2015) is a working group established by the IEEE (Institute of Electrical and Electronic Engineers) in the USA specializing in developing WLAN standards and amendments. Therefore, the IEEE 802.11 is a set of standards for implementing wireless local area network (WLAN) computer communication in the 2.4 GHz, 3.6 GHz and 5 GHz frequency bands. They are created and maintained by the IEEE LAN/MAN Standards Committee (IEEE 802). The base version of the standard IEEE 802.11-2007 has had subsequent amendments. These standards provide the basis for wireless network products using the Wi-Fi brand name.

2.2.4.3 The Wireless Broadband Alliance

The Wireless Broadband Alliance (2015) was established in 2003 by a group of telecom operators and is integrated by representatives from international broadband, cellular and integrated operators. The founders of this Alliance viewed Wi-Fi as an integral and strategic complement to other wireless and broadband networks such as 3G/UMTS, WiMAX, DSL, Cable and more.

WBA was created to support and further the vision of ubiquitous and seamless wireless broadband services. The mission of WBA is to facilitate adoption of Wi-Fi enabled services through improvements in user experience, interoperability and service delivery across technologies, devices and networks. By leveraging on the strengths of its unique membership mix and strong operator heritage, WBA seeks to engage the ecosystem for enabling a seamless Wi-Fi experience with the benefit of the end-user in mind.

Table 2.1 Wi-Fi standards.

Standard	Modulation	Maximum data rate	Spectrum bands	Bandwidth	MIMO
802.11-1997	DSSS-FHSS	1–2 Mbps	2.4 GHz	20 MHz	No
802.11a	OFDM	54 Mbps	5.25, 5.6 and 5.8 GHz	20 MHz	No
802.11b	CCK	11 Mbps	2.4 GHz	20 MHz	No
802.11g	OFDM	54 Mbps	2.4 GHz	20 MHz	No
802.11-2007	OFDM	54 Mbps	2.4 GHz	20 MHz	No
802.11n	BPSK/QPSK/QAM	54–600 Mbps	2.4 GHz	20/40 MHz	Yes

2.2.5   Wi-Fi Standards

Within the IEEE 802.11 Working Group, a series of IEEE Standards exist, as depicted in Table 2.1. These are explained in this section.

2.2.5.1   IEEE 802.11-1997
The WLAN standard was originally 1 Mbit/s and 2 Mbit/s (Mbps), 2.4 GHz RF and infrared (IR) standard (1997). Although the IR was considered in the standard, it was never implemented in commercial products. This standard is now a legacy only, as it is obsolete.

2.2.5.2   IEEE 802.11a
54 Mbps, 5.25, 5.6 and 5.8 GHz frequency bands, uses OFDM as its modulation scheme, having a bandwidth of 20 MHz (1999, shipping products in 2001).

2.2.5.3   IEEE 802.11b
Enhancements to 802.11 to support 5.5 and 11 Mbps, uses CCK modulation and operates in the 2.4 GHz ISM band, also having a bandwidth of 20 MHz (1999).

2.2.5.4   IEEE 802.11g
Enhanced 802.11b standard, totally compatible, 54 Mbps, 2.4 GHz standard, also uses OFDM modulation with a 20 MHz bandwidth (backwards compatible with b) (2003).

2.2.5.5   IEEE 802.11-2007
A new release of the standard that includes amendments a, b, d, e, g, h, i and j (July 2007).

2.2.5.6   IEEE 802.11n
Higher throughput improvements using MIMO (multiple-input, multiple-output antennas), BPSK/QPSK/QAM modulation combinations, 54–600 Mbps, working also in the 2.4 GHz band, with a bandwidth of 20/40 MHz (2010).

2.2.6 Spectrum

Wi-Fi has been allocated portions of the spectrum that can be used depending on the regions of the world and also whether newer standards are deployed. In the beginning, the ISM band in the 2.4 GHz was allocated for 802.11b/g usage, having 76 MHz of spectrum divided in 14 channels with 5 MHz separation, each of 22 MHz bandwidth. For the 802.11a/n standards, the 5 GHz band is used. The unlicensed 3.6 GHz band is only used in the USA for 802.11y networks, having 14 channels of 5/10/20 MHz of bandwidth. Finally, the 60 GHz band is intended for use in the future, with expected very high data rates (802.11a/d).

In addition to these frequency allocations, not all the channels can be used in every country. For example, in North America, channels 1 to 11 are used for the 2.4 GHz band, whereas in Europe, channels 1 to 13 are used. On the other hand, power levels are restricted and should not be exceeded, as Wi-Fi operates in unlicensed bands, interference should be minimized. There is in fact lack of protection in the unlicensed spectrum, suffering from potential interference from other systems, network security issues and, as in the case of the ISM, lots of congestion from other wireless services and technologies.

2.2.6.1 2.4 GHz Band

There are 14 channels in the 2.4 GHz spectrum available for WLAN use, as shown in Figure 2.7. Note that depending on local and regional regulations, not all the channels can be used; for example, in the US and Canada, channels 1 to 11 can be used; in Europe, channels 1 to 13 can be used (in the early days of WLAN, Spain and France were restricted to only use channels 11 to 13, but this restriction has been removed).

The bandwidth of each channel is 22 MHz, although the channel separation is only 5 MHz. This channel spacing governs the use and allocation of channels in a multi-AP (access point) environment, such as an office or campus. APs are usually deployed in a cellular fashion within an enterprise, where adjacent APs are allocated non-overlapping channels.

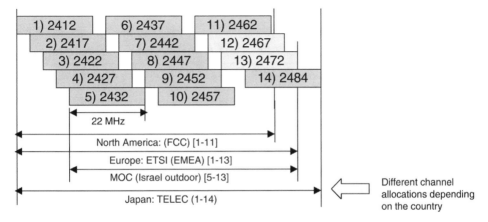

Figure 2.7 2.4 GHz channel allocations.

According to Figure 2.7, only three non-overlapping channels can be used at a time; therefore careful channel assignment should be done to avoid interference in WLAN networks.

As a sidenote, Japan is the only country in the world for which the fourteen channels are available in the spectrum – this is due to its very high utilization. According to the frequency map shown in Figure 2.7, Japan can thus have four non-overlapping channels for use.

2.2.6.2 5 GHz Band

Figure 2.8 shows the channel allocation for the 5 GHz frequency band, used mainly in the 802.11a and 802.11n standards. The centre frequency is shown for each case; the frequency of the channel is 10 MHz on either side of the dotted line. There is 5 MHz of separation between channels and thus special care should be taken when doing channel planning to avoid co-channel interference issues.

There are some restrictions and limitations in the use of the 5 GHz band for WLAN use, which are summarized as follows:

- Since other wireless systems such as radar operate in the same band, it is very likely that for some environments and regions, the use of Wi-Fi in the 5 GHz band could be very restricted.
- Path loss increases with frequency, so at 5 GHz there is substantially more path loss than at 2.4 or 3.6 GHz. This forces the use of more APs at 5 GHz than at lower frequency bands to cover the same area.
- Power control is not considered in the 802.11 standard, so it is required to implement transmit power control and dynamic frequency selection in the network to maximize performance.

Table 2.2 shows the frequency allocations in different parts of the world for the 5 GHz WLAN service.

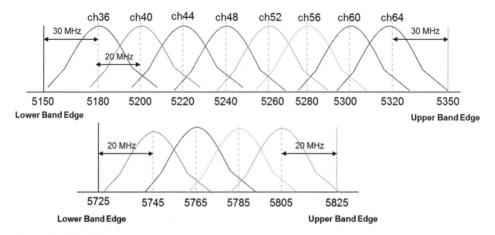

Figure 2.8 5 GHz channel plans.

Table 2.2 5 GHz band allocations around the world.

Regulatory domain	Frequency band	Channel number	Centre frequency
Japan	U-NII lower band	36	5.180 GHz
		40	5.200 GHz
		44	5.220 GHz
		48	5.240 GHz
Singapore	U-NII lower band	36	5.180 GHz
		40	5.200 GHz
		44	5.220 GHz
		48	5.240 GHz
Taiwan		52	5.260 GHz
		56	5.280 GHz
		60	5.300 GHz
		64	5.320 GHz
EMEA 1 – Australia and New Zealand	Same as USA	Same as USA	Same as USA
EMEA 2	U-NII lower band	36	5.180 GHz
		40	5.200 GHz
		44	5.220 GHz

2.2.7 Modulation Schemes Used in Wi-Fi

Modulation, which is a physical layer function, is a process in which the radio transceiver prepares the digital signal within the network interface card (NIC) for transmission over the channel. An analogue carrier is sent over the radio interface, for which its amplitude, frequency or phase is changed in proportion to the digital message. For example, for BPSK, the phase of the carrier is changed according to the digital message, using two symbols (0 and 1) – hence the name BPSK. If two bits are used to produce four symbols (00, 01, 10 and 11) and then these symbols are mapped to phase changes, then QPSK is produced. By combining some of these modulation techniques with coding it is possible to send more bits in the same timeframe, protecting the message against noise and interference. This has the benefit of increasing the efficient data rate.

2.2.8 Multiple Access (MA) Techniques

On the other hand, three MA techniques are often used in WLAN: direct sequence spread spectrum (DSSS), orthogonal frequency division multiplexing (OFDM) and frequency hopping spread spectrum (FHSS).

2.2.8.1 Frequency-Hopping Spread Spectrum (FHSS)
Frequency-hopping works very much as it name implies. The signal hops from frequency to frequency as a function of time over a wide band of frequencies, as shown in Figure 2.9. A hopping code determines the frequencies the radio will transmit and in which order. To receive the signal properly, the receiver must be set to the same hopping code and listen to the incoming signal at the right time and correct frequency.

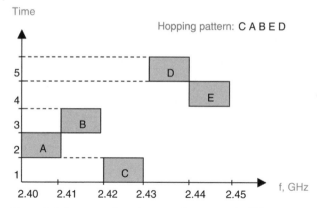

Figure 2.9 Frequency-hopping spread spectrum (FHSS).

The frequency-hopping technique reduces interference because an interfering signal from a narrowband system will affect the spread spectrum signal only if both are transmitting at the same frequency at the same time. Thus, the aggregate interference will be very low, resulting in few or no bit errors.

FHSS is rarely used nowadays for WLANs.

2.2.8.2 Direct Sequence Spread Spectrum (DSSS)

Spread spectrum techniques 'spreads' a signal's power over a wide band of frequencies, as seen in Figure 2.10, sacrificing bandwidth in order to gain signal-to-noise performance; that is process gain. This contradicts the desire to conserve frequency bandwidth, but the spreading process makes the data signal much less susceptible to electrical noise than otherwise would be achieved. Other transmission and electrical noise, typically narrow in bandwidth, will interfere with only a small portion of the spread spectrum signal, resulting in much less interference and fewer errors.

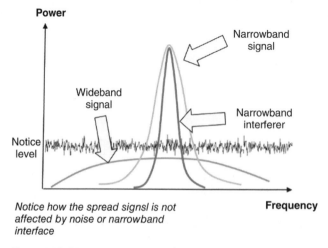

Figure 2.10 Direct sequence spread spectrum.

In particular, the direct sequence spread spectrum combines a data signal at the sending station with a higher data rate bit sequence, also known as the chipping code. A high processing gain increases the signal's resistance to interference. The direct sequence, having higher potential data rates than FHSS, would be best for bandwidth-intensive applications.

2.2.8.3 Orthogonal Frequency Division Multiplexing (OFDM)

Orthogonal frequency-division multiplexing (OFDM) is a frequency-division multiplexing (FDM) scheme used as a digital multicarrier modulation method. A large number of closely spaced orthogonal subcarriers are used to carry data, as shown in Figure 2.11. The data are divided into several parallel data streams or channels, one for each subcarrier. Each subcarrier is modulated with a conventional modulation scheme (such as quadrature amplitude modulation or phase-shift keying) at a low symbol rate, maintaining total data rates similar to conventional single-carrier modulation schemes in the same bandwidth.

OFDM has developed into a popular scheme for wideband digital communication, whether wireless or over copper wires, used in applications such as digital television and audio broadcasting, wireless networking and broadband Internet access.

The primary advantage of OFDM over single-carrier schemes is its ability to cope with severe channel conditions (for example, attenuation of high frequencies in a long copper wire, narrowband interference and frequency-selective fading due to a multipath) without complex equalization filters. Channel equalization is simplified because OFDM may be viewed as using many slowly modulated narrowband signals rather than one rapidly modulated wideband signal. The low symbol rate makes the use of a

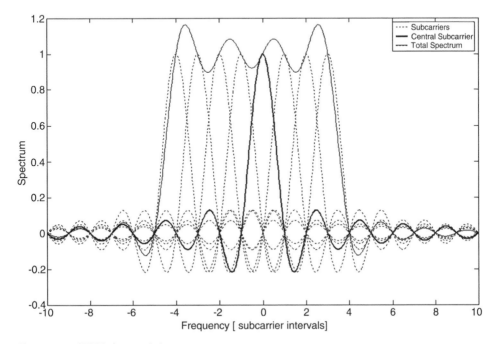

Figure 2.11 OFDM characteristics.

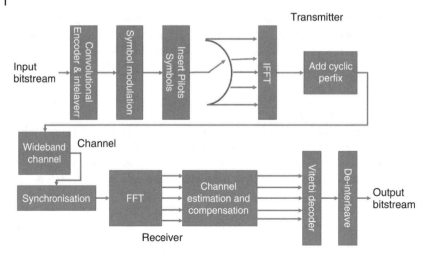

Figure 2.12 OFDM communication system architecture.

guard interval between symbols affordable, making it possible to eliminate intersymbol interference (ISI) and utilize echoes and time-spreading (that shows up as ghosting on analogue TV) to achieve a diversity gain, that is a signal-to-noise ratio improvement. This mechanism also facilitates the design of single-frequency networks (SFNs), where several adjacent transmitters send the same signal simultaneously at the same frequency, as the signals from multiple distant transmitters may be combined constructively, rather than interfering as would typically occur in a traditional single-carrier system.

OFDM was first introduced in the 802.11a standard and was then taken again for the 802.11g having improved data rates. Figure 2.12 shows an example of the architecture used in an OFDM system often employed in WLAN. A wideband channel is assumed for this example.

2.2.9 Power Levels

In terms of maximum power levels that can be transmitted per AP, there are regulations in terms of maximum EIRP (effective isotropic radiated power) limits per standard. Recall that EIRP refers to the maximum radiated power transmitted by the antenna taking into account antenna gain and cable losses (in dB):

$$\text{EIRP} = \text{AP transmit power} - \text{cable losses} + \text{antenna gain} \qquad (2.1)$$

Table 2.3 shows EIRP maximum levels for 802.11b/g and 802.11a standards. Special care should be taken here when choosing antennas for the APs, since according to the EIRP equation, higher antenna gain would be proportionally converted in an increase of the EIRP. This is the case, for example, when coverage is to be enhanced and highly directional antennas (such as Yagi-Uda) are used for 'bridging' two buildings.

There is another parameter that is often confused with EIRP, called ERP (*effective radiated power*). The difference is that EIRP takes as a reference the antenna gain with respect to an isotropic radiator (dBi) and ERP takes as a reference for antenna gain a half-wave dipole (dBi). In fact, ERP and EIRP are related by a factor of 2.15 (gain of a half-wave dipole in dBi).

Table 2.3 Power levels for EIRP around the world.

Standard	Regulatory domain	Maximum EIRP
802.11a	Americas (-A)	160 mW (channels 36–48)
		800 mW (channels 52–64)
	Japan (-J)	10 mW/MHz
	Singapore (-S)	100 mW
	Taiwan (-T)	800 mW (channels 52–64)
802.11b/g	Americas (-A)	4 W
	EMEA (-E)	100 mW
	Israel (-I)	100 mW
	China (-C)	10 mW
	Japan (-J)	10 mW/MHz

2.2.10 Performance Indicators

To assess Wi-Fi performance, curves of SINR (signal-to-interference and noise ratio) versus throughput for 802.11b/g are shown in Figure 2.13, with a bandwidth of 20 MHz. Recall that the SINR determines the performance of a Wi-Fi network since it establishes the minimum signal level above noise and interference for proper operation. Various modulation schemes are shown, starting with BPSK with low spectral efficiency (but high noise resistance) and ending with a high-level QAM modulation, which offers very high spectral efficiency at the expense of being more vulnerable to noise. Notice how as the number of symbols increase, more throughput is achievable but higher SINRs are required. These curves are considered for an OFDM system.

Similar curves to the ones shown in Figure 2.13 are presented in Figure 2.14, now for 802.11n having a bandwidth of 40 MHz. A similar behaviour is observed; as the throughput is increased and multilevel modulation is used, the SINR requirement also increases. Wi-Fi designers should take this into account when calculating the link budgets according to given throughput requirements.

2.2.11 Target Signal Levels and Link Budgets

As an example, assume there is a need to deploy a WLAN 802.11g network, having 60 Mbps for data (SINR of 18.8 dB) and a voice requirement of 25 dB, according to equipment vendor's specifications. Table 2.4 shows the calculations needed for this.

- An EIRP of 100 mW is used to account for maximum power level regulations (which are always specified in a maximum EIRP). Therefore, the transmit antenna gain is irrelevant – a word of caution, be careful not to use highly directive antennas without adjusting the transmit power from the AP, otherwise the maximum EIRP limits will be superseded!
- The maximum acceptable path loss (MAPL) is calculated depending on the SINR levels as given by the requirements.

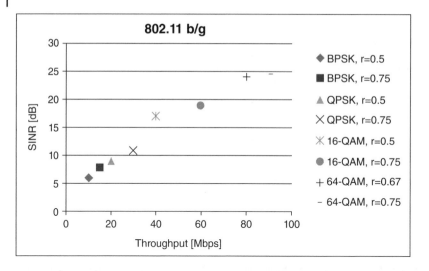

Modulation	Coding rate	MCS efficiency	Throughput (Mbps)	SINR (dB)
BPSK	0.5	0.5	10	6
BPSK	0.75	0.75	15	7.8
QPSK	0.5	1	20	9
QPSK	0.75	1.5	30	10.8
16-QAM	0.5	2	40	17
16-QAM	0.75	3	60	18.8
64-QAM	0.67	4	80	24
64-QAM	0.75	4.5	90	24.5

Figure 2.13 802.11b/g Wi-Fi performance.

- Noise contributions are taken from the noise floor level + noise figure of the client device, which in this case is assumed to be 20 dB.
- Interference is assumed to be −87 dBm from a co-channel nearby AP.
- A bandwidth of 20 MHz is taken for 802.11g channels.

A need to account for a sufficient interference margin in the design has to be considered, as well as a shadowing margin – the MAPL value obtained only represents the median path loss. This is especially critical for densely cluttered environments, where location variability tends to be quite large. For example, assuming a location variability of 12 dB and a 90% cell-edge confidence percentile, a fade margin for shadowing of 15.4 dB is to be considered, and hence the MAPL value with this margin is adjusted to only 96.6 dB.

Data rates affect AP coverage areas. Lower data rates, such as 1 Mbps, can extend the coverage area farther from the AP than higher data rates, such as 54 Mbps. Thus, the data rate and power level affects coverage and consequently the number of APs required. This is illustrated in Figure 2.15, where simulations have been performed to estimate the effects of throughput on coverage, assuming the ITU indoor model is used – refer to Chapter 4 for more details about the model.

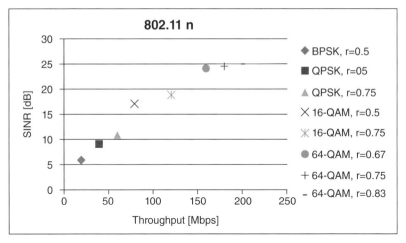

Modulation	Coding rate	MCS efficiency	Throughput (Mbps)	SINR (dB)
BPSK	0.5	0.5	20	6
QPSK	0.5	1	40	9
QPSK	0.75	1.5	60	10.8
16-QAM	0.5	2	80	17
16-QAM	0.75	3	120	18.8
64-QAM	0.67	4	160	24
64-QAM	0.75	4.5	180	24.5
64-QAM	0.83	5	200	25

Figure 2.14 802.11n Wi-Fi performance.

Table 2.4 Wi-Fi link budget example.

Parameter	Symbol	Value	Comments
EIRP	A	0.1 W	Usual EIRP
	B	20 dBm	$B = 10 \log(A/1000)$
Transmit power	E	20 dBm	Antenna gain B is irrelevant as max EIRP is set by regulations
Required SINR	F	18.8 dB	For 60 Mbps data throughput, 802.11g
Interference power	G	−87 dBm	Assumed from a co-channel AP nearby
Boltzmann's constant	H	1.38×10^{-23}	Units are $m^2kg\ s^{-2}\ K^{-2}$
Temperature	I	300 K	Room temperature, 27 °C
Noise density	J	−203.83 dB Hz^{-1}	$J = 10\log(HI)$
Noise figure	K	20 dB	Assumed performance of client device
Bandwidth	L	20 MHz	Applicable to 802.11g
Noise power	M	−111 dB Hz	$M = 10\log(L \times 10^6) + J + K$
Required signal power	N	−92 dBm	$N = M + F$, assuming 0 dBi receive antenna
Maximum acceptable path loss (median)	R	112 dB	$R = E - N$

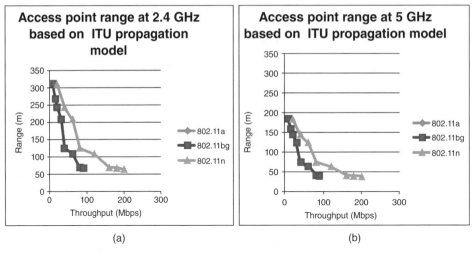

Figure 2.15 Impact of targets on range: (a) 2.4 GHz; (b) 5 GHz.

Different data rates are achieved by the AP sending a redundant signal on the wireless link, allowing data to be more easily recovered from noise. The number of symbols in a packet for lower data rates is greater than that for a higher data rate, so sending data at lower data rates takes longer. Therefore, clients working at lower data rates slow the network and thus it is preferable to limit the WLAN for higher data rates and force those clients further away to roam to closer APs. In a way, it seems better to have more APs at lower power than fewer APs transmitting at higher power, having the risk of slowing down the performance of the network.

Figure 2.16 shows a picture of the estimated ranges (not to scale) as the data throughput varies for 2.4 GHz and 5 GHz. The coverage is reduced as the throughput

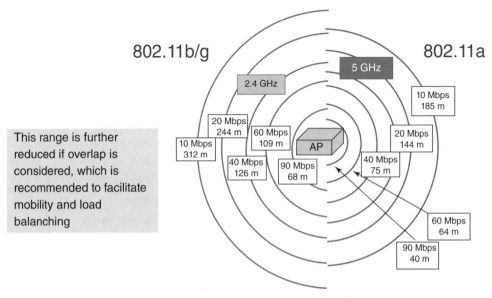

Figure 2.16 Estimated ranges versus data throughput for 2.4 GHz and 5 GHz.

is increased. If an overlap of 30% is considered between APs, this coverage range is further reduced – this should be considered when performing WLAN coverage design.

2.2.12 Interference Challenges

Interference is one of the key elements to take into consideration in WLAN networks, since it can degrade the performance of the system based on the MAC protocols used. These interference sources can be subdivided into three types:

- Co-channel interference. This comes as a result of either other Wi-Fi networks within range using the same channel or non-Wi-Fi systems transmitting in the same band. Examples of this include: microwave ovens, baby monitors, Bluetooth devices, cordless phones, etc. This kind of interference is seen as one of the worst ones, as uncontrolled emissions from in-band appliances can severely affect WLAN performance.
- Adjacent channel interference. This is caused by overlapping channels interfering with contiguous ones. According to the IEEE standards, for 802.11b/g, the adjacent channel rejection should be of at least 35 dB at 25 MHz from the centre frequency of the receiver.
- Out-of-band interference: caused by spurious emissions from out-of-band appliances. An out-of-band transmit power of −13 dBm/MHz is recommended according to IEEE standards.

For the CSMA/CD protocol (carrier-sense multiple access/collision detection), AP transmit only if the channel is clear; that is idle. If the channel is busy, transmission is not allowed. Since in the standard there is no channel reservation or scheduling, two APs can detect the signal clear and both transmit at the same time, hence producing collisions and reducing the throughput. Also, something else to notice is that carrier sensing can be triggered by non-WiFi signals (e.g. spurious emissions from other appliances), thus making the channel seem 'busy' and hence a reduction of throughput is experienced. Beacon packets can also stop nodes to transmit as the channel is seen 'busy', even when no user traffic is detected.

2.2.13 Channel Planning

Since Wi-Fi channels overlap, a careful selection should be made to avoid adjacent channel interference problems. As can be seen in Figure 2.17, for example, channels 1, 6 and 11 can be chosen in contiguous cells since these do not overlap; a partial overlap

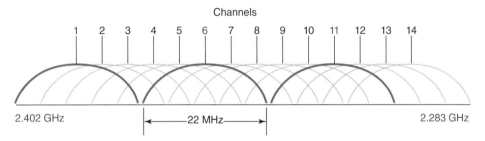

Figure 2.17 Wi-Fi channel planning.

channel plan selecting channels 3, 8 and 13 overlaying this can also be deployed, and can be used to reduce interference in the second tier. The patterns are to be repeated to guarantee a minimum SINR. For this, cells are deployed either in same-floor or vertical-floor configurations to make use of the propagation characteristics of the building and the high transmission losses associated with walls and floors.

For *data networks*, too much overlapping is not required; this is mainly determined by data rate specifications. In general, data links can accept less stringent latency requirements and can in principle reduce the data rate if coverage is low – this will only have the effect of reducing the throughput but the link will be kept operating. In fact, clients respond to low signal strength by stepping down data rate and taking longer to transmit.

For *voice networks*, more overlap is needed, since voice clients should roam to a better AP before dropping packets, in a similar way for handovers in cellular. Also, smaller cells are required, to reduce latency processor load in handheld and increase link stability.

2.2.13.1 Single-Floor and Vertical Channel Planning

For *single-floor channel planning*, sufficient overlap should be allowed for seamless connectivity, especially for voice deployments. The overlap can be 50% because of channel separation and the minimum recommended is 15%. Co-located non-overlapping channels should be chosen in contiguous AP footprints (cells) to minimize adjacent channel interference. The cell pattern should be repeated only when sufficient SINR levels are achieved; that is avoid reusing channel numbers in low SINR level regions.

Finally, as discussed earlier, the coverage area is not only dependent on transmit power, antenna gain and the environment but also on data rate. For example, at 2.4 GHz and transmitting at 100 mW for 2 Mbps, a coverage diameter of 132 m can be achieved. If the data rate is increased, this coverage is reduced to only 70 m having the same transmit power. Therefore, if the minimum overlap is accounted for, for 2 Mbps, four APs are needed to cover an area of 250×250 m^2. Likewise, for 11 Mbps, 16 APs would then be needed to cover the same area.

When using *vertical channel separation*, the idea is to co-locate non-overlapping channels in a vertical separation (different floors), as depicted in Figure 2.18, taking

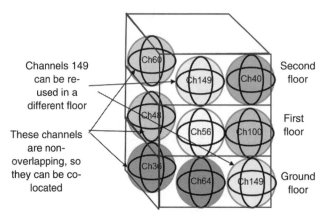

Figure 2.18 Vertical channel separation.

advantage of the high transmission losses due to the signal penetrating floors. This reduces the possibility of co-channel interference by using floor attenuation to separate channels, as indicated in Figure 2.17. The pattern can therefore be reused in a different floor as long as the SINR level is sufficiently high.

2.2.13.2 Multichannel Access Points

The idea of *multichannel access points* is to enhance throughput over a conventional AP by supporting multiple SSIDs and radios in the same unit. For example, an 802.11b PDA user at a moderate distance from an AP brings today's single-channel 802.11g network down from 54 Mbps to close to 1 Mbps. Therefore, WLAN systems must contend with an unpredictable and heterogeneous client environment, something that has been happening in recent years. At the same time, end users want to add services beyond just data, such as VoWIP. By having multiple channels operating from the same access point, we are able to partition the client traffic to dramatically enhance throughput. As end-users deploy new services on to their WLAN system, Engim enables these services to be partitioned intelligently and implemented on the same WLAN system.

2.2.13.3 Automated Planning

Radio resource management (RRM), also known as Auto-RF, can adjust the channel (dynamic channel assignment) and power (dynamic transmit power control) to maintain the RF coverage area. It adjusts the power level of the AP to maintain a baseline signal strength with neighbouring APs at −65 dBm (configurable). It adjusts the channel of the AP when it notices nearby interference sources on the channel on which the AP is currently located. It continues to optimize the RF coverage for the best reception and throughput for the wireless network.

RRM understands that the RF environment is not static. As different RF affecting variables change (people in the room, amount of devices stored in the facility, leaves on trees for outside deployment, interference from different RF sources, and so on), the RF coverage adjusts to these variables and changes with them. Because these variables change continuously, monitoring for the RF coverage and adjusting it periodically is necessary.

2.2.14 Mobility Issues

One of the key elements of mobility in any network has to do with roaming. *Roaming* in an enterprise 802.11 network can be described as when an 802.11 client changes its AP association from one AP within an ESS (extended service set) to another AP within the same ESS. Depending on the network features and configuration, a lot may occur between the clients, WLCs, and upstream hops in the network, but at the most basic level it is simply a change of association.

When a wireless client authenticates and associates with an AP, the WLC of the AP places an entry for that client in its client database. This entry includes the client MAC and IP addresses, security context and associations, QoS context, WLAN and associated AP. The WLC uses this information to forward frames and manage traffic to and from the wireless client. When the wireless client moves its association from one AP to another, the WLC simply updates the client database with the new associated AP. If necessary, a new security context and associations are established as well.

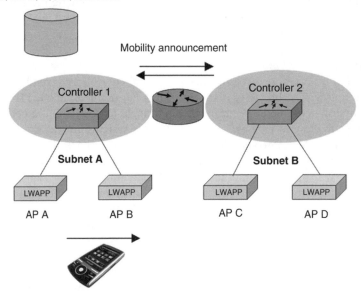

Figure 2.19 Layer 2 roam.

2.2.14.1 Layer 2 Roam

A *Layer 2 roam* occurs when a client roams from one AP and (re)associates to a new AP, providing the same client subnet. In most cases, the foreign AP can be on the same WLC as the home AP. This is a very simple roam because the WLC maintains a database with all the information of the client. All upstream network components from the WLC are unaffected by the client moving from home to foreign AP, as illustrated in Figure 2.19. In instances when there are multiple WLCs connected to the same subnet, and therefore a client can roam between WLCs but remain on the same subnet, mobility announcements are passed between the related WLCs to pass client context information between WLCs. This WLC then becomes the anchor WLC for that client.

2.2.14.2 Layer 3 Roam

In instances where the client roams between APs that are connected to different WLCs and the WLC WLAN is connected to a different subnet, a *Layer 3 roam* is performed and there is an update between the new WLC (foreign WLC) and the old WLC (anchor WLC) mobility databases.

 If this is the case, return traffic to the client still goes through its originating anchor WLC. The anchor WLC uses Ethernet over IP (EoIP) to forward the client traffic to the foreign WLC, to where the client has roamed. Traffic from the roaming client is forwarded out of the foreign WLC interface on which it resides; it is not tunnelled back. The client MAC address for its default gateway remains the same, with the WLC changing the MAC address to the local interface gateway MAC address when the client traffic is sent to the default gateway.

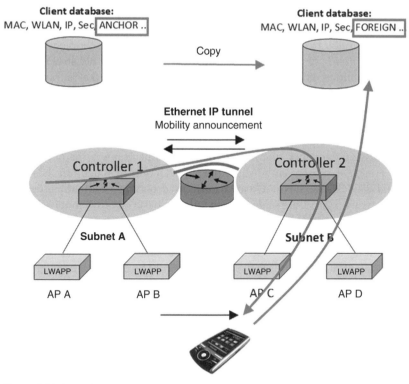

Figure 2.20 Layer 3 roam.

The example in Figure 2.20 describes a client Layer 3 roam with PMK (pair-wise master key). The client begins with a connection to AP B on WLC 1. This creates an ANCHOR entry in the WLC client database. As the client moves away from AP B and makes an association with AP C, WLC 2 sends a mobility announcement to peers in the mobility group looking for the WLC with the client MAC address. WLC 1 responds to the announcement, handshakes and ACKs. Next the client database entry for the roaming client is copied to WLC 2 and is marked as FOREIGN. Included PMK data (master key data from the RADIUS server) are also copied to WLC 2. This provides fast roam times for WPA2/802.11i clients because there is no need to reauthenticate to the RADIUS server. After a simple key exchange between the client and AP, the client is added to the WLC 2 database and is similar, except that it is marked as FOREIGN.

2.3 Bluetooth

Bluetooth (IEEE 802.16 standard, 2015) is a technology employed for short-range wireless communications, mainly in the ISM band around 2.4 GHz (same frequency as Wi-Fi) from fixed and mobile terminals, and building personal area networks (PANs). The standard was developed by the telecom vendor Ericsson in the mid-1990s and was originally conceived as a wireless alternative for RS-232 serial communications.

Bluetooth operates in the range of 2400–2483.5 MHz (including guard bands). This is in the globally unlicensed (but not unregulated) industrial, scientific and medical (ISM) 2.4 GHz short-range radio frequency band. Bluetooth uses a radio technology called a *frequency-hopping spread spectrum* (FHSS), similar to that used in old IS-95 cellular standards. The transmitted data are divided into packets and each packet is transmitted on one of the 79 designated Bluetooth channels. Each channel has a bandwidth of 1 MHz.

Bluetooth 4.0 uses 2 MHz spacing, which allows for 40 channels. The first channel starts at 2402 MHz and continues up to 2480 MHz in 1 MHz steps. It usually performs 1600 hops per second, with adaptive frequency-hopping (AFH) enabled.

Originally, *Gaussian frequency-shift keying* (GFSK) modulation was the only modulation scheme available; subsequently, since the introduction of Bluetooth 2.0+EDR, $\pi/4$-DQPSK and 8-DPSK modulation may also be used between compatible devices. Devices functioning with GFSK are said to be operating in basic rate (BR) mode where an instantaneous data rate of 1 Mbit/s is possible.

Bluetooth is a packet-based protocol with a master–slave structure. One master may communicate with up to seven slaves in a *piconet* (an ad hoc computer network using Bluetooth technology); all devices share the master's clock. The devices can switch roles, by agreement, and the slave can become the master (e.g. a headset initiating a connection to a phone will necessarily begin as master, as initiator of the connection, but may subsequently prefer to be slave).

Bluetooth is a standard wire-replacement communications protocol primarily designed for low power consumption, with a short range based on low-cost transceiver microchips in each device. Because the devices use a radio (broadcast) communications system, they do not have to be in visual line of sight of each other; however, a quasi-optical wireless path must be viable. Range is power-class-dependent, but effective ranges vary in practice. For example, class 1 devices that transmit at 100 mW (20 dBm) may reach up to 100 metres under clear propagation conditions (not cluttered environments). As discussed earlier in this chapter, obstructions may introduce additional losses to the wireless link and hence reduce the range considerably.

There is a range of applications for which Bluetooth has been used, as follows:

- Wireless control of and communication between a mobile phone and a hands-free headset. This was one of the earliest applications to become popular.
- Wireless control of and communication between a mobile phone and a Bluetooth compatible car stereo system.
- Wireless control of and communication with tablets and speakers such as iPad and Android devices.
- Wireless Bluetooth headset and Intercom. Idiomatically, a headset is sometimes called 'a Bluetooth'.
- Wireless networking between PCs in a confined space and where little bandwidth is required.
- Wireless communication with PC input and output devices, the most common being the mouse, keyboard and printer.
- Transfer of files, contact details, calendar appointments and reminders between devices with OBEX.
- Replacement of previous wired RS-232 serial communications in test equipment, GPS receivers, medical equipment, bar code scanners and traffic control devices.
- For controls where infrared was often used.

- For low bandwidth applications where a higher USB bandwidth is not required and cable-free connection is desired.
- Sending small advertisements from Bluetooth-enabled advertising hoardings to other, discoverable, Bluetooth devices.
- Wireless bridge between two Industrial Ethernet (e.g. PROFINET) networks.
- Three seventh and eighth generation game consoles, Nintendo's Wii and Sony's PlayStation 3, use Bluetooth for their respective wireless controllers.
- Dial-up Internet access on personal computers or PDAs using a data-capable mobile phone as a wireless modem.
- Short-range transmission of health sensor data from medical devices to mobile phone, set-top box or dedicated tele-health devices.
- Allowing a DECT phone to ring and answer calls on behalf of a nearby mobile phone.
- Real-time location systems (RTLSs) are used to track and identify the location of objects in real-time using 'Nodes' or 'tags' attached to or embedded in the objects tracked, and 'Readers' that receive and process the wireless signals from these tags to determine their locations.
- Personal security application on mobile phones for prevention of theft or loss of items. The protected item has a Bluetooth marker (e.g. a tag) that is in constant communication with the phone. If the connection is broken (the marker is out of range of the phone) an alarm is raised. This can also be used as a crew overboard alarm. A product using this technology has been available since 2009.

Many devices nowadays, especially for home, are provided with a Bluetooth connection. Various examples of Bluetooth devices or cards are shown in Figure 2.21. For a

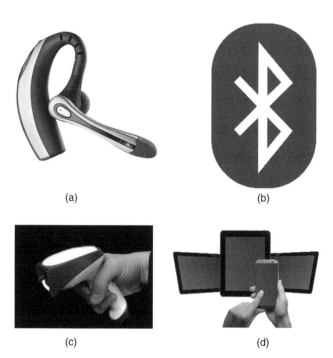

(a) (b)

(c) (d)

Figure 2.21 Bluetooth: (a) handset; (b) logo; (c) barcode scanner; (d) mobile phone equipped with Bluetooth.

product to be certified, the Bluetooth logo (also shown in Figure 2.21) should be displayed.

2.4 ZigBee

ZigBee is a protocol (IEEE 802.15 standard, 2015) that has been created for personal area networks (PANs) built from small, low-power digital radios. Even though the standard is designed to operate at low power, larger distances can be achieved by passing data through intermediate devices to reach more distant ones, thus creating a *mesh network*. A mesh network is one with no centralized control or high-power transceiver able to reach all the networked devices.

One of the characteristics of ZigBee is that it is mainly used to operate those applications that require a low data rate, long battery life and secure networking – ideal for industrial automation applications. The standard has a defined rate of 250 kbps and is best suited for periodic or intermittent data or single data transmission from a sensor or input device. The low cost allows the technology to be widely deployed in wireless control and monitoring applications. Low power usage allows longer life with smaller batteries. Mesh networking provides high reliability and a more extensive range. ZigBee chip vendors typically sell integrated radios and microcontrollers with a flash memory of between 60 kb and 256 kb.

Some typical applications that involve ZigBee include wireless light switches, electrical metres with in-home displays, traffic management systems and other consumer and industrial equipment that require short-range wireless transfer of data at relatively low rates. In summary:

- Home entertainment and control – home automation, smart lighting, advanced temperature control, safety and security, movies and music
- Wireless sensor networks
- Industrial control
- Embedded sensing
- Medical data collection
- Smoke and intruder warning
- Building automation.

The technology defined by the ZigBee specification is intended to be simpler and less expensive than other wireless personal area networks (WPANs), such as Bluetooth or Wi-Fi. ZigBee is not intended to support powerline networking but to interface with it at least for smart metering and smart appliance purposes. An example of a ZigBee module is shown in Figure 2.22.

One key characteristic of ZigBee is its robust security. ZigBee networks are secured by 128 bit symmetric encryption keys. In home automation applications, transmission distances range from 10 to 100 meters line-of-sight, depending on power output and environmental characteristics.

ZigBee operates on the industrial, scientific and medical (ISM) radio bands: 868 MHz in Europe, 915 MHz in the USA and Australia and 2.4 GHz in most jurisdictions worldwide. Data transmission rates vary from 20 kilobits/second in the 868 MHz frequency band to 250 kilobits/second in the 2.4 GHz frequency band.

Figure 2.22 ZigBee module.

Finally, there is a range of 'application profiles' that are either published or in development, as follows:

Released specifications:

- ZigBee Home Automation 1.2
- ZigBee Smart Energy 1.1b
- ZigBee Telecommunication Services 1.0
- ZigBee Health Care 1.0
- ZigBee RF4CE – Remote Control 1.0
- ZigBee RF4CE – Input Device 1.0
- ZigBee Light Link 1.0
- ZigBee IP 1.0
- ZigBee Building Automation 1.0
- ZigBee Gateway 1.0
- ZigBee Green Power 1.0 as optional feature of ZigBee 2012

Specifications under development:

- ZigBee Smart Energy 2.0
- ZigBee Retail Services
- ZigBee Smart Energy 1.2/1.3
- ZigBee Light Link 1.1
- ZigBee Home Automation 1.3

2.5 Radio Frequency Identification (RFID)

Radio frequency identification, or RFID, is becoming extremely popular in logic automation and in industry. Radio frequency identification (RFID) is a generic term for non-contacting technologies that use radio waves to automatically identify people or objects. There are several methods of identification, but the most common is to store a unique serial number that identifies a person or object on a microchip that is attached to an antenna. The combined antenna and microchip is called an 'RFID transponder' or 'RFID tag' and works in combination with an 'RFID reader' (sometimes called an 'RFID interrogator'). Examples of tags are given in Figure 2.23.

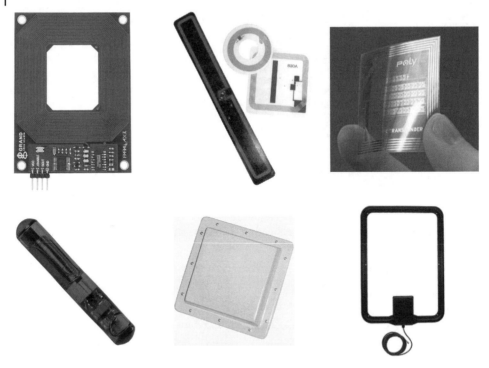

Figure 2.23 Examples of various RFID tags and readers.

An RFID system consists of a reader and one or more tags. The reader's antenna is used to transmit radio frequency (RF) energy. Depending on the tag type, the energy is 'harvested' by the tag's antenna and used to power-up the internal circuitry of the tag. The tag will then modulate the electromagnetic waves generated by the reader in order to transmit data back to the reader. The reader receives the modulated waves and converts them into digital data. This principle is called *backscattering* and is shown in Figure 2.24.

There are two major types of tag technologies. *Passive* tags are tags that do not contain their own power source or transmitter. When radio waves from the reader reach the chip's antenna, the energy is converted by the antenna into electricity that can power-up

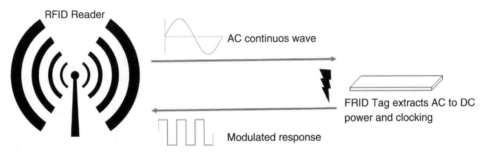

Figure 2.24 Backscattering principle used to power RFID passive systems.

the microchip in the tag (known as 'parasitic power'). The tag is then able to send back any information stored on the tag by reflecting the electromagnetic waves as described above. *Active* tags have their own power source and transmitter.

The power source, usually a battery, is used to run the microchip circuitry and to broadcast a signal to a reader. Due to the fact that passive tags do not have their own transmitter and must reflect their signal to the reader, the reading distance is much shorter than with active tags. However, active tags are typically larger, more expensive and require occasional service.

Frequency refers to the size of the radio waves used to communicate between the RFID system components. Just as you tune your radio to different frequencies in order to hear different radio stations, RFID tags and readers must be tuned to the same frequency in order to communicate effectively. RFID systems typically use one of the following frequency ranges: low frequency (or LF, around 125 kHz), high frequency (or HF, around 13.56 MHz), ultra-high frequency (or UHF, around 868 and 928 MHz) or microwave (around 2.45 and 5.8 GHz).

There really is no such thing as a 'typical' RFID tag. The read range of a tag ultimately depends on many factors: the frequency of RFID system operation, the power of the reader, environmental conditions, physical size of the tag antenna and interference from other RF devices.

2.6 Private Mobile Radio (PMR)

A private mobile radio (PMR) was developed for business users who need to keep in contact over relatively short distances with a central base station, such as taxi companies or emergency services. PMR networks consist of one or more base stations and a number of mobile terminals, as depicted in Figure 2.25. Such a system serves a closed user group that is normally owned and operated by the same organization as its users.

PMR systems use dedicated radio frequencies that are licensed by the regulator that manages the radio spectrum for a country. This ensures that the radio frequencies are available for the sole use of the PMR system and that other radios in the same frequency band do not interfere with the operation of the system.

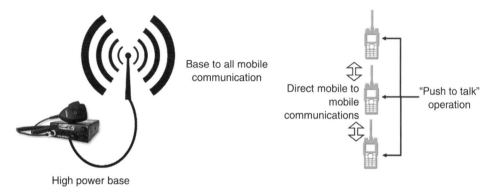

Figure 2.25 Private mobile radio.

From their early designs, PMR systems have developed into 'trunked' systems, the most notable of which is TETRA, a *terrestrial trunked radio*. Trunking is a technique where the resources of the communications network are shared, thus providing both flexibility and economy in the allocation of network resources. Typically, a communication channel is allocated for the duration of a call and then automatically released to allow it to be used for another call, perhaps between different users on the same system. The technique also enables multiple base stations to be connected and to provide coverage across a wider area than with a single base station.

PMR systems generally provide facilities for closed user groups, group call and push-to-talk, and have call setup times that are generally short compared with cellular systems. Many PMR systems allow direct mode operation in which terminals can communicate with one another directly when they are out of the coverage area of a network.

PMR systems may also be developed to allow public access (by subscription) and they are then known as a *public access mobile radio* (PAMR). The users of PAMR systems are usually not the same as the system's owner and operator. Traditionally, PMR systems have usually been based on European standards for the equipment, but operated under licence and subject to national frequency management plans. An exception is PMR 446, a consumer 'walkie-talkie', which has six analogue channels allocated in most European countries for use without a licence.

2.6.1 PMR Elements

PMR systems consist of two main elements – infrastructure and terminals. *Infrastructure* is provided by one or more radio base stations or repeaters which can be deployed on a single site or dispersed over a large geographical area to provide coverage over a wide area. *Terminals* usually refer to the hand portables carried by users or the mobile radios commonly installed into vehicles, but can also include applications such as voice and data dispatching.

2.6.2 Attributes

Some of the key attributes of PMR systems are:

- **Coverage.** A PMR system is designed to provide coverage for the entire operational environment so that radio signals are transmitted to all of the locations when communications are required by the customer. In addition to outside locations, this may include in-building coverage and underground locations such as tunnels.
- **Group communications.** Business operations quite often rely on group calls to facilitate efficient operational communications. A simple example of this can be provided by security staff in large sporting stadiums, where a member of staff can call for assistance from a number of colleagues by making a group call using a single button press.
- **Reliability and resilience.** PMR systems are often the first choice for communications where safety is paramount. This is why the public safety services of police, fire and ambulance use PMR as their primary means of voice communications. Due to the requirements of these organizations that the systems they use are ultimately reliable, PMR systems are often designed to continue to provide a service in the event of

equipment or power failure through the use of intelligent recovery mechanisms and power backup using batteries and/or generators.

- **Performance.** PMR systems use efficient signalling protocols to ensure very quick call establishment times. Often user organizations have requirements to be able to start speaking almost as soon as they decide to make a call and cannot wait for someone to answer a phone. For this reason, PMR systems provide a large range of call types from open channel calls where the user can quickly press and talk to emergency calls, which will always get through no matter how busy the system is.

- **Contention.** The size of a PMR system can be designed to deliver the quality of service required by the users. For non-essential communications it might be acceptable for users to queue or wait until resources are available and can therefore use a small system, but for higher end-users the system can be designed with a large amount of capacity to ensure that calls are always completed regardless of how busy the system becomes.

- **Security.** Privacy of communications can be very important to organizations to ensure that their conversations do remain private. PMR systems are able to offer varying levels of security from basic voice scrambling right up to complex military-grade encryption algorithms depending upon the needs of the encryption.

2.6.3 TETRA

A terrestrial trunked radio (TETRA) is a digital trunked mobile radio standard developed to meet the needs of traditional PMR user organizations such as: public safety, transportation, utilities, government, police forces, fire departments, ambulance, rail transport staff for train radios, transport services and the military. TETRA is a European Telecommunications Standards Institute (ETSI) standard and its first version was published 1995; it is mentioned by the European Radiocommunications Committee (ERC). An example of vehicle and portable TETRA radios is illustrated in Figure 2.26.

TETRA uses time division multiple access (TDMA) with four user channels on one radio carrier and 25 kHz spacing between carriers. Both point-to-point and point-to-multipoint transfer can be used. Digital data transmission is also included in the standard at a low data rate.

In terms of architecture, TETRA mobile stations (MS) can communicate using either direct-mode operation (DMO) or trunked-mode operation (TMO), employing

(a)   (b)

Figure 2.26 TETRA radios for: (a) vehicle and (b) portable.

switching and management infrastructure (SwMI) made of TETRA base stations (TBS). One of the advantages of using DMO is that it allows direct communications when network coverage is not available, using other TETRA terminals as relays – this is due to the very strict network reliability that TETRA needs to have. The base stations normally transmit continuously and (simultaneously) receive continuously from various mobiles on different carrier frequencies; hence the TETRA system is an FDD system. TETRA also uses FDMA/TDMA, similar to GSM.

TETRA also supports data communications, to send short messages over the system's control channels, whereas specific assigned channels are employed for packet-switched or circuit-switched data. For its modulation, TETRA uses $\pi/4$ DQPSK, a form of phase shift keying. The symbol (baud) rate is 18 000 symbols per second and each symbol maps to 2 bits, thus resulting in 36 000 bits/s gross.

Since TETRA needs to be well protected against eavesdropping, it is equipped with air interface encryption algorithms, which make it very robust.

The common mode of operation is in a group calling mode in which a single button push will connect the user to the users in a selected call group and/or a dispatcher. It is also possible for the terminal to act as a one-to-one 'walkie talkie', but without the normal range limitation since the call still uses the network. TETRA terminals can act as mobile phones, with a full-duplex direct connection to other TETRA users or the PSTN. Emergency buttons, provided on the terminals, enable the users to transmit emergency signals to the dispatcher, overriding any other activity taking place at the same time.

Although the allocated frequency band for TETRA is around 400 MHz, some countries employ slightly different channels for emergency and/or civilian use. For example, in France, 410–430 MHz is allocated for civilian/private use, with 380–400 MHz for emergency services. In Italy, the armed forces/emergency services employ the frequency pair 380–390 MHz and 462 MHz for civilian/private usage.

Finally, the TETRA Association, working with ETSI, developed the *TEDS standard*, a wideband data solution, which enhances TETRA with a much higher capacity and throughput for data. In addition to those provided by TETRA, TEDS uses a range of adaptive modulation schemes and a number of different carrier sizes from 25 kHz to 150 kHz, thus employing a similar strategy to the adaptive modulation schemes defined in cellular LTE. Initial implementations of TEDS will be in the existing TETRA radio spectrum and is likely to employ 50 kHz channel bandwidths as this enables an equivalent coverage footprint for voice and TEDS services. TEDS performance is optimized for wideband data rates, wide area coverage and spectrum efficiency.

2.7 Digital Enhanced Cordless Telecommunications (DECT)

Digital Enhanced Cordless Telecommunications (DECT™) is the ETSI standard for short-range cordless communications, which can be adapted for many applications and can be used over unlicensed frequency allocations world-wide. Although it was initially developed as a European standard, it was later adopted by many other countries (more than 110) and today has become a world-wide de-facto standard for cordless telephony applications. The standard is also used in electronic cash terminals, traffic lights and remote door openers.

The most common spectrum allocation is 1.88 GHz to 1.9 GHz used in Europe. This spectrum is unlicensed and technology exclusive, which ensures the operation of the standard with practically no interference, thus contributing to the very high spectral efficiency of the technology. The bands 1.9 GHz to 1.92 GHz and 1.91 GHz to 1.93 GHz are also very common in many countries outside Europe. In the United States the frequency allocation is 1920–1930 MHz, known as the UPCS band. In this case, the allocation is not technology exclusive, but is, in practice, safe enough to achieve similar interference-free operation.

Moreover, frequency allocations in Europe of 1.9–1.92 GHz (shared with UTRAN TDD), 1.92–1.98 GHz (shared with the uplink of UTRAN FDD) and 2.01–2.025 GHz have been foreseen by IMT-2000 for potential expansion of the standard, but actually they are not used yet.

DECT is suited to voice (including PSTN and VoIP telephony), data and networking applications with a range up to 500 metres. It dominates the cordless residential market and the enterprise PABX (private automatic branch exchange) market. DECT is also used in the wireless local loop (WLL) to replace copper in the 'last mile' for user premises.

DECT operates in the 1880–1900 MHz band and defines ten channels from 1881.792 MHz to 1897.344 MHz with a band guard of 1728 kHz. Each base station frame provides twelve duplex speech channels, with each time slot occupying any channel. The standard operates in the multicarrier/TDMA/TDD structure. DECT also provides a frequency-hopping spread spectrum over the TDMA/TDD structure. If frequency-hopping is avoided, each base station can provide up to 120 channels in the DECT spectrum before frequency reuse. Each timeslot can be assigned to a different channel in order to exploit advantages of frequency hopping and to avoid interference from other users in an asynchronous fashion.

The main technical characteristics of DECT include (European Telecommunications Standards Institute, 2015):

General

- Multicarrier TDMA technology with TDD (time division duplex)
- Frame time 10 ms
- Number of slots per frame: 24 (2 × 12)
- Modulation: GFSK, 4PSK, 8PSK, 16QAM and 64QAM
- Average transmission power: 10 mW (250 mW peak) in Europe, 4 mW (100 mW peak) in the US
- Seamless handover and full mobility management capabilities with authentication and ciphering.

Audio and telephony

- Supported audio codecs: G.726, G.711, G.722 (wideband), G.729.1 (wideband) and MPEG-4 ER LD AAC (wideband and super-wideband)
- Detailed audio specifications for narrowband (3.1 kHz) telephony and wideband (7 kHz) telephony
- Complete set of signalling and NWK procedures for PSTN/ISDN telephony, VoIP telephony (SIP or H.323), mixed scenarios (FP with PSTN and VoIP network connectivity), PABX environments, WLL (wireless local loop) applications and CTM (Cordless Terminal Mobility) applications.

Data

- DECT packet radio service (DPRS) providing packet data up to 840 kbit/s (GFSK modulation) or up to 5 Mbit/s (high-level modulation) with support for multibearer, asymmetric connections and efficient packet data handling
- Channel access time (from suspend state) = 15 ms (first bearer), 25 ms (additional bearers)
- Connection oriented and full mobility management procedures (similar to a cellular system)
- Simplified modes for Wireless LAN operation and low data rate applications
- Dual ARQ architecture with ARQ at MAC and at DLC layer
- Optional channel encoding (based on Turbo coding) for use with high-level modulation modes
- Additional data profiles for circuit mode transmission and for ISDN emulation.

Channel selection procedures

- Automatic frequency planning based on a distributed algorithm (dynamic channel allocation, or DCA) executed by the portable parts
- No need for any frequency planning at all: all DECT devices may access the whole DECT spectrum
- DECT DCA can be considered a preliminary implementation of what today is called 'cognitive radio'. It implements the paradigm 'spectrum sensing cognitive radio.

Miscellaneous

- Low power consumption due to technology architecture
- Low cost of DECT chipsets and radio parts due to mass production.

References

European Telecommunications Standards Institute (2015) DECT: Digital Enhanced Cordless Telecommunications. URL: http://www.etsi.org/technologies-clusters/technologies/dect.

IEEE 802.11 (2015) Wireless Local Area Networks Working Group. URL: http://www.ieee802.org/11.

IEEE 802.15 (2015) Wireless Personal Area Networks (WPAN) Working Group. URL: http://www.ieee802.org/15.

IEEE 802.16 (2015) Broadband Wireless Access Working Group. URL: http://www.ieee802.org/16.

European Telecommunications Standards Institute (2015) DECT: Digital Enhanced Cordless Telecommunications. URL: http://www.etsi.org/technologies-clusters/technologies/dect.

Tolstrup, M. (2011) *Indoor Radio Planning: A Practical Guide for GSM, DCS, UMTS, HSPA and LTE*, 2nd edition, John Wiley and Sons, Ltd, Chichester, ISBN 0-470-710-708.

Wireless Broadband Alliance (WBA) (2015) URL: http://www.wballiance.com.

Wi-Fi Alliance (WFA) (2015) URL: http://www.wi-fi.org.

3

System Requirements

An indoor wireless system is based on a set of requirements that should be met if the system is to deliver the required quality-of-service (QoS) for each of the offered services. These requirements, for example, could be specified in signal strength or path loss levels above or below a certain threshold, at various places inside and outside the building, which are taken as targets at which the design should aim for. The definition of such requirements is strongly related to the type of technology that is being deployed in the building and also depends on specific facility needs.

This chapter describes the definition of such requirements in a general context for any wireless communication system. Specific requirements depending on technology may vary and differences will be highlighted and explained.

3.1 Environments

The mobile propagation channel is particularly characterized by a strong influence of surrounding clutter and objects around transmitters and receivers in a wireless system. These obstructions mainly determine the way in which the signal propagates along a specific path, and therefore should not be underestimated. Furthermore, indoor scenarios have stronger effects as they are more densely populated by clutter, walls and partitions of different materials and open spaces such as windows, doors and atriums. Thus, it is very important to identify the characteristics of such obstructions and their distribution, in order to account for these changes and variations in field strength.

Indeed, indoor wireless design is very building-specific; that is each building has unique characteristics that make radio propagation different, and therefore the expected performance of the system varies. Radio propagation determines to a great extent the characteristics of the wireless system that will be deployed in terms of coverage, interference and leakage, but there are other performance requirements that can also be affected by the type of buildings and facilities, such as capacity, or the number of wireless resources/channels required to satisfy users' demands.

Despite this, it is possible to make a classification of buildings based on the type of facility the design is aimed for, and therefore premises can be categorized based on this criterion. For our purpose, these categories are to be called *environments*, and buildings that can fall within each category will be explained in each particular section. At this stage, it is important to point out that this classification is mainly based on the specific

Indoor Wireless Communications: From Theory to Implementation, First Edition. Alejandro Aragón-Zavala.
© 2017 John Wiley & Sons Ltd. Published 2017 by John Wiley & Sons Ltd.

building characteristics that affect radio requirements and therefore make a type of building fall within one or another category.

3.1.1 Corporate Buildings

Corporate buildings are facilities that are normally used by enterprises to have their headquarters or branches across a country. They include offices, meeting rooms, lounges, warehouses and, for very specific firms, labs and leisure venues. They vary in size from single-floor to multistorey buildings, as well as in land extension. Figure 3.1 shows examples of these buildings.

Corporate buildings often have *multiple floors*, which make the need to provide coverage in upper and lower floors using the minimum number of resources, to minimize cost. There are also radio propagation characteristics associated with multiple floors that need to be taken into account for leakage outside the building as well as penetration from outdoor wireless systems.

In some cases, a *uniform layout of floors* can be observed in many types of corporate buildings, which facilitates the design of the wireless coverage solution as propagation effects can be assumed to be the same for similar floors. This is normally done in the design of cellular systems, especially for building towers having a dozen floors.

Corporate building blocks are normally made of a wide variety of materials, but in most cases it is concrete and brick for the external walls and plasterboard and similar construction materials for the indoor partitions. Metallized windows, which do not allow macrocell penetration inside the building, are also common; this has the benefit of

Figure 3.1 Corporate buildings for various enterprises showing a variety of architectural designs.

Figure 3.2 *Technology Park* office building, Tecnológico de Monterrey Campus Querétaro, Mexico.

providing better levels of isolation between the indoor and the outdoor systems, as it is the case for cellular networks (Tolstrup, 2011). Nevertheless, leakage levels should be carefully controlled outside the building, for security reasons.

Some modern buildings exhibit a peculiar architecture, being surrounded by windows and for which the area dedicated to walls tends to be quite reduced, as shown in Figure 3.2. Most of these windows are made of materials that attenuate the radio signal strength and hence provide some form of 'isolation'. However, this is not the case for all scenarios and there are occasions in which isolation is very hard to achieve between an indoor dedicated cell and an outdoor cell. We will examine various techniques and methods later on in this book to account for these issues.

It is common for many wireless services to coexist in corporate buildings, for example wireless local area networks, cellular services, VMTS, VoWLAN, mobile computing, paging, two-way radio and fire/life/safety systems. Also, it has been a recent tendency for some companies to prefer mobile phone usage rather than fixed telephone lines amongst their employees. Thus, high wireless connectivity to enhance staff productivity is pursued, with the use of many wireless devices (laptop, PDAs, mobile phones, wireless printers, etc.). This has also pushed wireless network providers and operators to offer attractive cost tariffs for enterprises willing to replace the use of fixed telephone lines by mobile telephones.

3.1.2 Airports

An *airport* is a location where aircraft, such as commercial and private airplanes, helicopters and airships (free-flying aircrafts such as a Zeppelin) take off and land, according to modern definitions. Throughout the years, airport design and layout has evolved drastically, providing more comfort and facilities to airport staff and passengers.

Figure 3.3 Airports are large and busy facilities with a large variety of spaces that are relevant to wireless coverage: (a) airport access; (b) departures hall in the terminal building; (c) baggage reclaim area; (d) airport piers; (e) arrivals hall; (f) transfer and waiting rooms.

This includes wireless connectivity and access to the Internet, 'wherever you are and whenever you want'. It is therefore not surprising that an airport is one of the most important ports of entry for wireless operators and service providers, and thus special care should be taken when providing indoor wireless coverage there.

Although airports across the world vary in size and passenger traffic, most consist of runways, hangars and terminal buildings, as illustrated in Figure 3.3. Architecture and style is also very varied across the world, ranging from very simple one-terminal small airports to very large facilities, with many terminals and multistorey buildings, tunnels and even underground stations inside the airport. Large airports have also included dedicated areas for passengers, such as: restaurant and cafe areas, VIP rooms, waiting lounges, casinos, duty-free shops, hotels, etc.

Airports are mainly characterized for being highly changeable environments, where entire areas and zones are often refurbished and/or removed to relocate lounges, shops or open areas. This is especially applicable to large airports, such as Heathrow in London, UK; Charles de Gaulle in Paris, France; Schiphol in Amsterdam, Netherlands; Tokyo International in Tokyo, Japan; O'Hare International in Chicago, IL, USA; Hartsfield-Jackson in Atlanta, GA, USA; just to mention some of the busiest airports in the world.

Wireless coverage inside airports is a very important issue for operators, service providers, airport owners and authorities. Many passengers use the airport every day with various specific demands for wireless services, which span from getting in touch with friends and family; downloading music or video streams while waiting for their flight to depart or someone to arrive; working remotely and connecting their mobile devices to remote servers and accessing the Internet; and downloading and uploading emails. In addition to this, airport staff also make use of wireless services, such as: private mobile radio (PMR) and emergency services, and radio frequency identification (RFID) for baggage handling and tagging, amongst others. The fact that many wireless technologies are thus present within the airport imposes a further challenge for indoor wireless designers, since interference needs to be minimized and, in some cases, infrastructure should be reused whenever possible, which makes multioperator and multitechnology solutions more widespread nowadays; for example, many cellular operators sharing the same distribution system, Wi-Fi and cellular sharing antennas, etc. Interoperability should be guaranteed amongst the various wireless technologies (cellular, radar, public safety, PMR, Wi-Fi, etc.)

Airport terminal buildings have a large variety of spaces and areas where coverage is required, as described above; these places are normally referred to as *hot spots* since traffic is more likely to congregate there. Examples of this include all waiting lounges, food court, immigration, arrivals and departures halls. The characteristics of some airport terminal buildings, which include open spaces and large atriums, make propagation predictions challenging.

In particular, airport piers are a special case for indoor wireless design, since they are normally long corridors for which *waveguiding* effects (which will be addressed in detail in Chapter 4) are observed (Saunders and Aragón-Zavala, 2007), similar to those seen in tunnels. Not only are propagation characteristics peculiar here but also *bursty* traffic is observed, since passengers arrive in bursts. More specifically, for cellular systems, arriving passengers normally switch on their mobile phones as soon as they land and operators are keen to catch as many roamers as possible by having the best signal strength and quality in the airport piers.

To provide cellular coverage to airports, a common technique is to rely on *macrocell penetration* from surrounding outdoor base stations (Tolstrup, 2011), especially if airport size and traffic is reasonably small. If this technique is used, special care should be taken to assess coverage levels inside and outside the airport to control handovers efficiently, especially for cellular 3G networks where soft handover could make excessive use of resources if handover limits are not carefully chosen and controlled. For larger airports with higher traffic demands, dedicated picocells are often employed. For other wireless technologies, dedicated base stations are used. Optical fibre links are also widely used for data networking inside airports, and sometimes radio-over-fibre technologies are used to transport radio signals over long distances, using the existing fibre infrastructure.

In addition to terminal buildings and piers, wireless coverage is also required in car parks and surrounding areas and locations near the airport, as passengers and airport users require seamless coverage for personal and professional use. This includes emergency and safety systems, for which very high reliability is needed.

In terms of capacity dimensioning, various traffic user profiles are present in an airport for which various traffic demands can be identified: arriving passengers, departing passengers, visitors, airport staff, on-transit passengers (Aragón-Zavala *et al.*, 2009). This implies special techniques for capacity dimensioning, which also depend on the areas where people are more likely to congregate, and thus network resources need to be distributed accordingly. Moreover, broadband demands in airport hot spots make data traffic so high that nowadays it is common to perform data offloading to other wireless networks such as Wi-Fi, to avoid congestion on 3G and LTE networks.

Finally, the exclusion of wireless coverage in runways for safety reasons has been a topic for debate for a long time. Since it is claimed that wireless coverage should be avoided on the runways to minimize interference with aircraft electronic navigation equipment, airport authorities are pushing to ban wireless coverage in these areas. On the other hand, some operators argue that coverage is desired as soon as the aircraft lands and it is there inside the aircraft and even before getting off the plane where passengers switch on their mobiles phones – a fair amount of roamers can be caught here with proper cellular coverage design!

3.1.3 Trains and Railway Stations

A *railway station* is a train facility where trains regularly stop to load or unload passengers or freight, such as the ones shown in Figure 3.4. It consists of a platform next to the tracks and a building providing related services such as ticket sales and

(a)   (b)

Figure 3.4 Examples of railway stations: (a) terminus; (b) halt.

waiting rooms. Connections may be available to intersecting rail lines or other transport modes such as buses or the underground (tube).

In European countries, trains are one of the most popular choices for passengers travelling from their homes to their workplaces (commuters) or tourists travelling from one city to another, sometimes in different countries. In most of the world, cargo trains are employed to transport a wide variety of items. Either way, train stations have specific needs for wireless coverage. For instance, terminus and halts (small stations) need a different design approach, since they vary in size and passenger demands; waiting rooms and lounges may require sufficient coverage for wireless Internet services for passengers; coverage is desired in both terminals and platforms; and hot spots of traffic for large stations are also identified, which need to be properly addressed. Modern trains offer on-board wireless connectivity, which allows passengers to access Internet and other broadband services on the move. The main challenge here is not only the provision of coverage inside the train, but the backhaul connection to another network that guarantees a reliable connectivity during the entire journey.

As far as radio propagation is concerned, train stations are commonly ancient buildings, with thick wall materials that produce high propagation losses. More modern stations are made of less-attenuating materials, which allow a signal to propagate better. Underground platforms are a challenge since radio signals cannot penetrate walls to reach these areas and therefore dedicated base stations are required. Large open spaces and atriums also make train stations a challenging environment for accurate propagation predictions. In terms of propagation along the train coaches, this is normally confined to multiple reflections and some absorption in the seating areas.

In recent times, the provision of coverage on board trains has become more popular, giving commuters the facility to have wireless connectivity whilst travelling to work or coming back home. Most rail companies in Europe provide this service to their customers and thus special requirements should be specified for on-board passengers, some of which may clearly differ if at the terminal. Figure 3.5 shows an example of the interior of a European train coach.

Figure 3.5 Train coach for which on-board coverage is provided.

Figure 3.6 Typical contemporary shopping mall.

3.1.4 Shopping Centres

Since the development of suburban shopping centres in the United States in the 1950s, shopping malls have become the most popular type of retail venues throughout the world. The first malls were open as air retail centres, but very quickly a need to shelter shops and customers from the elements was recognized, and the concept of an enclosed shopping centre was born. An example of a contemporary shopping mall is shown in Figure 3.6.

A *shopping centre* or *mall* is one or more buildings forming a complex of shops representing merchandisers, with interconnecting walkways enabling visitors to easily walk from unit to unit, along with a car park area, as seen in Figure 3.7.

Some shopping malls have large exterior windows that provide plenty of natural light. For malls that have solid external walls, it is a glass-covered opening in the roof also known as a *soonroof* that provides the natural light. Some shopping malls have a multilevel parking structure attached to it. However, most malls only have an open air surface carpark next to the mall that provides no cover from the elements. Inside an enclosed mall, small retail shops are lined against the perimeter while leaving the middle area open to pedestrian traffic. An example of this 'open floor plan' interior mall architecture is shown in Figure 3.8.

There is not a great deal of difference in terms of construction materials and building layout of a shopping centre and other public facilities such as airport terminal buildings

(a)   (b)

Figure 3.7 Examples of shopping centres: (a) modern mall; (b) marketplace.

Figure 3.8 Inside view of a shopping centre.

or train stations. Perhaps one of the most distinctive features of a shopping centre is the fact that requirements can be drastically different to those for airports or train stations. For example, some areas and shops in the mall may have restricted access and even coverage should be kept to a minimum level. Antennas should be hid as much as possible and aesthetics becomes a high priority. Special large chain stores (departmental stores) may have specific coverage requirements for their customers, offering a range of wireless services inside their stores. A food court is especially important for the deployment of various wireless services to customers, with different user profiles and characteristics to those of airport or train passengers.

Maybe it is worth mentioning that there are a couple of characteristics that should not be overseen for shopping centres. Modern malls are multistorey buildings, having large atriums and open spaces combined with small rooms, where the shops are located, and at least a couple of large spaces per floor for departmental stores.

Car par coverage is especially of relevance, and in most of the cases it is quite difficult to achieve if dedicated base stations or antennas are not used, as car parks are often found in underground locations. Multistorey car parks can also be found, for which case wireless coverage is often achieved by surrounding outdoor base stations.

3.1.5 Hospitals

A *hospital* is an institution for health care providing patient treatment by specialized staff and equipment, and often, but not always, providing for longer-term patient stays. This definition has evolved over the years, as in ancient times a hospital was 'a place for hospitality'.

Modern hospital buildings are designed to minimize the effort of medical personnel and the possibility of contamination while maximizing the efficiency of the whole system. Travel time for personnel within the hospital and the transportation of patients between units is facilitated and minimized. The building also should be built to accommodate heavy departments such as radiology and operating rooms while space for special wiring, plumbing and waste disposal must be allowed for in the design.

Although these characteristics are desired for most hospitals, the reality is that many hospitals are the product of continual and often badly managed growth over decades or even centuries, with utilitarian new sections added on as needs and finances dictate. This obviously has an impact on wireless systems design and dimensioning, often pushing the wireless infrastructure to the limit to accommodate these new additions.

Hospitals have distinctive areas with different coverage requirements, which are summarized here:

- Isolation areas may be required for some critical zones, especially those containing specialized medical electronic equipment.
- Patient rooms and wards, where visitors are welcome, should have sufficient coverage to allow patients and family to communicate effectively.
- Interference is a major concern for hospitals and good isolation is a high priority.

In terms of equipment, short-range wireless devices using technologies such as Bluetooth may be used to interconnect devices in patients to monitor vital signs and other variables. They can also be used for interconnectivity between caregivers, consultants, specialists and hospital staff within the facility.

An important factor that should not be underestimated is that of *scalability*, to activate new wireless services quickly and extend services to new buildings is needed, including a high-mobility profile for users and caregivers, with a wireless services portfolio wherever they go.

Hospitals are made of a variety of building materials, basically depending of the specific architecture and style; for example, there is a clear difference in construction materials between a modern hospital building and an ancient facility. For the former, plasterboard and concrete walls are often employed and in some cases brick walls are used in various areas of the hospital. Multistorey hospital buildings are commonly found, with floors having a very similar layout. On the other hand, for the latter, old buildings are made of stone and high loss materials, with very thick walls and not many windows or doors allowing signals to penetrate the building.

3.1.6 Arenas and Stadiums

An *arena* is an enclosed area, often circular or oval-shaped, designed to showcase theatre, musical performances or sporting events. It is composed of a large open space surrounded on most or all sides by tiered seating for spectators. The key feature of an arena is that the event space is the lowest point, allowing for maximum visibility. Usually, an arena is designed to accommodate a fairly large number of spectators, as shown in Figure 3.9(a).

A *stadium* is a place, or venue, for mostly outdoor sports, concerts or other events, consisting of a field or stage partly or completely surrounded by a structure designed to allow spectators to stand or sit and view the event. See Figure 3.9(b) for an example of a football stadium.

Sports such as ice hockey, basketball, volleyball, gymnastics, boxing or martial arts are normally played in an arena; other sports such as football, baseball, rugby or cricket are played in a stadium. Music concerts and other events can take place in either venue.

| (a) | (b) |

Figure 3.9 Two types of sports venue: (a) sports arena; (b) football stadium.

For both cases, traffic is limited to the schedule of the events that will take place in the venue; otherwise it is rarely used at its full capacity. For example, in the early days of cellular systems, it was not rare to see deployments of temporary cellular sites in stadiums, to account for the very high peaks in traffic that could overload the serving macrocells. This strategy is still in use nowadays in many parts of the world, although others have decided to have dedicated coverage and resources at all times.

Stadiums are characterized by having very large open spaces, which could even be classified as outdoor and therefore arguably a stadium could be treated as an indoor venue. However, due to the nature of the construction, behind the spectators' area, there are areas in which people congregate in half-term periods or before the match starts, which are indeed part of a building and therefore indoor propagation principles apply. In the worst case scenario, wireless coverage to the spectators' area and seating can be classified as leakage from the inside part of the stadium. Finally, heavy concrete and steel construction used to build the facility produces high radio signals loss and therefore antennas should be carefully placed to guarantee sufficient coverage in all areas.

Arenas have clearly differences in terms of propagation characteristics to stadiums. First, arenas are enclosed spaces with a similar layout for spectators' areas, surrounding the large open space where events take place. Reflections can occur in the ceiling and therefore radio signals propagate better. Second, arenas are normally smaller than stadiums, and therefore less power is required from a base station to provide sufficient coverage.

User profiles in stadiums and arenas are quite distinctive from the other indoor venues we have discussed so far. A spectator will normally be paying close attention to the event rather than using his or her wireless device for long periods of time. However, there might be occasions in which a replay of the last minute event in the match may be required, or getting in touch with someone else to inform what the score is can be a priority. Also, users are not mobile most of the time, except for those times where people can go for a break in the event or to restrooms. The major issue here is to do with the very high volume of users needing to access network resources at the same time. Hence, there is a strong requirement for careful and efficient capacity dimensioning.

For arenas and stadiums, there is a need for Wi-Fi coverage for staff interconnectivity and wireless distribution for a multicast system providing supporters with a unique multimedia service for live game replays, highlights and statistics. Thus, multiple wireless voice and data operators and services are required.

Due to the fairly large distances inside these venues, a distribution system capable of accounting for large distances should be employed to distribute the signals inside the building appropriately. This should be a discrete, robust and durable system.

3.1.6.1 What Makes a Stadium so Special?

Stadiums are venues that, unlike other in-building scenarios, have particular special needs and challenges that any design engineer should know and understand in order to create a suitable design. The following aims at presenting an overview of what these challenges are that make a stadium 'so special'.

3.1.6.2 Mix of Communities with Different Needs

People with different backgrounds, interests, expectations, duties, etc., form the community in a stadium that has a need to have sufficient wireless connectivity and quality-of-service. However, the way 'sufficient' is achieved may vary significantly from one type of user to another, in various dimensions such as: coverage area, connection speed, service availability, grade of service necessities, etc. A few examples are now given to illustrate this better.

- **Visitors** come to the stadium for either an event, a match or simply as tourists wandering around the stadium. They normally make use of seating area spaces or, in the case of concerts, use of the playfield. During half time periods or breaks, they might access food and hospitality areas, bars and toilets. Often, some of them may want to share the latest goal with their friends or relatives who could not go to the stadium, and make use of their mobile devices to send streaming data, pictures or SMSs. VIP visitors have designated areas in rooms that have to be covered, as it is a 'very important area to be covered'. Finally, some visitors may want to get access to the Internet while in the stadium, so sufficient coverage and capacity should be available for this. Not only voice but also heavy data usage is anticipated for mostly all types of visitors coming to the stadium.
- **Stadium operations** refer to the personnel who work for the stadium on a daily basis, needing to have coverage and connectivity especially inside their offices. While people in front of desks may require Internet connection and perhaps cellular coverage, those who work for maintenance may need to have their PMR service, and perhaps security guards may need to have coverage in larger areas than those for the other personnel.
- **Event personnel** refers to the people who work for the specific event that is taking place at the stadium. For example, if it is a football match, the players may need coverage and connectivity in the dressing rooms, located normally in the basement. On the other hand, coaches and managers need coverage in the playfield, as well as the referees. If it is a music concert, the coverage requirements change drastically, and it might be that the bowl is the most important area for which coverage should be guaranteed.
- **Press and media** clearly require a press and conference room with enough coverage and capacity for press conferences, as well as connectivity in the playfield to cover the event. Heavy data usage is expected here, as well as voice.
- **Security** refers to police services that patrol the event, having strict coverage requirements with high service availability. An example of this is TETRA in Europe,

a private network used by the police. The areas where coverage is needed for security include even the most unexpected spaces in the stadium.

- **Health and emergency** services also require strict coverage requirements to meet their specific demands, especially in the event of an emergency.

Stadiums are challenging venues for which special radio design techniques are required. First of all, a mix of playfield area along with the seating and concession areas represent challenging environments for the prediction of radio waves. Inclined surfaces on these areas have special effects on the way radio signals propagate at different frequencies, and for which innovative models are required. In addition, stadium materials also make the penetration of signals difficult, and since there is a large mix of these materials in the venue, propagation mechanisms are expected to behave differently in each area – corridors, open space areas, enclosed spaces surrounded by heavy penetration loss walls, etc. Heavy duty construction, for example fire-resistant stairwells, numerous lift shafts, mix of very old and brand new buildings, and different radio requirements for non-event and event days, for example crowd management radios, are examples of other construction challenges that have an effect on radio design. Structural and architectural features, which make stadiums so special and distinct, also represent challenges in order to predict radio waves. Trade-offs between coverage in designated areas (e.g. sectors) and signal spillage to neighbouring ones (leakage) have to be made. In summary, demanding coverage requirements in areas such as VIP rooms, basement (dressing rooms) and car parks also require accurate radio planning and modelling, in order to create both a cost-effective and working solution.

There is a very relevant aspect that should not be overseen: interference. Around the stadium, surrounding macrocells provide cellular coverage and some of this signal penetrates into the premises. If interference management is not considered, a degradation in the signal quality is expected, which will be much more critical for 3G systems. On the other hand, interference between wireless technologies should also be taken into account. Thus, these interference issues should be considered and resolved, especially for broadcasters using wireless microphones, for example, of Wi-Fi networks operating in the stadium. The breadth of different systems and requirements to be deployed makes stadiums particularly very special venues in terms of radio considerations.

Finally, a large variety of wireless technologies and systems is often deployed in a stadium, such as cellular, broadcasting, Wi-Fi. TETRA, etc., all with different requirements and that have to coexist and minimize interference amongst them. Site sharing is a common practice for some stadiums, which must have special considerations. On the business side, if the radio coverage is not satisfactory according to acceptance criteria for some entities, especially security and emergency services, the opening and operation of the stadium can be severely compromised.

3.1.7 University Campuses

Medium to large university campuses often have multiple buildings that are somehow interconnected for different purposes. Perhaps one of the most common types of connectivity is for Wi-Fi, where highly directional antennas are used as *bridges* to communicate two campus buildings. Wireless coverage is therefore highly demanded in many spaces and areas around the campus, as depicted in Figure 3.10.

Figure 3.10 University campus areas where wireless coverage is often required: (a) classrooms; (b) conference rooms; (c) administrative offices; (d) labs; (e) computer rooms; (f) library.

Universities are characterized by a high variability of building layouts, structures and materials, thus having multiple propagation scenarios that need to be accounted for. For ancient universities, historic concrete structures are typically found, normally having thick walls. More modern universities use less attenuating materials and radio signals can penetrate much better.

It is often a requirement to allow coverage inside and outside university campus buildings for students, visitors and staff to access wireless services in the vicinity of the campus and be able to handover to dedicated cells (if it is the case) while inside the campus. Wireless coverage is desired in classrooms, labs, faculty offices, restaurants, dormitories and in outdoor areas inside the campus.

On the other hand, in many universities there are joint university–wireless providers' educational projects, where wireless devices can be used as part of university courses to assist students in their lectures and assignments. This fact, in conjunction with heavy data usage and high mobility, produce higher data rate demands (growing bandwidth requirements) with newer challenges in terms of quality-of-service to maintain those services.

Finally, university authorities normally ask wireless operators to provide mobile services without impacting the visual or historical integrity of the environment, something that has a great impact of the choice of antennas, distribution systems and system architectures.

3.1.8 Underground Stations

Train and underground stations have similar layouts: train platforms and a terminal building where tickets are issued, including waiting rooms, shops and restaurants. However, there is a major difference: for train stations, platforms are normally located at the same ground level as the terminal building, whereas for underground stations, platforms are always in tunnels well below the ground, as seen in Figure 3.11.

Underground tunnels exhibit very peculiar propagation behaviour and sometimes signal strength predictions using mathematical models can be challenging (Saunders and Aragón-Zavala, 2007). In addition, external base station penetration is very difficult and dedicated transmitters should be used to provide sufficient coverage, as is the case for cellular systems. For other technologies this is not such a problem as base stations are often located in close proximity to users; for example Wi-Fi access points.

Traffic behaviour tends to be bursty in the platforms, as it also is the case of train stations. There is a major constraint for the provision of wireless coverage in the underground tunnels: power sources to supply the electronic distribution or radio equipment are scarce and therefore alternative methods, such as the use of radiating cables or leaky feeders, are employed. Also, it is likely that interference with the operation of highly sensitive electronic equipment can be produced and therefore special care should be taken for cable or antenna placement in the tunnels. CCTV monitoring and security cameras could also potentially interfere with cables for wireless signal distribution.

In addition, special characteristics of underground environments such as curves, changes in elevation, humidity and surface texture of a tunnel wall can be non-RF friendly for radio propagation.

(a)   (b)

Figure 3.11 Underground stations critical zones: (a) platforms; (b) transit areas.

For maintenance purposes, there are space limitations as subways and tunnels can be quite crowded and access to them is limited; that is it can be difficult to install and maintain radio infrastructure there. Dirt and dust present in these environments can impose further challenges; thus radio equipment should be properly enclosed.

Wireless service providers (WSPs) often deliver multiple services across multiple licensed RF bands in underground stations, with the possibility of sharing a common radio distribution infrastructure to minimize costs and space. Thus, wireless systems installed there should have sufficient frequency and protocol flexibility.

3.1.9 Cinemas and Theatres

Also known as *movie theatres*, these are commercial operations catering to the general public, who attend to watch a film or movie. These are environments that can potentially congregate people who normally switch their devices off while the show, play or film is on, but who, on the other hand, use their wireless devices while in halls, lounges or other spaces within the theatre.

For cinemas or theatres, coverage is desired in lounges, toilets and waiting rooms, but should be excluded in the main auditorium; therefore, good isolation should exist between the auditorium and the other spaces.

Materials and layouts vary from country to country; some are ancient buildings with concrete and stone walls, producing high propagation losses and blocking radio signals; others are more modern and constructed of lighter materials that allow RF signals to penetrate.

As wireless technology develops, there might be the possibility in the near future that viewers may access remote locations with the use of a portable terminal (for example, to monitor their house via TV cameras).

Figure 3.12 shows typical areas for which wireless coverage is desired in a cinema.

3.1.10 Hotels

Hotels are a very important type of building for any wireless service provider and hotel owner, as connectivity is offered to guests during their stay as part of the hotel facilities. The most popular wireless services offered to guests are Wi-Fi for wireless Internet and cellular. Many hotels also have RFID tags for access control, car parks or guest identification.

(a)

(b)

Figure 3.12 Hot spot areas in a cinema.

For hospitality venues, reliable wireless coverage throughout the facility is required for guest convenience and staff productivity, making sure key areas are properly covered; for example lifts, guest rooms, lounges, offices, hallways, restaurants and meeting rooms, which are the areas where it is more likely that guests will use their wireless devices. High reliability should be guaranteed, as a bad wireless service can drastically impact the guest's impression for 'service quality' (not necessarily related to the wireless service provider but to the hotel!).

Also, special care should be taken as the indoor wireless solution must be aesthetically pleasing and non-disruptive to the guest's experience.

Leakage should be controlled to contain coverage inside the hotel, for security reasons as well as to minimize interference to external cells. Isolation between indoor and outdoor cells should be properly accounted for, especially in floors well above ground level, where macrocell penetration tends to dominate over a potential installed indoor cell.

Multioperator and multitechnology solution for all types of guests using all sorts of wireless network providers is nowadays becoming the preferred choice for hotels; give guests 'whatever they want whenever they want it' for wireless connectivity. This implies very good frequency planning as well as shared infrastructure negotiations.

For propagation considerations, the layout and materials for each floor are very similar, especially those where guest rooms are located. This makes propagation predictions and measurements heavily simplified.

An example of various views for different areas in a hotel is shown in Figure 3.13.

(a)

(b)

(c)

(d)

Figure 3.13 Relevant coverage areas for a hotel: (a) rooms; (b) toilets; (c) lobby; (d) restaurants.

3.1.11 Cruise Ships

A cruise chip can be defined as a 'moving building'; it is a transport system used by tourists where wireless services need to be available anywhere inside the ship, where wireless should work wherever passengers expect it to work. Another major issue is that passengers often want to stay in touch with family or friends and hence wireless connectivity becomes a major requirement.

There are a few important considerations in terms of requirements that need to be taken into account for indoor coverage on board cruise ships. Perhaps one of the most significant is that there is no equivalent land terrestrial radio infrastructure for the sea and thus all possible communications are established via satellite links. Therefore, there must be a way to distribute satellite-based signals inside the ship efficiently.

The usage and operation of mobile devices on board a cruise ship may be subject to strict regulations. A very good example is cellular mobile phone reception, which normally is to be restricted in zones designated for relaxation and quietness for passengers. Indoor wireless designers should take this into consideration. Later on, we will see that these areas are referred as 'exclusion' zones.

The materials of which cruise ships are made are somehow a big impairment for the propagation of radio signals. Vessels are constructed of dense, heavy metal materials that highly attenuate wireless signals. Thus, sufficient resources should be dedicated in terms of equipment (antennas, access points, etc.) to provide the required QoS. However, infrastructure installation and maintenance need to be simple, so as not to affect cruise schedules or displace passengers. Figure 3.14 shows the interior of a cruise ship showing various areas where wireless coverage is desired and others regarded as exclusion areas.

3.2 Coverage

Coverage is defined as the amount of signal strength level that should be present in a wireless system to meet a certain QoS for a particular service. This is applicable for wireless systems, since coverage could have different meanings for other technologies. For example, cellular operators often define indoor coverage requirements as a minimum signal strength level within at least a percentile of an area of the building. This area is normally termed the *coverage requirement area*, as seen in Figure 3.15.

In general, coverage is one of the most important requirements for any wireless system. Without sufficient coverage, access to the radio services cannot be achieved, or if levels are not within specified minimum limits, the quality can be seriously degraded.

As seen in Figure 3.15, there might be some areas within the building in which coverage is not desired and should be excluded. These areas are defined as *exclusion* areas, and the design may or may not attempt to minimize the signal strength there. This is especially true, for example, in the case of prisons, where coverage should be minimized inside the building, or for test laboratories, which need to be isolated from external electromagnetic interference.

3.2.1 Cellular

For cellular systems, coverage requirements are normally specified in terms of a minimum level over a certain percentile in a specified area within a building. Each

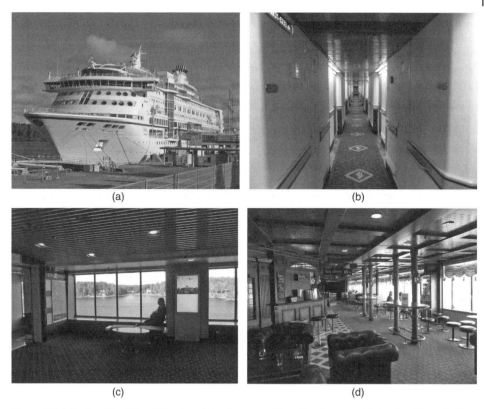

Figure 3.14 Cruise ship: (a) outside view; (b) cabin area; (c) resting lounges; (d) food and hospitality.

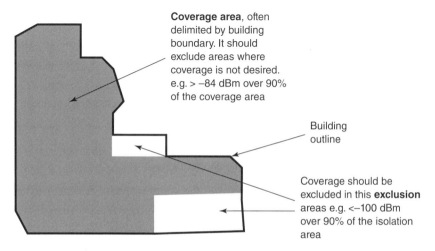

Figure 3.15 Indoor coverage area concept.

specific standard specifies minimum requirements according to well-established recommendations. The way in which the requirement is specified varies from standard to standard, but, overall, all refer to a minimum signal level that must be met in the desired area. Operators normally take these recommendations and build up from those figures to meet their specific coverage needs.

For GSM systems, the RxLev minimum coverage threshold for in-building systems is specified as −85 dBm or 5 to 7 dB greater than the residual macropilot, whichever is higher. Mobile operators often specify this coverage threshold over a certain percentile area; for example 90%.

For W-CDMA or 3G systems, this requirement is specified in terms of pilot strength/ CPICH RSCP greater than −85 dBm or 5–7 dB greater than the residual macropilot, whichever is higher. This is turn is converted by system designers in a minimum path loss to meet from transmit to receive antennas, over a certain percentile of coverage area.

For an LTE wireless systems, target design values for indoor systems are in terms of the RSRP (reference signal receive power), which should be greater than −100 dBm for a 5 MHz channel if the residual macrosignal is very low. It should be 5–7 dB higher if the residual macrosignal is high.

3.2.2 Wi-Fi

Wireless local area networks have slightly different coverage requirements than those specified in cellular systems. According to Geier (2001), coverage is specified in terms of the signal-to-interference and noise ratio (SINR) rather than a signal strength threshold. For example, a signal level of −53 dBm measured near an access point and a typical noise plus interference level of −90 dBm yields an SINR of 37 dB, a reasonable and 'healthy' value for wireless LANs.

An SINR directly impacts the performance of a wireless LAN connection. A higher SINR value means that the signal strength is stronger in relation to the noise levels, which allows higher data rates and fewer retransmissions – all of which offers better throughput and therefore faster connections. On the other hand, a lower SINR requires wireless LAN devices to operate at lower data rates, which decreases throughput. An SINR of 30 dB, for example, may allow an 802.11g client radio and access point to communicate at 24 Mbps, whereas an SINR of 15 dB may only provide for 6 Mbps.

From measurements reported in Geier (2008), user-oriented tests were conducted to determine the impacts of SINR values on the ability for a user with a typical client radio (set to 30 mW) to associate with an 802.11b/g access point and load a particular webpage. The following is reported for SINR:

- Greater than 40 dB = excellent signal (5 bars); always associated; fastest connection.
- 25 dB to 40 dB = very good signal (3–4 bars); always associated; very fast.
- 15 dB to 25 dB = low signal (2 bars); always associated; usually fast.
- 10 dB to 15 dB = very low signal (1 bar); mostly associated; mostly slow.
- 5 dB to 10 dB = no signal; not associated; no go.

Target SINRs are therefore specified by each specific WISP (wireless Internet service provider) when deploying large Wi-Fi installations, and so, based on this, the number of access points required to cover a certain area can be specified. For the home, access points are deployed more empirically, without taking the SINR values too much into consideration – as long as there is enough signal in the desired coverage areas this is enough!

There is an important limiting factor that should be taken into consideration when designing for coverage for Wi-Fi: Access points have a maximum allowed transmit power, which has been defined by the standards and varies depending on the specific standard and region. These regulations are specified in terms of EIRP (effective isotropic radiated power), which take into account cable losses and antenna gains. Table 2.3 shows EIRP maximum levels for 802.11b/g and 802.11a standards. Special care should be taken here when choosing antennas for the APs, since according to the EIRP definition, higher antenna gain would be proportionally converted in an increase of the EIRP. This is the case, for example, when coverage is to be enhanced, and highly directional antennas (such as Yagis) are used for 'bridging' two buildings.

Section 2.2 summarizes some system requirements and characteristics that are relevant to Wi-Fi and which should be considered for designing a Wi-Fi system.

3.2.3 Wireless Personal Area Networks (WPAN)

Technologies such as Bluetooth, ZigBee, IrDA (Infrared Data Association), wireless USB, Z-Wave and BAN (body area network) fall within this category in which the term 'coverage' is more referred to the capability of a device being within the range of another one willing to establish communications (or associate with). In other words, coverage requirements are more specified in terms of receiver sensitivity of the devices themselves, which is clearly regulated by the standard. Some examples for this are given below.

- For Bluetooth, a receiver sensitivity of −70 dBm or better is required (IEEE 802.16, 2015).
- For ZigBee, the IEEE 802.15.4 standard (2003) sets the requirements for the −85 dBm one percent packet ratio (PER) minimum sensitivity. In practice, and according to Le (2010), sensitivity depends of a minimum SNR that guarantees the target error rate as well as a given receiver noise figure. Table 3.1 shows typical sensitivity data for the different data rates specified there.

Thus, in order to meet such received power demands, these devices should have transmitters capable of providing enough power, according to the link budget calculations and the type of propagation environment, as will be discussed later in this book.

Table 3.1 Typical ZigBee receiver sensitivity performance.

Data rate (kbps)	SNR (dB)	Sensitivity (dBm)	Noise figure (dB)
250	3	−99	11.5
500	6	−96	11.5
1000	9	−93	11.5
2000	12	−90	11.5

3.3 Isolation

In the context of indoor wireless communications, *isolation* is referred to as the ability of a system to divide or isolate different areas of a wireless network. For example, for cellular systems, isolation is a property that is related to how well the outdoor cell is maintained isolated from the indoor cell, to avoid interference or excessively frequent soft handovers, for 3G and LTE systems. An example of isolation from the indoor cell to the outdoor macrocell in a stadium is shown in Figure 3.16. The really hard issue here is to guarantee that the coverage from the indoor cell is dominant *inside* the stadium. Sometimes this can also be controlled by adjusting the macroantennas downtilt as well as the transmit power.

In RF terms, isolation is related to how well a port is isolated from another port in a device. For example, the ratio (in dB) of the power level applied at one port of a mixer to the resulting power level at the same frequency appearing at another port is treated as isolation, and can be seen as an attenuation. For multitechnology or multioperator systems, co-located antennas that operate at various frequency bands should have some degree of isolation to guarantee that there is not adjacent channel interference that could degrade system performance.

The requirements for isolation vary from system to system and have a stronger influence on cellular systems, specifically 3G. Users in office buildings, for example, are typically close to windows; thus the dominance of the indoor system must be maintained throughout the building, even right next to the windows (Tolstrup, 2011). Also, for Wi-Fi systems, isolation is important as the indoor network should be prevented from hackers willing to access it from outside.

For GSM systems, a level of isolation of at least 11 dB is required, plus a margin of around 6–10 dB to account for fading (Tolstrup, 2011). For UMTS, a good guideline is to design the indoor cell to be 10–15 dB more powerful inside the building than any outside macrocell signal, just being careful not to leak excessive signal outside the building, as discussed in Section 2.4.

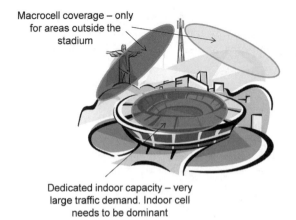

Macrocell coverage – only for areas outside the stadium

Dedicated indoor capacity – very large traffic demand. Indoor cell needs to be dominant

Figure 3.16 Isolation of an indoor cell from a macrocell in a stadium.

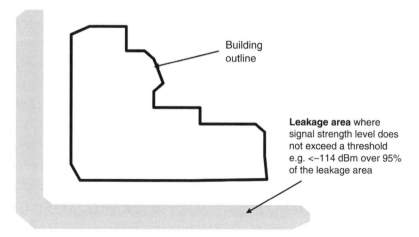

Figure 3.17 Leakage concept.

3.4 Leakage

In addition to coverage and isolation requirements, signal strength should be maintained at restricted levels to avoid signal spillage outside the building. This signal spillage is known as *leakage*, and is often desired to be kept at a minimum level over a percentile of an area outside the building. This area is defined as the *leakage requirement area*, as shown in Figure 3.17.

Although leakage is a serious issue for GSM networks, it can be counterbalanced with antenna downtilting from the external macrocell to make sure the macro dominates in the external areas of the building. For 3G systems, it becomes more critical due to soft handovers: if the coverage area of the indoor cell overlaps excessively out of the building with the macro cell, excessive soft handovers may occur, which can result in an overload of network resources, hence affecting system performance.

For Wi-Fi networks, leakage can represent a threat to the security of the network, as potential hackers can have access to the network if coverage is not controlled out of the building or corporate premises. For personal area networks, such as Bluetooth or ZigBee, leakage is not an issue, as devices communicate with each other.

3.5 Capacity

As the demand for wireless services increases, the number of channels assigned to a cell eventually becomes insufficient to support the required number of users. At this point, wireless design techniques are needed to provide more system resources per unit coverage area, that is to expand the capacity of wireless systems. Therefore, wireless service operators aim to maximize the efficiency of their networks in terms of increasing the number of users accommodated per coverage area. For example, for cellular, this can be achieved by increasing the number of available channels in the cell or by using other

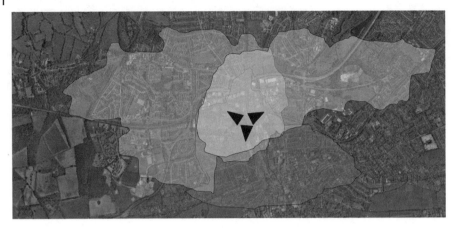

Figure 3.18 Three-sector configuration employed to increase capacity in an urban area.

techniques such as sectorization (outdoor), zoning (indoor) or cell splitting (Saunders and Aragón-Zavala, 2007). These are explained briefly as follows:

- Sectorization is a technique used in cellular systems to increase the number of users connecting to the network over a given area. This enhancement is achieved by replacing an omnidirectional configuration by a combination of two, three or six sectors using directional antennas, pointing to different locations and thus minimizing co-channel interference, as depicted in Figure 3.18.
- Zoning is the term often employed for sectorization if done indoors and zones are in essence similar to sectors in a macrocell outdoor environment. The distinction is that sectors are delimited using antenna radiation patterns primarily, whilst zones are separated by choosing groups of antennas that are isolated by virtue of in-building penetration losses. An example of the selection of these zones is given in Figure 3.19.

The capacity of a network is then often expressed as a percentile of maximum blocking probability or minimum grade of service. Blocking occurs as a result of a user attempting to make a phone call without available channels in the cell. On the other hand, the grade

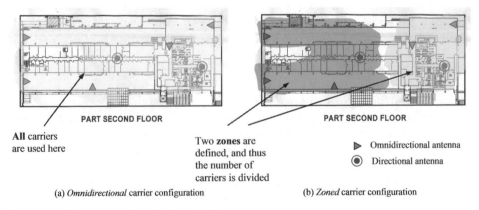

All carriers
are used here

(a) *Omnidirectional* carrier configuration

Two **zones** are
defined, and thus
the number of
carriers is divided

▷ Omnidirectional antenna

◉ Directional antenna

(b) *Zoned* carrier configuration

Figure 3.19 The concept of zoning in a building.

of service (GoS) is a measure of the ability of a user to access a trunked system during the busiest hour. The busy hour is based upon customer demand at the busiest hour during a week, month or year. The GOS is therefore a benchmark used to define the desired performance of a particular trunked system by specifying a desired likelihood of a user obtaining channel access given a specific number of channels available in the system. It is part of the system design process to estimate the maximum required capacity and to allocate the proper number of channels in order to meet the GoS. This is often the case for GSM cellular systems.

For 3G, capacity requirements are treated slightly differently. It is often said that 3G cellular systems have a 'soft capacity' in that the maximum number of users the network can grant access to does not necessarily depend on the number of available channels in the pool – we are in a CDMA system where all users can access the network using the same channel at the same time, but just having different codes. However, as more users access the network, the levels of interference increase and this reduces the maximum capacity of the network (Holma and Toskala, 2010). This is, of course, the case for noise rise as well, where special care should be taken not to overload the noise, especially in the uplink.

For LTE networks, capacity dimensioning is not too different from the other standards. The number of subscribers in a specific coverage area needs to be determined, along with an estimated forecast for the following years – remember that capacity is often installed for present and future needs. Traffic profiles are defined by each operator, from which traffic mix and models are used. Once this is completed, a rough estimate of the cell size and site count are performed to verify whether, with a given site density, the system can carry the specified load or new sites have to be added. In LTE, the main indicator of capacity is SINR (signal-to-interference and noise ratio) distribution in the cell. Thus, SINR can be directly mapped into system capacity (data rate) or cell throughput.

Other wireless technologies follow different approaches. For instance, for Wi-Fi networks, throughput is strongly influenced by both SINR levels and the number of users associated with the same AP. Since more users increase the number of packet collisions, the standard prevents this and reduces the data rate as more lost packets are detected. On the other side, coverage is strongly related to data rates and system capacity for a Wi-Fi network. Data rates affect AP coverage areas. Lower data rates, such as 1 Mbps, can extend the coverage area farther from the AP than higher data rates, such as 54 Mbps. Thus, the data rate and power level affects coverage and consequently the number of APs required, which in turn determines the maximum number of users in a Wi-Fi system. Therefore, a careful channel planning should be performed, especially taking into consideration that Wi-Fi contiguous channels overlap and that co-channel interference can affect the system performance considerably, as it has a direct impact on SINR levels.

For wireless local area networks (WLANs), capacity is normally closely related to the number of simultaneous users accessing the network, which in turn can severely degrade the system throughput during peak traffic hours. For some densely populated buildings, more APs are placed to dedicate more channels to the same coverage area and therefore satisfy the required demand. Special care should be taken in selecting the channels to be used to avoid co-channel interference. To illustrate how to dimension a WLAN deployment, let us assume a three-storey office building of 220 × 220 m, as shown

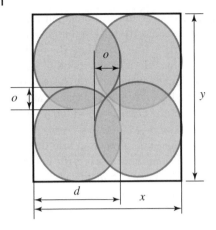

Figure 3.20 Capacity dimensioning example of a WLAN.

Table 3.2 Estimated number of APs required for 2.4 GHz and 5 GHz WLAN systems.

Frequency (GHz)	Data rate (Mbps)	Range (m)	Overlap (m)	AP per floor	Total AP
2.4	40	126	38	4	12
	90	68	20	16	48
5	40	75	22.5	14	42
	90	40	12	55	165

in Figure 3.20, for which an estimation of the number of required APs for 2.4 GHz and 5 GHz is to be performed. Two throughputs were selected: 40 Mbps and 90 Mbps. Thus, the required number of AP per floor N is given by

$$N \approx \left(\frac{x-d}{d-o}+1\right)\left(\frac{y-d}{d-o}+1\right) \tag{3.1}$$

For this example, x and y represent the coordinates of the single floor, d is the range at a given throughput and o is the overlap, taken as $0.3d$ for this example, or 30%. The total number of required APs in the building is given by multiplying $F \times N$, and estimations for 2.4 GHz and 5 GHz are given in Table 3.2.

For more detailed capacity calculations, refer to Chapter 8.

3.6 Interference

Interference is the major limiting factor in the performance of cellular radio systems, as has been mentioned for 3G cellular, LTE and Wi-Fi. Sufficient received signal levels are not sufficient to guarantee a certain quality of service; interference levels should also be taken into account, especially for broadband wireless systems.

Sources of interference include another mobile in the same cell, a call in progress in a neighbouring cell, other base stations operating in the same frequency band or any non-cellular system that inadvertently leaks energy into the cellular frequency band.

The two major types of system-generated cellular interference are co-channel interference and adjacent-channel interference. *Co-channel interference* comes as a result of several cells using the same set of frequencies and hence users from distant cells interfering with those of the wanted cell. On the other hand, interference resulting from signals that are adjacent in frequency to the desired signal is called adjacent-channel interference. *Adjacent-channel interference* results from imperfect receiver filters that allow nearby frequencies to leak into the passband.

For either type of interference, a system requirement is necessary to ensure that interference levels are kept to a minimum level to achieve certain QoS. For GSM, the *signal-to-interference ratio* (SIR) is often used, in dB, and represents the level of signal exceeding interference. For instance, a GSM voice system may require at least an SIR level of 15 dB to guarantee a certain quality of speech. More specifically, for indoor systems, a CIR (carrier-to-interference ratio) of 9 dB is needed for co-channel interference, whereas at least 5 dB of CIR are left for adjacent channel interference. For UMTS systems, the *signal energy-per-chip to noise and interference density ratio* is used, for which the noise is often not specified in the acronym since it is much less than the interference levels, yielding E_c/I_0, in dB. A pilot $E_c/I_0 > -5$ dB at 50% cell loading is often specified by operators for in-building. LTE networks use the SINR in dB to specify minimum levels of received signal above noise and interference to guarantee a cell throughput, given a certain scenario and MCS (modulation and coding scheme).

For licence-free bands, such as the ISM which is used by Wi-Fi and Bluetooth, interference can become a more serious threat. In fact, WLAN networks have to perform careful frequency planning to avoid overlapping channels to interfere with each other, since the use of contiguous channels is not an option for the designer. Other devices such as microwave ovens, baby monitors, some cordless phones and Bluetooth operate in the same frequency band as Wi-Fi, and therefore protocols should be robust enough to support fairly high levels of interference – as is the case of Bluetooth. According to the IEEE standards, for 802.11b/g, the adjacent channel rejection should be of at least 35 dB at 25 MHz from the centre frequency of the receiver. In the case of out-of-band interference, caused by spurious emissions from out-of-band appliances, an out-of-band transmit power of -13 dBm/MHz is recommended according to IEEE standards.

In HSPA (high-speed packet access) networks, a very important target design KPI (key performance indicator) is the SINR, as for LTE, which must be greater than 10 dB to guarantee good performance. As SINR and related data throughput are vendor specific, different operators may have distinct target values; for example, SINR of 15 dB for a data throughput of 7.2 Mbps.

3.7 Signal Quality

So far, we have described various requirements that are essential for an indoor wireless system for proper operation. Many of these requirements are established on the basis that a minimum signal quality is expected in the wireless system, for sound performance. Here are a few examples:

- Traditionally, cellular operators have measured and benchmarked their network performance in GSM using mostly the bit error rate (BER) and the dropped call rate (DCR) to quantify the speech quality and the rate of lost connections, respectively. There are special venues, such as stadiums, in which the DCR will give an indication of how well handover areas have been designed.

- In GSM, BER is mapped into RxQual, where values range from 0 (BER = 0.14%) to 7 (BER = 18.1%). Additionally, the call success rate (CSR) and the handover success rate (HSR) have been used to measure the performance of the signalling channels associated with call originations and handovers – these KPIs are also very relevant for some buildings since *sectorization* is a common practice to provide radio coverage and therefore excessive handovers should be minimized. However, on the other hand, successful handovers must be guaranteed in well-delimited areas. For data KPIs and for packet-switched (PS) GPRS, the most important parameters are reliability, throughput and delay.
- For LTE networks, RSRQ (reference signal receive quality) should be greater than 5 dB for 50% cell loading.

3.8 Technology

A careful assessment of the technology available to design an indoor wireless system should not be overlooked, especially in a high demanding and dynamic market such as the one in personal communications. For example, cellular operators may be aiming to be at the vanguard in the latest technology, but this does not necessarily ensure that system performance has been fully optimized. The use of state-of-the-art technology does not guarantee that the indoor cellular system is well designed, but nevertheless represents a valuable infrastructure from which system performance can be enhanced and optimized.

There is no "recipe" to choose the best technology for a project. It is often the case of evaluating all the alternatives that are available for a given set of operational requirements. In general, this tends to be the main driving force for the choice of technology. However, technology cannot be seen as a standalone factor, as it is normally closely related to other factors, such as cost and complexity. A careful balance of all these factors should be made for a successful design.

Finally, technology is perhaps one of the fastest changing requirements of all for wireless communications. Newer possibilities may exist within the timelines of a project, which can provide additional benefits to the system. Permanent research on the state-of-the-art is therefore highly recommended, in order to select the best possible technology at all times.

3.9 Cost

Costs should also be maintained within the budget of the project, and therefore key decisions about the selection of the most appropriate technology are essential. These decisions cannot be accurately made if a limited knowledge of what these options are exists. There is, however, a cost requirement that is taken as universally accepted: design the best solution with the lowest possible budget!

There are many factors that can affect the cost of a project and which should be taken into account when specifying requirements. These are summarized as follows:

- **Equipment and technology costs.** These are perhaps the most obvious expenditures, having all the required 'bits and bobs' to make the system go up and running.
- **Development costs.** Even if a wireless service provider or operator has its own engineering group, this represents a cost in man-hour terms that should not be overlooked and must therefore be budgeted appropriately.
- **Maintenance costs.** Once the indoor wireless system is installed, there is a need to keep it 'healthy' and thus sufficient maintenance costs should be accounted for in the planning process.
- **Installation costs.** The costs of installing the system may have an impact on the choice of technology; for example, if the budget is limited, a passive distribution system for cellular may not be an option if optical fibre is already available in the building and therefore installation costs can be significantly reduced.
- **Operational costs.** These are related with the operation of the system and are normally budgeted as a post-installation cost. These may also have an impact on the design of the wireless system, as the choice of a specific technology can represent much higher operational costs; for example a larger network.

Specific details associated with various types of costs and financial issues to be taken into account when deploying and operating an indoor wireless network are covered in Chapter 12.

3.10 Upgradeability

Many wireless technologies used inside buildings have fast developments and upgrades that should be considered before a new system is to be deployed. Some room should therefore be left to account for upgrades. A very clear example of this is in the Wi-Fi world, where for years the only approved standard was 802.11b. Recently, higher speeds can be achieved using the same frequency (2.4 GHz) in the 802.11g standard. Thus, the originally installed WLAN infrastructure should be able to migrate from 802.11b to 802.11g. Another example is the evolution in cellular from 2G to 3G, for instance, from GSM to UMTS. Albeit the differences in technologies, the possibility of upgrading a GSM indoor system to a UMTS indoor system planned since the beginning can drastically reduce migration time and costs.

It is fair to say that sometimes it is difficult to foresee the characteristics of emerging technologies in order to account for upgrades and design totally flexible systems. There are cases, however, where most of the existing infrastructure should be replaced; if this is the case, a sufficient budget should therefore be allowed for this migration, which does not only take financial but also human and time resources.

One key aspect to take into consideration when performing system upgrades in an indoor system is the type of traffic that needs to be handled. In the early days of mobile communications, voice was the main driver and all efforts were around the provision of good voice quality to customers. Nowadays, this requirement has been superseded by the enormous increase of mobile data traffic and the creation of newer and more sophisticated mobile devices that allow the use of mobile Internet virtually anywhere at any time. If an upgrade is required to accommodate data traffic and allow users to access the Internet, for example, in an IP-based system, special care should be taken when considering that voice needs also to be delivered. This may seem a straightforward consideration but, believe it or not, sometimes is not carefully assessed and no matter how good data service you may have, the voice is quite not there!

3.11 System Expansion

A good example where system expansion can play a decisive role is an indoor network deployment in an airport. Airports normally increase their number of passengers per year, and therefore are subject to expansions in their wireless infrastructure. Thus, careful planning should be in place to allow system expansion in terms of capacity and possible additional areas to provide wireless coverage. This also includes the possibility of adding more wireless services to be offered to passengers, visitors and airport staff.

There are special venues for which system expansion can result in a critical variable: stadiums. As we all know, stadiums not only hold sports events but for off-season periods concerts and massive events are taking place here, with a substantial increase in the need of wireless coverage and capacity. Due to cost constraints, it is often hard to deploy all the maximum network capacity in a stadium since the beginning, and thus this needs to be done periodically. Meanwhile, the capacity has to be shared with outdoor networks, resulting sometimes in blockage and excessive delays in data traffic that degrade system performance. At the earliest convenience it is very important to deploy the required capacity to accommodate all voice and data traffic, especially for in-building systems.

3.12 Conclusion

Every good indoor wireless system design should take as a reference a well-established set of requirements. The more ambiguous these requirements are, the less likely the system has optimum performance. It is therefore worth spending enough time to define these requirements from many different perspectives: technical, operational, financial and even physical. These requirements will dictate the characteristics of the design and will determine the technology to be used. It is often the case that design compliance standards need to be checked against these requirements, as will be discussed later.

References

Aragón-Zavala, A., Cuevas-Ruiz, J.L., Castañón, G. and Saunders, S.R. (2009) Mobility model and traffic mapping for in-building radio design, in *Proceedings of the 19th International Conference on Electrical, Communications and Computers, CONIELECOMP 2009*, Cholula, pp. 46–51.

Geier, J. (2001) *Wireless LANs, Implementing High Performance 802.11 Networks*, 2nd edition, SAMS, Carmel, IN, ISBN 0-672-32058-4.

Geier, J. (2008) *Wi-Fi: Define Minimum SNR Values for Signal Coverage.* URL: http://www.enterprisenetworkingplanet.com/netsp/article.php/3747656/WiFi-Define-Minimum-SNR-Values-for-Signal-Coverage.htm.

Holma, H. and Toskala, A. (2010) *WCDMA for UMTS: HSPA Evolution and LTE,* 5th edition, John Wiley and Sons, Ltd, Chichester, ISBN 0-470-68646-1.

IEEE standard 802.15.4, Part 15.4 (2003) *Wireless Medium Access Control (MAC) and Physical Layer (PHY) Specifications for Low-Rate Wireless Personal Area Networks (LR_WPANs),* IEEE Computer Society. ISBN 0-7381-3677-5.

IEEE 802.16 (2015) Broadband Wireless Access Working Group. URL: http://www.ieee802.org/16.

Le, K.T. (2010) Transceiver design for IEEE 802.15.4 and ZigBee-compliant systems, *Microwave Journal,* October 2010.

Saunders, S. and Aragón-Zavala, A. (2007) *Antennas and Propagation for Wireless Communication Systems,* 2nd edition, John Wiley and Sons, Ltd, Chichester. ISBN 0-470-84879-1.

Tolstrup, M. (2011) *Indoor Radio Planning: A Practical Guide for GSM, DCS, UMTS, HSPA and LTE,* 2nd edition, John Wiley and Sons, Ltd, Chichester. ISBN 0-470-71070-8, 2011.

4

Radio Propagation

Radio wave propagation is a key topic for an in-building wireless system designer. The underlying theory that gives rise to the production of radio waves is reviewed here, which (alongside light) are examples of electromagnetic waves. The simplest form of wave is a *plane wave*, which illustrates many of the ways in which more complex waves interact with their surroundings (e.g. a building). The properties of materials (e.g. walls, windows, roofs and trees) are also explained, which are relevant to their impact on waves.

There is theoretically an infinite number of ways in which waves interact with materials, which can be broken down into a small number of mechanisms. By understanding each of these mechanisms in turn an appreciation of the nature of the effects that are observed in practical in-building design is developed. These mechanisms, combined with the complicated structure of real buildings, make the practical prediction of in-building propagation very challenging, as stated for the various environments in the previous chapter.

Propagation effects are so important to indoor radio design that an entire chapter is dedicated to a deeper study and understanding of such mechanisms. In fact, radio waves interact in many ways with building walls, furniture, people, etc., and depend on aspects such as construction materials so these should be carefully considered for more accurate work.

4.1 Maxwell's Equations

Radio wave propagation has its foundations in electromagnetic theory principles. Electromagnetism is a very mature subject, which established its foundations many centuries ago, and it is expected that you have a basic understanding and background of its basic principles. It is out of the scope of this book to provide a detailed analysis of electromagnetic theory and principles that form the basis of Maxwell's equations. However, interested readers willing to expand their knowledge or review such principles can refer to Sadiku (2007), Kraus and Fleisch (1999) and Hayt and Buck (2011).

The existence of propagating electromagnetic waves can be predicted as a direct consequence of *Maxwell's equations*. These equations are a set of four partial differential equations, which specify the relationships between the variations of the vector electric field $\vec{E}$ and the vector magnetic field $\vec{H}$ in time and space within a medium. The $\vec{E}$ field strength is measured in volts per meter and is generated by either a time-varying

Indoor Wireless Communications: From Theory to Implementation, First Edition. Alejandro Aragón-Zavala.
© 2017 John Wiley & Sons Ltd. Published 2017 by John Wiley & Sons Ltd.

Figure 4.1 Four scientists who contributed with their observations, experiments and derivations to the formulation of Maxwell's equations: (a) James Clerk Maxwell; (b) Carl Friedrich Gauss; (c) Michael Faraday; (d) André-Marie Ampère.

magnetic field or by a free charge. The $\vec{H}$ field is measured in amperes per meter and is generated by either a time-varying electric field or by a current.

James Clerk Maxwell (1851–1879), shown in Figure 4.1(a), a Scottish physicist and mathematician, synthesized previously unrelated observations, experiments and equations of electricity, magnetism and optics into a consistent theory. Maxwell's equations take their name after J.C. Maxwell and can be found in his paper 'On physical lines of force' (Maxwell, 1861). He then added the term *displacement current* to one of the equations in Maxwell (1865).

Maxwell's four equations are summarized in this section, where a brief explanation is given to each.

4.1.1 Gauss's Law for Electricity

Carl Friedrich Gauss (1777–1855), shown in Figure 4.1(b), a German mathematician and scientist, contributed extensively to magnetism (hence the unit *Gauss* for magnetic flux density after his name) and in 1835 developed his famous Gauss's law, which in fact was not published until 1867. Gauss's law states that: *the electric flux through any closed surface is proportional to the enclosed electric charge.* In differential (4.1) and integral (4.2) forms, respectively, Gauss's law can be expressed as

$$\vec{\nabla} \cdot \vec{D} = \rho_v \tag{4.1}$$

$$\oint_S \vec{D} \cdot d\vec{S} = \iiint_{vol} \rho_v dV \tag{4.2}$$

The operator $\vec{\nabla} \cdot \vec{D}$ indicates the divergence of an electric flux displacement vector $\vec{D}$ and ρ_v is the volumetric charge density in coulombs per square metre. Recall that $\vec{D}$ is related to the electric field vector $\vec{E}$ by a constant known as *permittivity*, ϵ, as follows:

$$\vec{D} = \epsilon \vec{E} \tag{4.3}$$

In (4.2), the left-hand side of the equation is a surface integral denoting the electric flux through a closed surface $\mathbf{S}$ and the right-hand side of the equation is the total charge enclosed by $\mathbf{S}$, obtained by integrating the volumetric charge density, ρ_v.

Free-space permittivity is $\varepsilon_0 = 8.85 \times 10^{-12}$ F/m, and, in practice, wireless links are assumed to have free-space permittivity. However, should the radio wave propagate in a different medium, a different constant should be considered. Recall that the permittivity of a material is related to its electrical properties and each material is given a certain *relative permittivity* ε_r. This is discussed later in this chapter.

The mathematics under Equations (4.1) and (4.2) are rather complex, but its physical interpretation is much more important and relevant to the understanding of the phenomena behind indoor radio propagation. In summary, Gauss's law demonstrates that electric field lines may either start/end on charges or are continuous, as seen in Figure 4.2. Electric charges are therefore sources of electric field lines and can exist as monopoles.

Gauss's law is also known as *Gauss's flux theorem* and can be used to derive Coulomb's law, which is said to be its *dual* equation in magnetism.

4.1.2 Gauss's Law for Magnetism

Electricity and magnetism always have dual laws and equations. For Gauss's law, there is its counterpart for magnetism, which states that the divergence of any magnetic field $\vec{\nabla} \cdot \vec{B}$ does not exist; that is magnetic flux lines cannot be generated by magnetic poles or

Electric charge

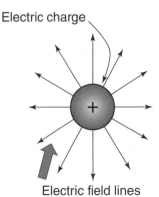

Electric field lines

Figure 4.2 Electric field force lines.

'charges, as is the case with electric fields. Therefore, as a consequence of this, magnetic monopoles cannot exist. In mathematical form:

$$\vec{\nabla} \cdot \mathbf{B} = 0 \tag{4.4}$$

$$\oint_S \vec{B} \cdot d\vec{S} = 0 \tag{4.5}$$

In integral form, as stated in (4.5), the net flux of magnetic field lines will always be zero as the same number of lines that emanate from the magnetic field source return back to it, as shown in Figure 4.3. This implies the existence of magnetic *dipoles* in the universe, seen as sources of magnetic fields, where magnetic field lines are always continuous.

There is also a correspondence between the magnetic flux density vector $\vec{\mathbf{B}}$ and the magnetic field intensity vector $\vec{\mathbf{H}}$, which is given by a parameter known as *permeability*, μ, as follows:

$$\vec{\mathbf{B}} = \mu\vec{\mathbf{H}} \tag{4.6}$$

Permittivity and permeability will be discussed in more detail later on in this chapter.

Magnetic field lines

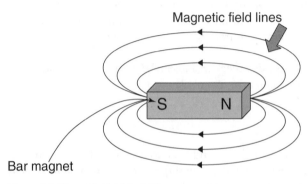

Bar magnet

Figure 4.3 Magnetic field force lines.

4.1.3 Faraday's Law of Induction

Electromagnetic induction was discovered and first published in 1831 by Michael Faraday (1791–1867), shown in Figure 4.1(c), an English chemist and physicist, by discovering that a time-changing magnetic flux creates a proportional electromotive force. Faraday also studied the magnetic field around a conductor carrying a continuous electric current and discovered diamagnetism and the laws of electrolysis. He was also a very successful inventor or various electromagnetic rotary devices, which in turn form the foundation of modern electric motors.

The discovery of electromagnetic induction had been in fact the subject of debate and controversy, as Joseph Henry (1797–1878), an American scientist, also discovered mutual inductance independently of Michael Faraday. However, it was demonstrated that Faraday was the first to publish his results. Nevertheless, the unit of inductance, Henry, is named in his honour.

Faraday's law of induction states: *the induced electromotive force or EMF in any closed circuit is equal to the time rate of change of the magnetic flux through the circuit* (Sadiku, 2007). In its original form, Faraday's law was expressed as a single equation describing two different phenomena: the motional electromotive force generated by a magnetic force on a moving wire and the transformer electromotive force generated by an electric force due to a changing magnetic field. It was not until 1861 when J.C. Maxwell drew attention to this fact and in Maxwell (1961) he gave a separate physical explanation for each of the two phenomena. Thus, Faraday's law of induction in the form of a partial differential equation is often known as the *Maxwell–Faraday equation*. In integral and differential forms, it can be expressed as

$$\vec{\nabla} \times \vec{E} = -\frac{\partial \vec{B}}{\partial t} \tag{4.7}$$

$$\oint_C \vec{E} \cdot d\vec{L} = -\frac{\partial}{\partial t} \iint_S \vec{B} \cdot d\vec{S} \tag{4.8}$$

In (4.7), a new concept was introduced in Maxwell's equations, known as *curl*; thus, the last two of Maxwell's equations are called curl equations due to this. By looking at (4.7) carefully, a time-changing magnetic field $\vec{B}$ can produce an electric field, which is orthogonal to the original magnetic field. Equation (4.8) is a direct consequence of *Green's theorem* and represents the same induction principle (Hayt and Buck, 2011).

4.1.4 Ampère's Circuital Law

The fourth Mawxell equation is known as *Ampère's circuital law*, after André-Marie Ampère (1775–1836), shown in Figure 4.1(d), a French physicist and mathematician, who is often referred as one of the main discoverers of electromagnetism. In fact, the unit of measurement for electric current, the ampere, is named after him.

Ampère's law was discovered in 1826 and states that magnetic fields can be generated in two ways: by an electric current circulating in a wire and by changing electric fields. In other words, it integrates the magnetic field around a closed loop to the electric current passing through the loop. In mathematical form:

$$\nabla \times \vec{H} = \vec{J} + \frac{\partial \vec{D}}{\partial t} \tag{4.9}$$

$$\oint_C \vec{\mathbf{H}} \cdot d\vec{\mathbf{L}} = \iint_S \left(\vec{\mathbf{J}} + \frac{\partial \vec{\mathbf{D}}}{\partial t} \right) \cdot d\vec{\mathbf{S}} \tag{4.10}$$

A new term is introduced in both (4.9) and (4.10), called *current density vector* $\vec{\mathbf{J}}$, and is used to denote direction and magnitude of electric current circulating in a wire. Equation (4.9), in particular, indicates that a changing electric field or a current can create a magnetic field, orthogonal to the original electric field.

4.1.5 Consequence of Maxwell's Equations

The last two equations, Maxwell's *curl* equations, contain constants of proportionality, which dictate the strengths of the fields. These are the *permeability* of the medium μ in henrys per meter and the *permittivity* of the medium ε in farads per meter. They are normally expressed relative to the values in free space, μ_0, ε_0:

$$\mu = \mu_0 \mu_r \tag{4.11}$$

$$\varepsilon = \varepsilon_0 \varepsilon_r \tag{4.12}$$

The relative permeability and permittivity represent intrinsic properties of materialsand therefore are taken as multiplicative factors to obtain the actual values of permeability and permittivity for any material, having the free-space values as reference. The permeability in free space is $4\pi \times 10^{-7}$ H/m, whereas the free-space permittivity is 8.85×10^{-12} F/m.

Free space strictly indicates a vacuum, but the same values can be used as good approximations in dry air at typical temperatures and pressures.

Maxwell's equations do not only anticipate the possibility of using electromagnetic waves for wireless communications but also have set a full range of applications where electric and magnetic fields are somehow involved, as is the case of machines that make use of induction principles to operate (transformers, motors, alternators, generators, etc.). Our focus, however, is on the wireless communications world, and thus all the remaining references to Maxwell's equations legacy will be made on that basis.

4.2 Plane Waves

There is a striking result that can be obtained from the solution to the curl equations. Since one of them suggests that a time-varying magnetic field produces an electric field whereas the other dictates that a time-varying electric field or a current produce a changing magnetic field, thus a *travelling electromagnetic wave* is generated, as a result of the interactions of both fields over space and time.

In free space, many types of waves can exist, which satisfy Maxwell's equations; that is constitute valid solutions. Amongst the most popular ones are spherical, cylindrical and plane waves, as shown in Figure 4.4. However, any wave type can be represented as a collection of *plane waves*; thus these are the fundamental types of waves that need to be examined. They are typically simpler to analyse than the other types, so they provide a good way to understand the properties that the more complex wave types exhibit.

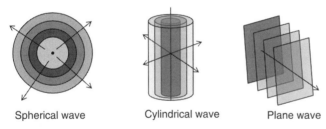

Spherical wave   Cylindrical wave   Plane wave

Figure 4.4 Types of electromagnetic waves.

Figure 4.5 Radiating cable, which produces cylindrical waves.

Note that all of these wave shapes can exist at the same time in a given area of space and they pass through each other without interacting (at least in the 'linear' materials, which are normal in buildings and at low power levels). A very good example of cylindrical waves is that produced by *radiating cables*, as shown in Figure 4.5, which will be analysed in more detail later on in this book.

4.2.1 Wave Equation

Suppose the electric and magnetic fields are in free space, so both permeability and permittivity are those of free space. Thus, as they are in a non-conducting medium, the vector current density **J** is zero. Note that Maxwell's equations state that electric and magnetic fields are always orthogonal to each other, so the assumption here is that **E** and **H** have components as suggested in Figure 4.6.

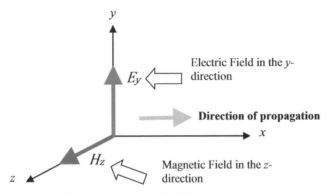

Figure 4.6 Electric and magnetic fields in free space.

Notice that it is assumed the generated waves will be travelling in the *x*-direction. Thus, by rearranging both curl equations, *wave equations* for the electric and magnetic fields can be obtained (Hayt and Buck, 2011):

$$\frac{\partial^2 E_y}{\partial t^2} = \frac{1}{\mu_0 \varepsilon_0} \frac{\partial^2 E_y}{\partial x^2} \tag{4.13}$$

$$\frac{\partial^2 H_z}{\partial t^2} = \frac{1}{\mu_0 \varepsilon_0} \frac{\partial^2 H_z}{\partial x^2} \tag{4.14}$$

The above equations describe the motion of an electric E_y or magnetic field H_z as a function of time and space. If a velocity parameter *v* is defined as

$$v^2 = \frac{1}{\mu_0 \varepsilon_0} \tag{4.15}$$

then, rearranging equation (4.13) in terms of velocity and using (4.15) gives

$$\frac{\partial^2 E_y}{\partial t^2} = v^2 \frac{\partial^2 E_y}{\partial x^2} \tag{4.16}$$

By examining (4.16) carefully, the dimensions of *v* truly represent a velocity, which in turn shows that the wave travels at the speed of light in free space (3×10^8 m/s).

From (4.16), the electric field wave equation, many solutions exist, all of which represent fields that could actually be produced in practice. However, they can all be represented as a sum of *plane waves*, which represent the simplest possible time-varying solution. A general solution can then be found if

$$E_y = E_0 \cos(\omega t + kx) + E_0 \cos(\omega t - kx) \tag{4.17}$$

The wavenumber *k* represents the number of wavelengths λ per unit distance, $2\pi/\lambda$, and ω is the angular frequency in radians per second, which is related to linear frequency as $\omega = 2\pi f$. Notice that (4.17) shows that travelling plane waves in both directions represent a solution to the wave equation, both in time and space, having a maximum electric field of E_0. More generally, using vector notation, the electric field can be expressed as

$$\vec{E} = E_0 \cos(\omega t \mp kx) \, a_y \tag{4.18}$$

Similarly, since **E** and **H** are perpendicular to each other, an expression for the magnetic field can be obtained:

$$\vec{H} = H_0 \cos(\omega t \mp kx) \, a_z \tag{4.19}$$

In (4.19), H_0 is the maximum magnetic field intensity, in A/m.

4.2.2 Plane Wave Properties

As seen in Figure 4.6, the direction of propagation of the wave is along the *x*-axis. The vector in this direction is the propagation vector or the *Poynting vector*. The two fields are in phase at any point in time or in space. Their magnitude is constant in the *yz* plane,

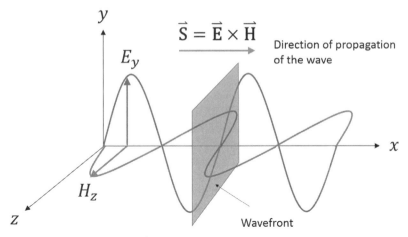

Figure 4.7 Electromagnetic plane wave.

and a surface of constant phase (a wavefront) forms a plane parallel to the yz plane – hence the term plane wave. This is better shown in Figure 4.7.

The oscillating electric field produces a magnetic field, which itself oscillates to recreate an electric field and so on, in accordance with Maxwell's curl equations. This interplay between the two fields stores energy and hence carries power along the Poynting vector **S**, which in fact describes the magnitude and direction of the power flow carried by the wave per square metre of area parallel to the yz plane, that is the power density of the wave

$$\vec{S} = \vec{E} \times \vec{H}^* \tag{4.20}$$

Equations (4.18) and (4.19) satisfy Maxwell's equations, provided the ratio of the field amplitudes is a constant for a given medium, known as *wave impedance* Z_w:

$$Z_w = \frac{|\vec{E}|}{|\vec{H}|} = \sqrt{\frac{\mu}{\varepsilon}} \tag{4.21}$$

In free space, the wave impedance is around 377 Ω.

Finally, the velocity of a point of constant phase on the wave, the *phase velocity* v at which wavefronts advance in the **S** direction, is given by

$$v = \frac{\omega}{k} = \frac{1}{\sqrt{\mu\varepsilon}} \tag{4.22}$$

The wavelength is therefore

$$\lambda = \frac{v}{f} \tag{4.23}$$

4.2.3 Wave Polarization

The alignment of the electric field vector of a plane wave relative to the direction of propagation defines the *polarization* of the wave. For example, if the electric field is parallel to the *x*-axis, this wave is *x*-polarized and could be generated by a straight wire antenna parallel to the *x*-axis. An entirely distinct *y*-polarized plane wave could be generated with the same direction of propagation and recovered independently of the other wave using pairs of transmit and receive antennas with perpendicular polarization. This principle is sometimes used in satellite communications to provide two independent communication channels on the same Earth satellite link. If the wave is generated by a vertical wire antenna (H-field horizontal) then the wave is said to be vertically polarized; a wire antenna parallel to the ground (E-field horizontal) primarily generates waves that are horizontally polarized.

The waves described so far have been linearly polarized, since the electric field vector has a single direction along the whole of the propagation axis. If two plane waves of equal amplitude and orthogonal polarization are combined with a 90° phase difference, the resulting wave will be circularly polarized (CP), in that the motion of the electric field vector will describe a circle centred on the propagation vector. The field vector will rotate by 360° for every wavelength travelled. Circularly polarized waves are most commonly used in satellite communications, since they can be generated and received using antennas that are oriented in any direction around their axis without loss of power. They may be generated as either right-hand circularly polarized (RHCP) or left-hand circularly polarized (LHCP); RHCP describes a wave with the electric field vector rotating clockwise when looking in the direction of propagation. In the most general case, the component waves could be of unequal amplitudes or at a phase angle other than 90°. The result is an elliptically polarized wave, where the electric field vector still rotates at the same rate but varies in amplitude with time, thereby describing an ellipse. This can be seen in Figure 4.8.

If there is a mismatch between the wave and receive antenna polarization, this represents a loss that has to be taken into account when planning and designing a wireless system. The polarization of standard mobile phones tends to be pretty random, both due to movement by the user and the antenna characteristics itself, so significant

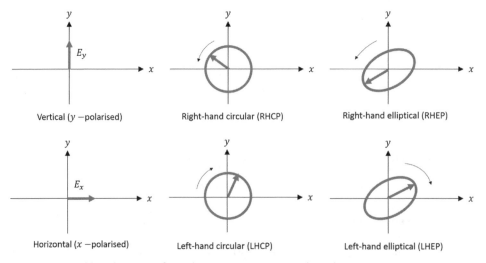

Vertical (*y* −polarised)　　Right-hand circular (RHCP)　　Right-hand elliptical (RHEP)

Horizontal (*x* −polarised)　　Left-hand circular (LHCP)　　Left-hand elliptical (LHEP)

Figure 4.8 Possible polarizations for a plane wave propagating along the *z*-axis.

polarization loss is often encountered. This is particularly enhanced inside buildings, where multiple reflections and other depolarizing effects take place.

For indoor wireless communications, linear polarization is often the preferred method of those described.

4.2.4 Wave Propagation in Lossy Media

So far only lossless media have been considered in our previous examples. This has been done in order to introduce the properties of electromagnetic waves more simply. When the medium has significant conductivity σ, the amplitude of the wave diminishes with distance travelled through the medium as energy is removed from the wave and converted to heat (losses). Thus, new expressions for (4.18) and (4.19) need to be derived, as follows:

$$\vec{E} = E_0 e^{-\alpha x} \cos(\omega t \mp kx)\mathbf{a}_y \tag{4.24}$$

$$\vec{H} = H_0 e^{-\alpha x} \cos(\omega t \mp kx)\mathbf{a}_z \tag{4.25}$$

The constant α is known as the *attenuation constant*, with units of per metre (m^{-1}), which depends on the permeability and permittivity of the medium, the frequency of the wave and the conductivity of the medium, measured in siemens per metre, or per-ohm-metre (Ωm^{-1}). Conductivity, permittivity and permeability are known as *constitutive parameters* of the medium.

Depending on the constitutive parameters of materials, a classification can be made as good insulators or good conductors, for which the attenuation constant, wavenumber, wave impedance, wavelength and phase velocity can be computed. From this, very relevant information can be obtained from building construction materials, which will have a direct impact on the way radio waves propagate inside a building, as we will see in the next sections. More details on the equations used to derive wave parameters for lossy media can be found in Saunders and Aragón-Zavala (2007).

4.3 Propagation Mechanisms

So far, the media in which waves were propagating consisted of regions within which the constitutive parameters (i.e. properties of materials) did not vary in space, being infinite in extent in all directions, regardless of whether the materials were considered as good conductors or insulators. In practice, however, the boundaries between media must be considered for accurate radio propagation work; for example between air and the ground, between buildings and the air, from Earth to space, etc. These boundary effects give rise to changes in the amplitude, phase and direction of propagating waves. Almost all of these effects can be understood in terms of combinations of simple *propagation mechanisms* operating on plane waves. These mechanisms are now described and will later be used to analyse wave propagation in the real world.

4.3.1 Is Electromagnetic Theory Wrong Inside Buildings?

A good start for this section could be a reflection on whether all the theory we have reviewed so far is applicable to all environments. More specifically, we are concerned in this book on wireless communications inside buildings. These principles have proven right for hundreds of years, yet are almost applicable to all scenarios.

One can easily predict (and indeed measure) the behaviour of various materials in isolation and determine that losses increase with frequency. For the moment, let us leave this assumption right until we formally explain the mechanism that makes this loss increase with frequency.

However, in the last decades, radio measurements inside buildings have proven a different outcome to that expected by the theory. If losses are examined in practical buildings, the building penetration loss has been found to decrease with frequency in some studies but to increase in others. Clearly the mechanisms involved are rather complex and hence need to be fully examined. In fact, such mechanisms tend to be more sophisticated than those related to outdoor environments. This is precisely one of the reasons why indoor radio propagation is considered much more difficult and challenging than outdoor radio propagation, and needs a good amount of understanding and effort.

The constitutive parameters of materials are themselves frequency-dependent, even for relatively uniform walls, due to the specific molecular structure of the materials used. For example, the relative permittivity for brick changes from 4.62 at 1.7 GHz down to 4.11 at 18 GHz. Commercial glass changes more drastically, from a permittivity of 4 at VHF to a relative permittivity of 9 at microwave frequencies.

In order to understand how these constitutive parameters further affect wave propagation, several propagation mechanisms need to be introduced, which are relevant to in-building systems.

In summary, we can conclude that the legacy from electromagnetic theory still applies well for practical buildings. However, we should also understand that the propagation mechanisms inside buildings are more complex and make the signal behave quite differently, as building geometries, shapes, roughness, etc., have an effect on the resulting field strength at a receiver.

4.3.2 Loss and Skin Effect

When the medium has significant conductivity, the amplitude of the wave decreases with distance travelled through the medium. This decrease depends also on the constitutive parameters of the materials, and is accounted for by the attenuation constant. In consequence, the field strength for $\vec{E}$ and $\vec{H}$ decreases exponentially as the wave travels through the medium, as shown in Figure 4.9. The distance through which the wave

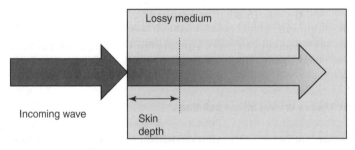

Figure 4.9 Skin effect.

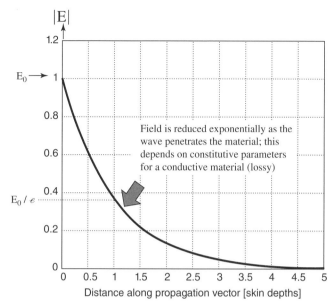

Figure 4.10 Calculation of skin depth.

travels before its field strength reduces to $e^{-1} = 0.368 = 36.8\%$ of its initial value is its *skin depth* δ (Figure 4.10), given by

$$\delta = \frac{1}{\alpha} \tag{4.26}$$

Thus, when a plane wave interacts with conductive material, mobile charges within the material are made to oscillate back and forth with the same frequency as the impinging fields. The movement of these charges constitutes a very rapidly alternating current, the magnitude of which is greatest at the conductor's surface, hence originating this decline in field strength.

Having the skin depth, it is possible to calculate the total field at a penetration distance x, as follows:

$$E(x) = E_0 e^{-x/\delta} \tag{4.27}$$

E_0 is the field strength at the boundary, x is the penetration distance and $E(x)$ is the field at the penetration distance. This equation reaffirms our initial statement made in Section 4.3.1: if it were for the skin effect in isolation, penetration losses for practical buildings will always have to increase with frequency! It is time now to examine in detail all the other mechanisms affecting these losses.

4.3.3 Reflection

Figure 4.11 shows a plane wave incident on to a plane boundary between two media with different permeability and permittivity. Both media are assumed lossless for the moment. The electric field vector may be in any direction perpendicular to the propagation vector. The propagation vector is at an angle θ_i to the surface normal at the point of incidence.

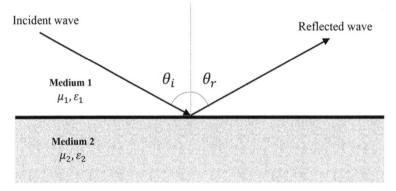

Figure 4.11 Reflection.

If Maxwell's equations are solved for this situation, the result is that two new waves are produced, each with the same frequency as the incident wave. Both the waves have their Poynting vectors in the plane, which contains both the incident propagation vector and the normal to the surface (i.e. normal to the plane of the paper in Figure 4.11). This is called the *scattering plane*.

The first wave propagates within medium 1 but moves away from the boundary. It makes an angle θ_r to the normal and is called the *reflected wave.*

When analysing reflection and refraction, it is convenient to work in terms of rays; in a homogeneous medium rays are drawn parallel to the Poynting vector of the wave at the point of incidence. They are always perpendicular to the wavefronts.

In the context of indoor radio propagation, reflections occur in confined spaces, hence producing multipath effects, which create deep fades (fast fading) in the received signal strength. These reflections can be so severe, especially in corridors and tunnels, that they can create other propagation effects; for example waveguiding.

4.3.4 Refraction (Transmission)

From Figure 4.12, the second wave produced from the incident wave travels on into medium 2, making an angle θ_t to the surface normal. This is the *transmitted* wave (also known as the *refracted* wave), which results from the mechanism of refraction.

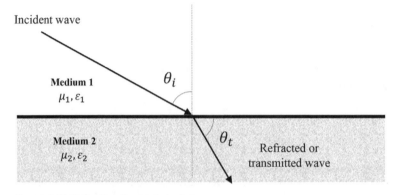

Figure 4.12 Refraction (transmission).

Figure 4.13 Refraction of light, which is observed when light changes from air to water.

Refraction depends on the angle of incidence as well as on constitutive parameters of materials. A very good example of how refraction affects a wave can be easily seen in a glass of water towards which a beam of light is directed, as seen in Figure 4.13. It will be seen that the ray of light 'bends' as it enters the water, which will mean that the medium has changed.

The energy has to come from somewhere: greater transmission through a medium implies less reflection and vice versa. Therefore a material that contains energy well (good reflector) will also not allow much energy to pass through it (good shielding).

For in-building systems, signal penetration through walls and floors is possible due to refraction. Wall and floor materials have a strong influence on the propagation (transmission) losses and thus in the amount of signal that can be received after going deeper into a building. As we will examine later in Chapter 4, many propagation models rely on wall and floor penetration losses at various frequencies to improve path loss prediction accuracy. *Building penetration loss* is the transmission loss experienced by a signal as it penetrates an external wall of a building. This parameter is used when coverage is to be provided from an external macrocell inside a building. Other mechanisms should also be taken into account for better predictions, such as diffraction, which is explained next.

4.3.5 Diffraction

This chapter has described ways in which the interactions between plane waves and infinite plane surfaces can be calculated with high accuracy, provided that the constitutive parameters are known. In practice, however, these conditions are rarely fulfilled and approximations must be made. For example, in a typical in-building scenario, the source

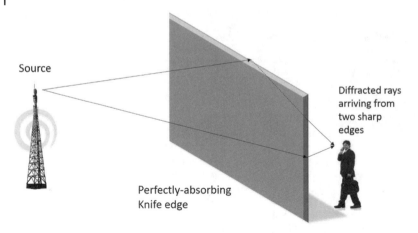

Figure 4.14 Diffracted rays occurring behind an obstructing obstacle.

or transmit antenna could be located inside the building in one floor and a receive antenna could be in another floor. Various reflections on walls and furniture, refractions through walls as the wave is transmitted, some scattered energy and other effects may be present before the radio wave reaches the receiver. These effects strongly depend on frequency and specific building materials, plus the characteristics of the furniture.

One of the effects that often can be either neglected or underestimated is when the path from the transmitter to the receiver is blocked, probably assuming that no energy is present in the *shadow region* behind an absorbing obstruction. This suggests that there is an infinitely sharp transition from the shadow region to the illuminated region outside. In practice, however, shadows are never completely sharp and some energy does propagate into the shadow region, as depicted in Figure 4.14.

The effect due to which waves can propagate even into the shadow region behind an obstruction is known as *diffraction* and can most easily be understood by using Huygen's principle:

- Each element of a wavefront at a point in time may be regarded as the centre of a secondary disturbance, which gives rise to spherical wavelets.
- The position of the wavefront at any later time is the envelope of all such wavelets.

Diffraction makes non-line-of-sight (NLOS) propagation possible inside buildings. For example, the signal diffracts into a room through an open door or windows, even for walls with high transmission losses. As long as the obstructions or 'sharp edges' are within the 60% of the first Fresnel zone radius, diffraction occurs and causes the signal to reach areas that otherwise would not be possible.

4.3.6 Scattering

The reflection processes discussed so far have been applicable to smooth surfaces only; this is termed *specular reflection*. When the surface is made progressively rougher, the reflected wave becomes scattered from a large number of positions on the surface, broadening the scattered energy. This reduces the energy in the specular direction and increases the energy radiated in other directions, as shown in Figure 4.15.

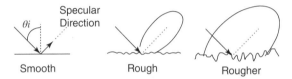

Figure 4.15 Rough surface scattering.

The degree of scattering depends on the angle of incidence and on the roughness of the surface in comparison to the wavelength. The apparent roughness of the surface is reduced as the incidence angle comes closer to the grazing incidence ($\theta_i \approx 90°$) and as the wavelength is made larger. This means that some surfaces can be considered 'smooth' at certain frequencies, but exhibit 'roughness' at higher frequencies.

If a surface is to be considered smooth, then waves reflected from the surface must be only very slightly shifted in phase with respect to each other. A reasonable criterion for considering a surface smooth is if this phase shift is less than 90°, which leads to the *Rayleigh criterion*:

$$\Delta h < \frac{\lambda}{8 \cos \theta_i} \qquad (4.28)$$

The parameter Δh is known as the *surface height difference*, as shown in Figure 4.16. All surfaces are considered smooth if the grazing incidence is considered.

For various building architectures, it is common that the inside walls have a rough finishing that could lead to scattering, which needs to be taken into account for extra losses. Thus, it is necessary to compute these losses.

As can be seen from Figure 4.16, the surface height difference is not a deterministic parameter that can be calculated straightforwardly; this difference varies randomly for every wall and thus we need to use probability to describe it.

It is correct to assume that Δh follows a Gaussian or normal distribution (Figure 4.17), for which an expected value or mean of the surface height $\mu_{\Delta h}$ and a standard deviation $\sigma_{\Delta h}$ can be defined.

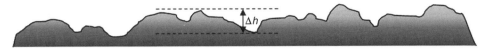

Figure 4.16 Surface height difference Δh.

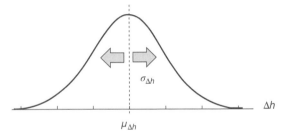

Figure 4.17 Gaussian distribution for surface height difference.

Once a surface has been determined to be 'rough' according to the Rayleigh criterion, the amount of scattered energy needs to be computed. When the surface is rough, the reduction in the amplitude of the specular component may be accounted for by multiplying the corresponding value of the Fresnel specular reflection coefficient R by a roughness factor f, which depends on the angle of incidence and on the standard deviation of the surface height $\sigma_{\Delta h}$. The formulation for this factor is:

$$f(\sigma_{\Delta h}) = e^{-0.5\left(\frac{4\pi\sigma_{\Delta h}\cos\theta_i}{\lambda}\right)^2} \tag{4.29}$$

Therefore, an effective reflection coefficient R_{rough} needs to be considered:

$$R_{rough} = Rf(\sigma_{\Delta h}) \tag{4.30}$$

Effective reflection coefficient

Fresnel reflection coefficient

Roughness factor

Finally, the total reflected field $\vec{\mathbf{E}}_r$ (specular + scattered) is

$$\vec{\mathbf{E}}_r = R_{rough}\vec{\mathbf{E}}_i \tag{4.31}$$

Note that in equation (4.31), the incident $\vec{\mathbf{E}}_i$ and reflected electric fields are vectors, since they can contain components in either parallel, perpendicular or both directions; for example the latter will be the case for circular polarization. Details on how to compute the Fresnel specular reflection coefficient R can be found in Saunders and Aragón-Zavala (2007).

4.3.7 Waveguiding

Some corridors, airport piers and long enclosed spaces exhibit reflections and refractions similar to those of a waveguide or tunnel. The waves reflect off the walls many times and this has the effect of carrying the wave along the waveguide 'further' than would normally be the case. The only provision is that the width of the waveguide must be much greater than a wavelength, so unlike many other propagation effects the propagation loss in waveguides actually reduces with frequency. This can be seen in Figure 4.18.

To understand waveguiding better, let us examine how path loss changes with distance in an enclosed space, but with the transmitting antenna outside of the waveguide itself (e.g. imagine a transmitter in a large hallway with a corridor leading off it). Notice how the field drops rapidly in the first few metres following entrance to the space. This drop is dependent on the distance of the transmitting antenna from the entrance, the angle of arrival of waves into the space relative to its axis, its cross-sectional area and the frequency of operation. Propagation within the space is further influenced by the space's shape and by the construction materials (see Figure 4.19).

Waveguiding is so important for enclosed spaces such as corridors, for example, that they should be understood and taken into account. Failure to account for waveguiding may result in being too pessimistic in signal coverage along this enclosed space, which may result in exceeding the anticipated signal levels and possibly interfering with other systems.

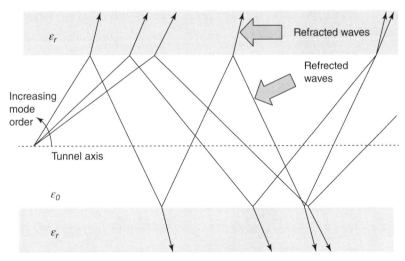

Figure 4.18 Tunnel propagation, similar to what happens with waveguiding.

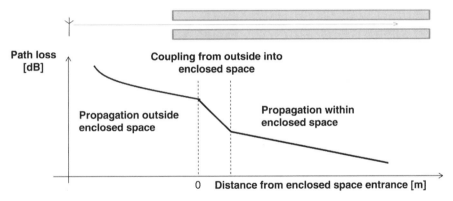

Figure 4.19 Waveguiding effect.

Waveguiding accounts for reflections and refractions given in areas such as corridors or other narrow gaps between walls, which are characteristic of many buildings. This effect depends on angles, material types and distances between walls, and produces what is known as a *waveguiding gain*, increasing the penetration depth along corridors. Because of this, the path loss exponent in such environments appears to be less than 2, which is the one for free-space loss! See Figure 4.20 for an example of how waveguiding can affect signal strength along a corridor – notice how the signal strength is maintained along corridors, despite the large distances to the transmit antenna.

4.4 Effects of Materials

Maxwell's curl equations described in Section 4.1 also dictate special behaviour when impinging on a material. Such behaviour depends on many factors, which can be described depending on specific characteristics of the material. Some of these factors are:

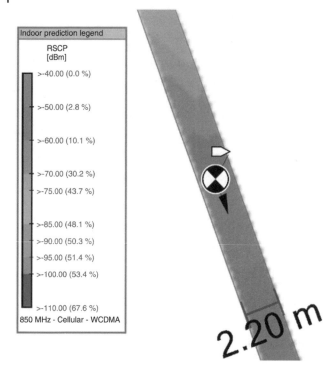

Figure 4.20 Strength of waveguiding inside a building.

- The material properties, also known as constitutive parameters; that is conductivity, permittivity and permeability. These characteristics dictate whether a material is to be considered as either a good conductor or a good insulator and therefore how a wave propagates through them varies (phase velocity, impedance, etc.). For practical cases, conductivity and permittivity are the most important.
- The shape of the material and its roughness. As described in Section 4.3, reflections and scattering depend on these properties, thus affecting the amount of loss that is produced by the materials of the walls upon which the radio wave is impinging. Diffraction is also affected, as 'sharp edges' accentuate this effect.
- The frequency and wavelength of the wave, which in turn makes a radio wave of a specific wavelength exhibit different behaviour upon impinging or going through the same material; that is propagation effects depend on frequency and hence wavelength (Saunders and Aragón-Zavala, 2007).
- The type of wave; that is plane, spherical, cylindrical, etc.

Some properties of electromagnetic waves are changed depending on the constitutive parameters of materials to which they propagate, such as phase velocity and wavelength. However, materials also change waves in many other ways. For example, they change direction and phase, as well as losing energy (attenuation) for those cases where waves propagate through lossy materials.

The constitutive parameters of materials are themselves frequency dependent, even for relatively uniform walls, due to the specific molecular structure of the materials used. For example, the relative permittivity μ_r for brick changes from 4.62 at 1.7 GHz down to

4.11 at 18 GHz. The commercial glass change more drastically, from a permittivity of 4 at VHF to a relative permittivity of 9 at microwave frequencies.

The building wall structure frequently has several layers, setting up multipath interference and associated resonances within the structure. These can be analysed by treating each layer as a section of a transmission line, with a characteristic impedance determined by the wave impedance, the frequency and the angle of incidence. At each boundary reflections are created leading to a set of multiple reflections with interference between each contribution. For a deeper study on these effects, refer to Saunders and Aragón-Zavala (2007).

Another good example of the effects of materials in radio propagation is that of a metallized window. For a very lossy (highly conductive) medium or for very high frequencies, the skin depth is small, so most of the current stays on the surface of the material and the penetration depth is small. Therefore a window with metallization will have a high current on its surface near an antenna but will allow little wave penetration.

Conversely, even a highly conductive material that is thinner than its skin depth will still allow energy to pass through.

4.5 Path Loss

An average decrease of received field strength as the mobile moves away from the base station (macrocell if outdoors, picocell/femtocell if indoors) is often used to determine the range of a wireless system. Most of the indoor path loss models, which will be described in Chapter 5, aim to predict this average decrease, or path loss, which is dependent on distance at given frequencies. A formal definition of path loss is therefore required here.

The *path loss* between a pair of antennas is the ratio of the transmitted power to the received power, usually expressed in decibels. It includes all of the possible elements of loss associated with interactions between the propagating wave and any objects between transmit and receive antennas.

4.5.1 Median Path Loss

In the case of channels with large amounts of fast fading (deep signal fades as the mobile receiver moves away from the transmitter), such as mobile channels, the path loss applies to the power averaged over several fading cycles (the local median path loss). This path loss is hard to measure directly, since various losses and gains in the radio system also have to be considered. These are best accounted for by constructing a *link budget*, which is usually the first step in the analysis of a wireless communication system and is discussed in detail in Section 4.5.2. In fact, for in-building systems, multipath effects are quite severe, thus producing strong fast fading and the median path loss is used for system design.

In order to define the path loss properly, the losses and gains in the system must be considered. The elements of a simple wireless link are shown in Figure 4.21 and are as follows:

- Transmit power, P_T. The amount of available power taken at the output of the transmitter. This is expressed either in watts, milliwatts or picowatts, or in decibels: dBm, dBW.

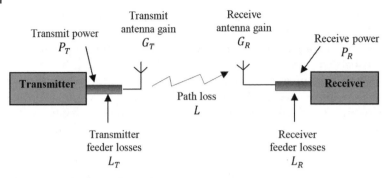

Figure 4.21 Elements of a wireless communication system.

- Feeder loss, L_T. These losses account for cable and connector losses in the transmitter. They are unitless or expressed in dB.
- Transmit antenna gain, G_T. Gains are expressed either in dBi or dBd, if in decibels, or are unitless.
- Path loss, L. The path loss units are dB.
- Receiver antenna gain, G_R. Also expressed in dBi, dBd or unitless; this is the antenna gain at the receiver end.
- Feeder loss, L_R. These losses are referred to the receiver end and are taken between the antenna and receiver; expressed in dB.
- Received power, P_R. Expressed in milliwatts, watts or if in decibels: dBm, dBW.
- Effective isotropic transmitted power, P_{TI}. This parameter refers to the power level that is radiated from the antenna, in dBm or dBW, taking into account transmit antenna gain and cable losses, as follows:

$$P_{TI} = P_T + G_T - L_T \tag{4.32}$$

Note that all parameters in uppercase are expressed in decibels. Another way to call P_{TI} is as an effective isotropic radiated power, or EIRP. This term is more popular for radio communications and is extensively used for terrestrial and satellite systems.

- Effective isotropic received power, P_{RI}. Power received at the antenna, before taking into consideration receive antenna gain or cable losses:

$$P_{RI} = P_R + L_R - G_R \tag{4.33}$$

Having the effective isotropic radiated and received powers, it is possible to obtain the total path loss from P_{TI} and P_{RI} as follows:

$$L = P_{TI} - P_{RI} \tag{4.34}$$

The elements presented in Figure 4.21 are generic to any wireless system, including in-building radio systems.

4.5.2 Link Budgets

A *link budget* is a procedure that is done to obtain the estimated received signal strength in a wireless link, taking into account transmit power, gains and losses in the system.

Thus, from Figure 4.21, the received power P_R can be calculated taking into account all gains and losses in the system:

$$\text{Received power} = \text{Transmitted power} + \text{Gains} - \text{Losses} \qquad (4.35)$$

When designing an indoor wireless system, one of the main requirements (as seen in Chapter 3) is to estimate its range for given system parameters (antenna gains, cable and feeder losses, transmit power, etc.). Since most of these parameters can be known, the path loss must be estimated as accurately as possible, as it depends on the propagation environment – reflections, refractions, diffraction, scattering and waveguiding affect the way in which the signal propagates inside the building. Therefore, to predict the path loss as accurately as possible is one of the main goals of an in-building radio designer, allowing the range of a radio system to be determined before installation. The path loss can be simply computed by rearranging the link budget equation, depicted in equation (4.35), to determine the path loss corresponding to a given received power:

$$L = P_T - L_T + G_T - P_R + G_R - L_R \qquad (4.36)$$

For indoor scenarios, link budgets are often constructed based on the specific distribution system in use. Losses are accounted for in a different way, as active distribution systems usually compensate for such losses with the use of amplification. Examples of indoor link budgets will be given in Chapter 11, where case studies are examined, since distribution systems need to be introduced, which is done until Chapter 9. For now, let us focus on the basics first to understand the 'full picture' and be able to use this knowledge later on.

4.5.3 Receiver Sensitivity

When the received power falls beyond a certain threshold level, which provides just acceptable communication quality, the wireless link is said to be right at the edge of coverage. This minimum received signal strength is known as *receiver sensitivity* and is used in link budgets to estimate coverage range based on parameters such as: frequency, base station antenna height, transmit power, antenna gains, etc.

For example, for GSM cellular systems, this receiver sensitivity is around −104 dBm. This means that a dropped call due to a low signal strength level may occur if the received signal goes beyond this threshold; for example −106 dBm.

4.5.4 Maximum Acceptable Path Loss (MAPL)

A very useful parameter used to delimit the range of a wireless system is known as the *maximum acceptable path loss* (MAPL). The maximum range of such a system occurs when the received power falls beyond the given receiver sensitivity. Thus, the value of a path loss for which this power is received is the MAPL, expressed in decibels.

Note that as a consequence of the reciprocity theorem, the definition of the path loss is unaltered by swapping the roles of transmit and receive antennas, provided that the frequency is maintained and the medium does not vary with time. However, the maximum acceptable propagation loss may be different in the two directions, as the applicable losses and sensitivities may be different. For example, a base station receiver is usually designed to be more sensitive than the mobile, to compensate for the reduced transmit power available from the mobile.

The specific value of MAPL is very dependent on the technology used and the coverage and signal quality requirements of the particular project. However, it would be very unusual for it to be outside the range of $70 - 150$ dB.

4.5.5 Free-Space Loss

Free-space loss (FSL) is the minimum path loss experienced between two antennas, assuming free-space conditions are given between the two points. To understand the mechanism upon which free-space loss is defined, let us examine a point source (antenna) producing radio waves. Power radiates equally in all directions and is spread over the surface of a sphere, so the power available in an antenna of a fixed size decreases with the square of the distance. This is known as the *inverse square law* and is very easy to verify experimentally.

Additionally, the gain of a fixed size of antenna increases linearly with wavelength (or it is inversely proportional to frequency). Since this affects both transmit and receive antennas, the power also decreases with the square of frequency.

To compute free-space loss, the Friis equation is used, which takes into account the antenna gains. If the gains are not considered, a more useful expression in decibels for the free-space loss L_{FSL} is used, as follows:

$$L_{FSL} = 20 \log \left(\frac{4\pi r f}{c} \right) \tag{4.37}$$

In (4.37), f is the frequency in Hz and r is the distance between the transmitter and receiver in metres.

For in-building scenarios, free-space loss can only be assumed for very short distances between transmit and receive antennas, since clutter and objects are normally close to either side of the link and within the first Fresnel zone radius, thus causing multipath or diffraction. An example of this is when the transmit power of a Wi-Fi access point is to be calculated, for which a laptop is brought close to it (1.5 m) and the received signal is recorded; for example suppose -23 dBm. Since the antenna gains of transmit and receive antennas are known (2 dBi and 0 dBi, respectively), the free-space loss is calculated as

$$L_{FSL} = 20 \log \left(\frac{4\pi \times 2.4 \times 10^9 \times 1.5}{3 \times 10^8} \right) = 43.6 \, \text{dB}$$

Therefore, using the link budget equation (4.35) and rearranging for transmit power yields

$$P_T = P_R - G_T + L_{FSL} - G_R = -23 - 2 + 43.6 - 0 \approx 18.6 \, \text{dBm}$$

$$p_T = 10^{P_T/10} = 10^{18.6/10} = 72.4 \, \text{mW}$$

4.5.6 Excess Loss

As free-space loss is the minimum path loss experienced between two antennas under normal conditions, as defined in Section 4.5.5, it will be expected that for most conditions and situations, FSL could be taken as this true minimum. However, there are exceptions for effects like waveguiding (Section 4.3.7), where the total path loss is actually less than that of free-space loss – this effect can be explained as multiple reflections occurring, say, on the walls of corridors or as enclosed spaces adding coherently and contributing to the received signal strength, seeing the effect as a 'waveguiding gain'. This is a very special case and, until now, there is nothing else reported that may follow this behaviour.

All additional losses to the 'minimum' free-space loss are called 'excess losses'. Since free-space loss only takes into account frequency and distance, the excess loss considers other propagation effects, especially inside buildings, such as diffraction, wall and floor losses and scattering. In mathematical terms, the excess loss L_{ex} is given by

$$L_{ex} = L - L_{FSL} \tag{4.38}$$

In other words, the excess loss can be computed if the total path loss L is known, and information about transmit–receive distance as well as frequency is available. In Chapter 5 we will see that many of these propagation effects are taken into consideration in the indoor propagation models, and some of them add these factors to the free-space loss.

4.6 Fast Fading

In a wireless channel, the noise sources that affect it can be subdivided into multiplicative and additive effects. The additive noise arises from noise generated within the receiver itself, such as thermal and shot noise in passive and active components, and also from external sources, such as atmospheric effects, cosmic radiation and interference from other transmitters and electrical appliances. The multiplicative noise arises from the various processes encountered by transmitted waves on their way from the transmitter antenna to the receiver antenna.

It is conventional to further subdivide the multiplicative processes in the channel into three types of fading: path loss, shadowing (or slow fading) and fast fading (or multipath fading), which appear as time-varying processes between the antennas. All of these processes vary as the relative positions of the transmitter and receiver change and as any contributing objects or materials between the antennas are moved.

In Section 4.5.1 we discussed the median path loss, which is often used as the main parameter for wireless radio design. However, the two other multiplicative processes are very relevant and should not be underestimated. Fast fading is discussed here, leaving the slow fading definition for Section 4.7.

Fast fading is normally characterized by rapid signal variations on the scale of half-wavelength and frequently introduces variations as large as 35 to 40 dB. It results from the constructive and destructive interference between multiple waves reaching the mobile from the indoor base station, as depicted in Figure 4.22. This type of fading is also known as multipath fading.

For indoor scenarios, multipath strongly influences the behaviour of the signal strength, as the mobile moves away from the base station. An example of fast fading affecting propagation inside a building is shown in Figure 4.23, where thousands of signal strength samples were collected in a university library building at 872 MHz.

Figure 4.24 shows the environment where the samples were collected. Notice that this floor has many bookshelves and desks for students, which create strong multipath effects that produce variations of around 30 dB, as shown in Figure 4.23.

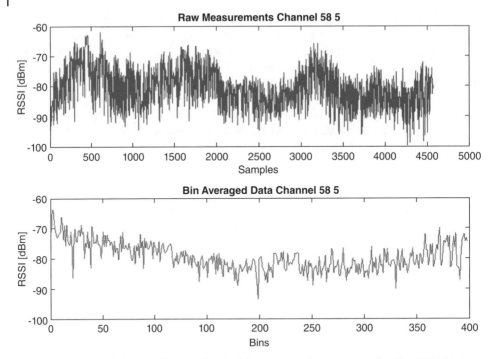

Figure 4.22 Multipath fading effects inside a building compared to averaged data for 1900 MHz in a university campus building.

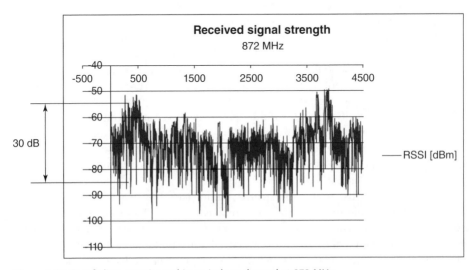

Figure 4.23 Fast fading experienced in an indoor channel at 872 MHz.

(a)   (b)

Figure 4.24 University library building for the measurements reported in Figure 4.23.

During system design, fast fading is often removed by filtering, as an additional allowance of system resources is made to overcome deep fades.

4.7 Shadowing (Slow Fading)

Median path loss does not consider the fact that the surrounding environmental clutter may be vastly different at two different locations having the same transmitter–receiver separation. This leads to measured signals that are vastly different than the median path loss of Section 4.5.1. *Shadowing* is a zero-mean log-normally distributed random variable (Gaussian in dB), which describes the random effects that occur over a large number of measurement locations that have the same transmitter–receiver separation but different levels of clutter on the propagation path.

For indoor environments, shadowing arises due to the varying nature of the particular obstructions between the base station and the mobile, such as particularly tall pieces of furniture or partitions within a room, or simply the specific layout of the building, as depicted in Figure 4.25.

Failure to take shadowing into account while predicting path loss can result in areas of a building wrongly predicted. Those areas with less predicted path loss would produce outage spots for which coverage levels will be less than the required levels. On the other hand, if path loss is predicted in excess, predictions would dictate that signal strength values will be less than in reality and thus interference to other cells or systems is prone to occur. Both scenarios are shown in Figure 4.26, in comparison with a full prediction of the same area considering shadowing.

The standard deviation of the shadowing is known as *location variability* and normally has higher reported values than those of macrocellular environments. It is usual to model shadowing in indoor environments as lognormal, just as in other environments. However, there is some evidence that the location variability is itself more environment-dependent. Reported values span from 3 to 15 dB, which depend on frequency and environment (Fiacco *et al.*, 2000).

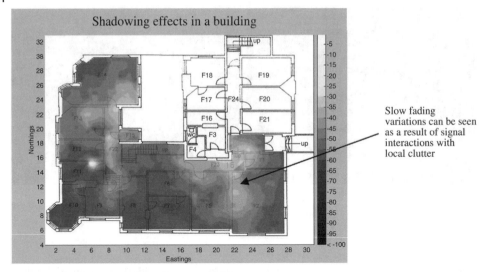

Figure 4.25 Indoor shadowing effects in a building.

4.8 Building Penetration Loss

Many efforts in characterizing the losses caused by building walls have been reported in the literature, having contradictory conclusions and substantially different methodologies, something that makes a comparison amongst studies an almost impossible task. One of the regulatory bodies in Europe that has performed a substantial amount of work in this field is Ofcom, the UK regulator, issuing reports and recommendations for penetration losses. Ofcom's findings and approach will be taken in this section to discuss how building penetration losses are characterized.

The Ofcom February 2009 mobile liberalization consultation (Ofcom, 2009) defined building penetration loss (BPL) as:

> The difference (in decibels) between the median of the location variability of the signal level at the building location, as predicted by the outdoor propagation model, and the signal level inside the building at the same height above ground, with multi-path fading spatially averaged for both signals.

The BPL is only one component of the overall propagation loss between a base station and an (indoor) mobile device and all need to be considered to determine the overall impact on network performance.

In practice, there are three key variables that determine the extent of BPL as follows (Ofcom, 2012):

- Frequency
- Depth (or consistency) of penetration
- Building (or clutter) type.

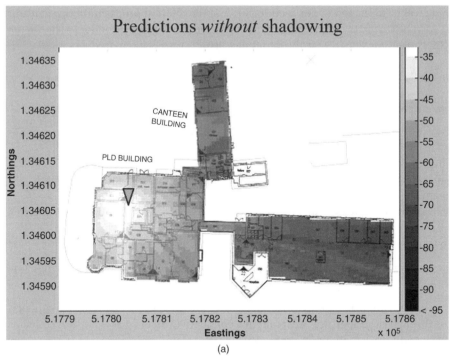

(a)

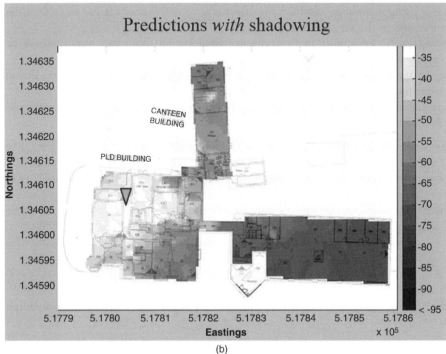

(b)

Figure 4.26 Consequences of not taking shadowing into account for a building: (a) predictions without shadowing; (b) predictions with shadowing.

These will be discussed in this section, but first the relevant propagation mechanisms involved in BPL need to be examined. An example of a consultation conducted by Ofcom in 2009 is used for illustrating the variations of BPL across many frequencies.

4.8.1 Radio Wave Propagation into Buildings

The main propagation mechanisms of relevance are:

- Absorption loss due to penetration through bulk materials (e.g. solid, homogeneous walls)
- Diffraction around edges (e.g. window frames, reinforcement rods, etc.)
- Scattering from rough surfaces and objects small compared with a wavelength (e.g. furniture and rough walls)
- Multipath effects (e.g. from the combination of multiple reflections from internal walls)
- Waveguide effects (e.g. along long, straight corridors).

These mechanisms exhibit different frequency dependencies, as follows (Ofcom, 2009):

- Loss through uniform external walls and metallized glass will increase with frequency.
- Loss through large apertures with minimal conductivity (plain glass windows) remains unchanged with variations in frequency.
- Loss in regions dominated by diffraction and no line-of-sight increases with frequency.
- Losses through smaller windows may decrease with frequency if the wavelength is comparable to the size; i.e. around 25 cm.
- Loss at particular points in the presence of a strong multipath may increase or decrease with frequency over limited frequency ranges.
- Loss through internal walls and through floors will increase with frequency, giving greater differences for a greater target in-building depths.
- Corridors may exhibit unusually low penetration loss, which either remains unaffected by frequency or tends to decrease.

The balance between the mechanisms – and hence the overall frequency dependency – will depend on the materials and geometry of the building in question as well as to a lesser extent the geometry of the surrounding buildings and the relative placement of the transmitter and receiver. Beyond this simple summary, it is important to recognize that resonance and multipath effects mean that losses over a given range can be positively or negatively dependent on frequency, even within a single mechanism.

4.8.2 Variations with Frequency

The Ofcom February 2009 mobile spectrum liberalization consultation (Ofcom, 2009) provided an overview of the relevant propagation mechanisms and effects, concluding that:

> Any given building will be a mixture of these effects, so a large spread of values is anticipated relative to the overall trend. Nevertheless, the general trend of increasing loss with frequency arising from the skin effect and other mechanisms

Table 4.1 BPL variations with depth and clutter. Taken from Ofcom (2009).

Depth 0	Increasing with Frequency (Rising) (Base case)		
Frequency	Dense urban	Urban	Suburban
900 MHz	7.0	5.0	3.0
1800 MHz	7.9	5.9	3.9
2.1 GHz	8.6	6.6	4.6

Depth 1	No Variation with Frequency (Constant)			Increasing with Frequency (Rising) (Base case)			Increasing with Frequency (Rising at higher rate)		
Frequency	Dense urban	Urban	Suburban	Dense urban	Urban	Suburban	Dense urban	Urban	Suburban
900 MHz	11.5	9.5	7.5	11.5	9.5	7.5	11.5	9.5	7.5
1300 MHZ	11.5	9.5	7.5	13.1	11.1	9.1	19.5	17.5	15.5
2.1 GHz	11.5	9.5	7.5	14.0	12.0	10.0	21.5	19.S	17.5

Depth 2	No Variation with Frequency (Constant)			Increasing with Frequency (Rising) (Base case)			Increasing with Frequency (Rising at higher rate)		
Frequency	Dense Urban	Urban	Suburban	Dense Urban	Urban	Suburban	Dense Urban	Urban	Suburban
900 MHz	14	12	10	14	12	10	14	12	10
1800 MHz	14	12	10	16	14	12	22	20	18
2.1 GHz	14	12	10	17	15	13	24	22	20

suggests an overall trend to increase with frequency, particularly when the penetration depth into a building is high. The frequency variation might thus take the form of Figure 23 for any given building.

This Figure 23 cited above from Ofcom (2009) shows that path loss decreases with frequency over a limited range, especially in a region termed *wall cavity*. At higher frequencies there is another loss decrease peak arising from the skin effect, within a region denoted as *materials matching*. Otherwise, the loss increase with frequency

Table 4.2 Comparison of BPL assumptions made by Ofcom (2012).

		Frequency (MHz)					Frequency exponent (dB/decade)
		800	900	1800	2100	2600	
Sept '07	Mean 8PL (dB) Sept 07		10	12	13		8.2
	Sept 07 Standard deviation		6	6	6		
Feb '09	Mean BPL (dB) Feb 2009 suburban base case depth 2		10	12	13		8.2
	Mean BPL (dB) Feb 2009 suburban base case depth 1		7.5	9.1	10		6.8
	Mean BPL (dB) Feb 2009 suburban base case depth 0		3	3.9	4.6		4.3
	Mean BPL (dB) Feb 2009 suburban no freq variation depth 2		10	10	10		0.0
Mar '11	Mean BPL (dB) Mar 2011 suburban Depth 1	7.2		9.3		11	7.4
	Mar 11 depth 1 standard deviation	6		6		6	
	Mean BPL (dB) Msr 2011 suburban Depth 2 +	9.6		14.8		19.1	18.6
	Mar 11 depth 2+ standard deviation	7		9		9	
Jan'12	Mean suburban Jan '12 lower bound 1m	1.85	1.85	1.85	1.85	1.85	0.0
	Mean suburban Jan '12 lower bound 15m	6.39	6.39	6.39	6.39	6.39	0.0
	Jan 12 lower bound standard deviation	4		5.4		6	
	Mean suburban Jan '12 upper bound 1m	3.71	3.99	5.63	6	6.5	5.5
	Mean suburban Jan '12 upper bound 5m	6.29	6.86	10.17	10.9	11.92	11.0
	Mean suburban Jan '12 upper bound 10m	9.55	10.46	15.86	17.06	18.72	17.9
	Mean suburban Jan '12 upper bound 15m	12.79	14.06	21.54	23.2	25.5	24.8
	Jan 12 upper bound standard deviation	8		10.8		12	

follows a linear behaviour, as would be expected simply by applying the free-space loss equation.

Despite this, there was a great effort to characterize the behaviour of BPL with frequency by conducting extensive measurement campaigns; the overall finding was that there was a clear trend for BPL to increase with frequency, but the variability of data amongst buildings at a given frequency was substantially larger than the variation of the mean with frequency. If some of this variation may also be due to methodological differences amongst authors, the studies concluded that it was difficult to assert a firm value for the variation of losses in the mean.

4.8.3 Variations with Depth and Clutter

A new set of assumptions were created in Ofcom (2009). In contrast to the previous work, these additionally varied with depth (to allow examination of shallower and deeper penetration cases) and clutter type, as shown in Table 4.1.

4.8.4 Comparison of Assumptions Made by Ofcom

A summary of key cases of variations of BPL with frequency, depth and clutter are illustrated in Table 4.2 (Ofcom, 2012). Note that the frequency exponent in this table is based on the difference between the loss values in the widest frequency range included in the assumptions, as follows:

$$\text{BPL frequency exponent(dB per decade)} = \frac{L_2 - L_1}{\log_{10}\left(\frac{f_2}{f_1}\right)} \tag{4.39}$$

4.9 Conclusion

Radio propagation inside buildings exhibits a more complex behaviour than in outdoor environments, thus making it more difficult to predict. We have started with the fundamentals of electromagnetism to enable the reader to understand and comprehend its origins. Maxwell's equations are much more than a frightening form of expressing electromagnetic effects and constitute the basis of many practical devices and technologies today!

It has been seen that if some propagation mechanisms are not taken into account while attempting to predict path loss, errors can be produced that affect system performance – this applies both ways, either if the path loss is underestimated or overestimated. Waveguiding is also typical of in-building scenarios and is often neglected or wrongly predicted.

An introduction to link budgets has also been presented, for which some modifications are to be done when used in buildings. However, this is presented later on in the book, when technologies to provide and distribute radio coverage inside buildings are revised.

A deep understanding of the propagation mechanisms that affect link performance is thus essential if path loss and/or signal strength is to be predicted inside buildings with reasonable accuracy. The application of this knowledge in the construction of indoor propagation models was discussed in Chapter 4.

Finally, the dependency of building penetration loss and in general the dependency of transmission loss with frequency has been the study of researchers for many years. However, there are relatively few studies that measure using the same measurement approach across the range of frequencies of interest and over a large sample of buildings, making it difficult to compare directly between studies. Each study of building penetration loss is approached in a different manner. Some studies look to derive theoretical models based on the physical properties of materials and then to tune these models using precise measurement data of individual building materials. Other studies are purely empirical, measuring a representative sample of real buildings and forming a conclusion from these.

There is a great deal of variability in the approach as well as differences in the specific buildings examined. As a consequence of this, the spread of values of BPL arising from these studies is large. Some of the factors that will affect the outcome of BPL studies are, for example:

- Cell type, for example macro versus micro
- Building types, for example office building versus residential home
- Building location, for example isolated versus built-up area
- Penetration depth into building, for example external wall or deep inside
- Number of buildings surveyed
- Materials, for example isolated or part of a wall or part of a building.

Also, it has been found throughout the years that propagation loss changes depending on the penetration depth inside the building. This is strongly related to the types of buildings and construction materials and thus needs to be further carefully inspected and researched.

References

Fiacco, M., Stavrou, S., Browne, J., Jones, S., and Saunders, S. (2000) Measurement and modelling of small-cell shadowing cross-correlation at 2 GHz and 5 GHz, in *Proceedings of the 7th International Symposium on Antennas and Propagation, ISAP 2000*, Fukoka, Japan, August 2000.

Hayt, W.H. and Buck, J.A. (2011) *Engineering Electromagnetics*, 8th edition, McGraw-Hill Higher Education, USA. ISBN 978-007338066-7.

Kraus, J.D. and Fleisch, D. (1999) *Electromagnetics*, 5th edition, McGraw-Hill Higher Education, USA. ISBN 978-007116429-0.

Maxwell, J.C. (1861) On physical lines of force, *The London, Edinburgh and Dublin Philosophical Magazine*, Fourth Series, pp. 162–195, March 1861.

Maxwell, J.C. (1865) A dynamical theory of the electromagnetic field, *Philosophical Transactions of the Royal Society of London*, pp. 459–512, 1865.

Ofcom (2009) Application of spectrum liberalisation and trading to the mobile sector – A further consultation. Ofcom, February 2009. URL: http://stakeholders.ofcom.org.uk/consultations/spectrumlib/.

Ofcom (2012) Propagation losses into and within buildings in the 800, 900, 1800, 2100 and 2600 MHz bands. Report for Ofcom by Real Wireless, 19 July 2012. Annex A within zip file available at: http://www.ofcom.org.uk/static/spectrum/ RW_investigation_of_combined_award_technical_issues.zip.

Sadiku, M.N.O. (2007) *Elements of Electromagnetics*, 4th edition, Oxford University Press, Oxford (UK) and New York (USA). ISBN 978-0-19530048-3.

Saunders, S. and Aragón-Zavala, A. (2007) *Antennas and Propagation for Wireless Communication Systems*, 2nd edition, John Wiley and Sons, Ltd, Chichester. ISBN 0-470-84879-1.

5

Channel Modelling

Radio signals propagate in real indoor environments in all sorts of ways, since the in-building channel is quite challenging in terms of specific objects affecting the electromagnetic wave, leading to the propagation effects presented in Chapter 4. For system design and planning, it is essential to predict signal strength levels inside the building, as well as other important parameters (e.g. power delay profile) to guarantee system performance and dimension required resources.

This chapter focuses on channel models and techniques often used for indoor wireless systems, in order to provide useful tools that assist the engineer in the design of such systems. Although most of the emphasis is given to narrowband models used to estimate the local mean, a section to wideband channel modelling is included, since this is important to determine to a great extent the maximum data rate that can be achieved in a network.

5.1 The Importance of Channel Modelling

The basic parameters of antennas and radiating cables can be used together with an understanding of propagation mechanisms to calculate the range of a wireless communication system. For this purpose, mathematical expressions or models are a useful way to characterize the propagation phenomena for various environments, which can then be applied in predicting signal strength for similar scenarios. Antennas and radiating cables will be the subject of study for Chapter 6; indoor propagation models are presented in this chapter.

Accurate models assist the system designer when performing predictions to assess the expected coverage inside a building, while ensuring system performance as these predictions are closely related to real values.

On the basis that the propagation models are correctly chosen, predictions can be made quickly, even before visiting a building, and determine at a first glance the expected range of antennas and their potential locations. This applies to many wireless technologies and venues, although it becomes more useful for large buildings, where hundreds of antennas need to be deployed (e.g. airports, shopping centres, stadiums, etc.).

There are other cases in which predictions can be useful; for example, for areas where measurements cannot be made – access restrictions to designated areas in many

Indoor Wireless Communications: From Theory to Implementation, First Edition. Alejandro Aragón-Zavala.
© 2017 John Wiley & Sons Ltd. Published 2017 by John Wiley & Sons Ltd.

buildings can result in having model predictions as the only choice for estimating radio coverage.

Propagation models can also be used to predict signal strength or path loss everywhere in a building and allow optimization of antenna locations without repeated measurements. For example, an antenna can be 'moved' to a different location and by performing a new prediction, on the basis that the model used is accurate enough; there is no need to revisit the building and perform measurements for this new antenna location (this can be quite handy especially if the building in consideration is 130 km away!).

5.2 Propagation Modelling Challenges

The modelling of any physical phenomenon in nature is in its own a challenging task. Scientists and researchers have spent fairly large amounts of time establishing mathematical models to describe how key physical mechanisms behave. Luckily for humanity, this has led to extraordinary developments in science and technology, thanks to the fact that once the model is available, experimentation is thus possible.

When a wireless system is to be deployed inside a building and modelling needs to be performed, further challenges are encountered:

- Propagation models exhibit a large variability with design parameters such as frequency, distance between transmitter and receiver, antenna heights, etc. This means that some models account for these effects explicitly in their mathematical expression and often are created having in mind parameter limitations; that is a specific model can only be used within a certain frequency range and antenna heights.
- There is a large variability of buildings, materials, layouts, etc., which make indoor models more difficult to be applicable in a 'general' form and thus some of its parameters need to be 'tunable'.
- Propagation effects inside buildings are quite complex and need to be taken into account to improve model accuracy – this can be difficult since in many cases it is hard to separate them.
- Radio measurements are used for both model tuning and validation, which can in principle be straightforward but in practice could represent further complications; for example how could we perform post-design validation measurements with a full stadium to account for spectators' body losses and reflections?

5.3 Model Classification

Indoor propagation models can be classified into many categories depending on channel characteristics, propagation environment, design approach, etc. These classifications are briefly presented and explained as follows.

5.3.1 Channel Bandwidth

Mobile radio systems for voice and low bit rate data applications can consider the channel as having purely narrowband characteristics, but the wideband mobile radio channel has assumed increasing importance in recent years as mobile data rates have increased to support multimedia services. In non-mobile applications, such as television and fixed links, wideband channel characteristics have been important for a considerable period.

In the narrowband channel, multipath fading comes about as a result of small path length differences between rays coming from scatterers in the near vicinity of the mobile. These differences, on the order of a few wavelengths, lead to significant phase differences. Nevertheless, the rays all arrive at essentially the same time, so all frequencies within a wide bandwidth are affected in the same way.

By contrast, if strong scatterers exist well off the great circle path between the base and mobile, the time differences may be significant. If the relative delays are large compared to the basic unit of information transmitted on the channel (usually a symbol or a bit), the signal will then experience significant distortion, which varies across the channel bandwidth. The channel is then a wideband channel and any models need to account for these effects.

A very first model classification that can be done is according to the nature of the radio channel: *narrowband*, if local mean and possibly shadowing is to be estimated, and *wideband*, for power delay profile and the characterisation of other relevant parameters.

5.3.2 Propagation Environment

In addition to the nature of the channel (wideband or narrowband), a common approach to classify propagation models is by taking into account the environment for which they have been designed: *outdoor*, *indoor* (in-building) and *outdoor-to-indoor*. Our focus in this book is mainly on indoor and outdoor-to-indoor models, which are extensively used while designing in-building radio networks.

5.3.3 Model Construction Approach

Another classification is made for models depending on the approach taken to construct them: *physical*, if the model is based on underlying physics; *empirical*, if based on fitting functions to measurements; *statistical*, if based on random distributions; etc. The decision as to which model should be used and where is based on a clear understanding of what each the categories represent, and thus apply the models for real building scenarios.

A comparison of advantages and disadvantages of empirical versus physical models is presented in Table 5.1. In summary, while empirical models are simple and fast, they can be used to estimate buildings in general and there is no need for detailed building data, cannot guarantee accurate results for details, as well as strongly depend on a large number of measurements. The building and environment classification is hard, and, if operating frequency is changed, there is a need for complete recalibration of the building propagation model. Physical models fill those deficiencies quite well, having a wide parameter range, as well as producing details of the relation between path loss and angle of arrival, multipath, etc. However, the electrical characteristics and detailed structure are hard to capture accurately, making it difficult to implement and taking a much longer computational time.

Table 5.1 Comparison of empirical versus physical models.

	Empirical models	Physical models
Advantages	• Simple and fast to calculate • No need for detailed building data • Good way to estimate for buildings in general rather than for a particular location	• Potentially highly accurate: account for the real mechanisms in a site-specific way • Wide parameter range: one model can work across a wide frequency range and for very diverse building types • Can produce detail of relation between path loss and angle of arrival, multipath, etc.
Disadvantages	• May be accurate 'on average' but cannot account for details • Require lots of accurate measurements • Hard to classify: are all 'open plan offices' the same? • Limited range of applicability: need to completely recalibrate for different frequencies	• Even when building geometry is available, the electrical characteristics and detailed structure are hard to capture accurately • Complicated to implement • Long computation time

5.4 Model Accuracy

Before going into depth to analyse the different in-building propagation models, it is useful to define how the accuracy of a model can be evaluated and, based on this, select the best model to be used or determine if the selected model was employed using appropriate parameters.

The accuracy of a propagation model can be evaluated by checking some statistics of the error results, which compares the predicted values given by the model with real measurements at these points. The *mean error* μ_{error} is often used for this, which is the difference between the measured $L_{i,\text{measured}}$ and the predicted $L_{i,\text{predicted}}$ path loss, taking into consideration all the samples N:

$$\mu_{\text{error}} = \frac{1}{N}\sum_{i=1}^{N}\left(L_{i,\text{measured}} - L_{i,\text{predicted}}\right) \tag{5.1}$$

For example, measured and predicted values of path loss are shown in Figure 5.1. Notice that the solid line indicates a 'best fit' of the measured data, hence having a mean of zero. This is classified as a 'perfect model' in that it has been tuned to measurements as closely as it can be. However, notice that a zero mean is not a measure of a perfect prediction; it is the standard deviation of the error, which clearly shows how accurate the model fits the measurements, as positive and negative errors cancel out, thus forcing the mean to be zero.

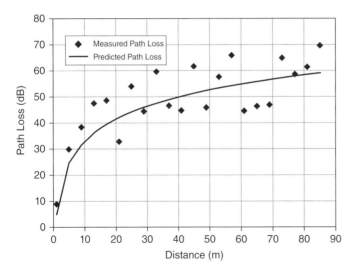

Figure 5.1 Measurements versus predictions.

The procedure employed to adjust model parameters with measurements so that first-order statistics of the prediction error (mean, standard deviation, correlation) are minimized is called *model tuning*. A well-tuned model should have a mean very close to zero, but careful interpretation should be made at this point since this does not guarantee a minimum error. The *standard deviation* of the error σ_{error} expresses the average deviation irrespective of whether it is positive or negative, thus being a better figure of merit for models:

$$\sigma_{\text{error}} = \frac{1}{N} \sum_{i=1}^{N} \sqrt{\left(L_{i,\text{measured}} - L_{i,\text{predicted}}\right)^2 - \mu_{\text{error}}^2} \tag{5.2}$$

Model tuning is described in more detail in Chapter 7, when radio measurements are introduced and discussed.

5.5 Empirical Models

An *empirical model* is based directly on measured results and it is from these measurements that model parameters of a simple equation are adjusted (tuned) to provide a good fit to results measured previously in similar environments. These models are relatively easy to develop and implement, but are strongly dependent on the accuracy of the measurements conducted to tune their parameters.

To create such a model, an extensive set of actual path loss measurements is made and an appropriate function is fitted to the measurements, with parameters derived for the particular environment, frequency and other variables, so as to minimize the error between the model and the measurements.

Empirical models provide no physical insight into the mechanisms by which propagation occurs. An example of an empirical model is shown in Figure 5.2, where a line fits

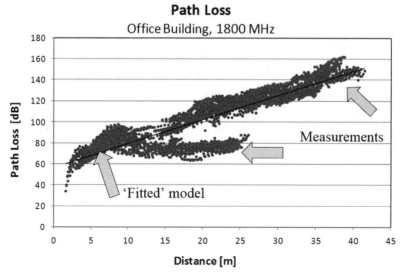

Figure 5.2 Example of measurements taken in an office building at 1800 MHz showing how a model can empirically be fitted to the measured data.

the measurements, which show a large variability (spread) given by the location variability of the environment.

This section provides a review of some of the most popular empirical indoor models used in practice, which have been developed throughout the years. Its use and applicability is explained in each section, as well as the in-building technologies for which they can be used.

5.5.1 Power Law Model

The *power law model* is considered to be the simplest form of an empirical path loss model; that is the only two model parameters n and K are tuned with measurements. The form of this model is

$$L = 10n \log r + K \tag{5.3}$$

The parameter K is also known as the *path loss intercept*, which represents the path loss taken at a reference distance, often taken as 1 m. Parameter n is known as the *path loss exponent*. These parameters depend on building materials, density of walls and frequency. The distance between the transmit and receive antennas is r.

The path loss intercept is found empirically, having a distinct value for every building and/or floor that is measured.

The path loss exponent has been found by measurement to depend on other system parameters, such as antenna heights and specific clutter surrounding the environment. For in-building systems and in dense clutter environments, it can take values of anywhere from a little less than 2 (in unobstructed waveguiding situations, such as corridors) to 5 or even more in very heavily obstructed environments. Values in the range 2–4 are most common. The path loss exponent is a critical factor in establishing

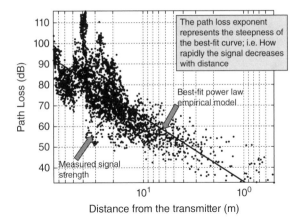

Figure 5.3 Power law model.

the coverage and capacity of a cellular system, since it strongly affects the range and therefore influences interference as well.

Figure 5.3 shows an example of a power law model fitted to some measurements. The straight line represents the best-fit power law empirical model curve. The dots are the measured data (after averaging out the fast fading). Note how the path loss exponent represents the steepness of the best-fit curve, showing how rapidly the signal decreases with distance.

This model is not very important for real predictions since research has shown that path loss behaviour inside buildings depends only on these two parameters. Nevertheless, it illustrates a number of points that will help in explaining the more practical indoor models later. It also accounts for why a different approach is taken for macrocells compared with in-building.

5.5.2 Keenan–Motley Model

The *Keenan and Motley* empirical model (Keenan and Motley, 1990) is usually taken as a reference for in-building propagation work, as the Okumura–Hata model (Okumura *et al.*, 1968; Hata, 1980) is for macrocell predictions. The model characterizes indoor path loss by a fixed path loss exponent of 2, just as in free space, plus additional loss factors relating to the number of floors n_f and walls n_w intersected by the straight-line distance r between the terminals. Thus:

$$L = L_1 + 20n \log r + n_f a_f + n_w a_w \tag{5.4}$$

where a_f and a_w are the attenuation factors (in decibels) per floor and per wall, respectively. L_1 is the loss at $r = 1$ m, also known as the path loss intercept, as defined in Section 5.5.1. Wall and floor factors are adjusted using measurements to minimize the error between the model predictions and measurements.

Figure 5.4 shows an example of a prediction using the Keenan and Motley model. Note that although some variability is obtained due to the use of wall factors and other

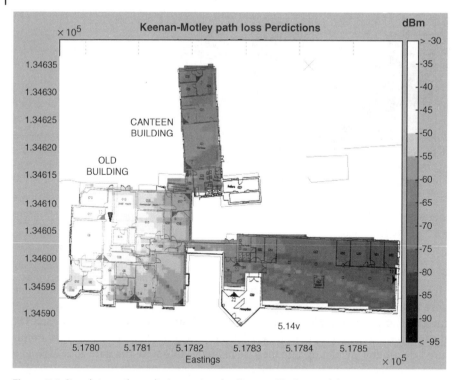

Figure 5.4 Signal strength predictions using the Keenan–Motley model.

material characteristics, the model does not account well for shadowing variations due to specific furniture around the site. Nevertheless, a reasonable accuracy can be obtained, with reported standard deviations of the errors (measured versus predicted values) between 5 and 8 dB, depending on the environment (Aragón-Zavala *et al.*, 2006).

Despite some of the limitations of the Keenan and Motley in-building propagation model, it is very simple to implement and is often taken as a reference to compare the accuracy of other more sophisticated models against – in a similar way that the Okumura–Hata model is taken as a reference for macrocell predictions.

One of the main limitations the Keenan and Motley model has is the fact that it averages all floor and wall losses, without being able to take into account individual partition losses coming from different materials. This assumption brings errors in the prediction that balance out when computing the mean of the error μ_{error}, but are shown when calculating the standard deviation of the error σ_{error}.

On the other hand, even though the model can yield good results (statistically), it fails to be accurate enough for some areas where path loss can be underestimated or overestimated, hence causing either leakage or outage problems.

Finally, as observed in Figure 5.4, sharp signal transitions between zones corresponding to different numbers of wall can be seen, which is clearly not accurate.

5.5.3 ITU-R Indoor Model

For this model, only the floor loss is accounted for explicitly and the loss between points on the same floor is included implicitly by changing the path loss exponent (ITU, 1997). The basic variation with frequency is assumed to be the same as in free space, producing the following total path loss model (in decibels):

$$L = 20 \log f_c + 10n \log r + L_f(n_f) - 28 \tag{5.5}$$

where n is the path loss exponent and $L_f(n_f)$ is the floor penetration loss, which varies with the number of penetrated floors n_f. Recommended values for path loss exponents are given in Table 5.2.

The 60 GHz figures apply only within a single room for distances less than around 100 m, since no wall transmission loss or gaseous absorption is included. Floor penetration factors for this model are presented in Table 5.3.

The ITU-R model is very simple to implement and key parameters can be tuned with measurements. Unlike the Keenan–Motley model, it attempts to model path loss taking into account floor penetration losses, which make it more suitable for 3-D scenarios.

Table 5.2 Path loss exponents n for the ITU-R model.

Frequency		Environment	
(GHz)	Residential	Office	Commercial
0.9	–	3.3	2.0
1.2–1.3	–	3.2	2.2
1.8–2.0	2.8	3.0	2.2
4.0	–	2.8	2.2
60.0	–	2.2	1.7

Table 5.3 Floor penetration factors, $L_f(n_f)$ (dB) for the ITU-R model.

Frequency		Environment	
(GHz)	Residential	Office	Commercial
0.9		9 (1 floor)	
	–	19 (2 floors)	–
		24 (3 floors)	
1.8–2.0	$4n_f$	$15 + 4(n_f - 1)$	$6 + 3(n_f - 1)$

On the other hand, the 'average' path loss exponents are supposed to take into account variations with wall penetration losses, but it is this averaging process that can make it very inaccurate for larger areas with multiple partitions made of a fairly large variety of materials. It is somehow an indoor power law model, which could work reasonably well for simple building geometries but not for more complex scenarios.

5.5.4   Siwiak–Bertoni–Yano (SBY) Multipath-Based Model

This model (Siwiak, Bertoni and Yano, 2003) is based on the assumption that, on average, as the wave expands in a cluttered environment, it attenuates due to several propagation mechanisms generating multipath components whose behaviour can be predicted. In addition to the inverse-square path loss, the wave additionally sheds energy into time in the form of multipath reflections. This shedding of energy into multipath was measured by the authors by delay spread, which increases with propagation distance. Thus the model can be expressed as free-space propagation modified by a multipath scattering term:

$$L = 10 \log \left[\left(\frac{c}{4\pi r f_c} \right)^2 \left(1 - e^{-(r_t/r)^{n-2}} \right) \right] \qquad (5.6)$$

For in-building scenarios, n is taken as 3. The SBY model predicts free-space propagation that transitions to the inverse n-power law beyond a breakpoint distance r_t. For $r \gg r_t$, the modifying term reduces to $1/r^{(n-2)}$, which, combined with the free-space term, yields an overall power law of n.

It seems to take into account multipath associated with the indoor propagation channel and it is relatively simple to implement. It can be applied for both a power delay profile and path loss propagation predictions.

It does neither take into account absorption due to lossy materials nor losses due to polarization cross-coupling in the propagation path. The model was created having in mind UWB propagation, and although the authors claim that it can be applied to narrowband systems, no evidence of this has been found in the literature.

5.5.5   Ericsson Multiple Breakpoint Model

The Ericsson radio system model was obtained by measurements in a multiple-floor office building (Akerberg, 1988). The model has four breakpoints and considers both an upper and a lower bound on the path loss. The model also assumes that there is 30 dB attenuation at 1 m, which can be shown to be accurate for a frequency of 900 MHz and unity-gain antennas. Rather than assuming a lognormal shadowing component, the Ericsson model provides a deterministic limit on the range of path loss at a particular distance.

Although intended to be used around 900 MHz, the model can be extended to 1800 MHz by the addition of 8.5 dB extra path loss at all distances. The model is shown in Table 5.4.

Implementation of this model is very simple and therefore predictions can be made very fast. Some of the model parameters can be tuned with measurements very easily if the range needs to be extended. On the other hand, the frequency range over which the

Table 5.4 Ericsson indoor propagation model.

Distance (m)	Path loss lower limit (dB)	Path loss upper limit (dB)
$1 < r < 10$	$30 + 20 \log r$	$30 + 40 \log r$
$10 \leq r < 20$	$20 + 30 \log r$	$40 + 30 \log r$
$20 \leq r < 40$	$-19 + 60 \log r$	$1 + 60 \log r$
$r \geq 40$	$-115 + 120 \log r$	$-95 + 120 \log r$

model is valid is fairly limited (900–1800 MHz) and since it is a power law model, it requires a very large amount of measurements to have reasonable accuracy.

5.5.6 Tuan Empirical Indoor Model: 900 MHz to 5.7 GHz

In one empirical model for use at Wi-Fi frequencies (Tuan *et al.*, 2003), the path loss has the general form:

$$L = k_1 + k_2 \log f_c + k_3 \log r + n_w(k_4 P_1 + k_5 P_2) + k_6 n_f \tag{5.7}$$

P_1 and P_2 are associated with the angle of incidence θ to a wall. Various forms of P_1 and P_2 were proposed in Tuan *et al.* (2003), and after validating the model with measurements, the path loss is given by

$$L = 19.07 + 37.3 \log f_c + 18.3 \log r + n_w[21 \sin \theta + 12.2(1 - \sin \theta)] + 8.6 n_f$$

$$\tag{5.8}$$

This model is given as valid for a frequency range between 900 MHz and 5.7 GHz and can be used in office environments, although the authors do not explicitly recommend other types of scenarios.

In general, a path loss equation of the form given here can be tuned with measurements conducted at the frequency of interest. The unknown coefficients k_1 to k_6 can be computed using linear regression from the measured data.

The model attempts to take into account key propagation mechanisms through measurements and key parameters can be tuned using measurements. There is a frequency dependency and wall and floor factors are considered as well. The frequency range over which the model is valid is also an advantage. It is also very easy to implement.

The authors do not show clear evidence of the validity of the model for frequencies below 2.4 GHz. Also, as diffraction and reflections are not explicitly considered (just through measurements), very inaccurate results could be produced at large distances or for specific buildings.

5.5.7 Attenuation Factor Model

An in-building site-specific propagation model that includes the effect of building type as well as the variations caused by obstacles was described in Seidel and Rappaport (1992).

The attenuation factor model is given by

$$L = L_{1m} + 10n_{SF}\log r + FAF + \sum PAF \qquad (5.9)$$

where n_{SF} represents the exponent value for the same floor measurement, FAF represents a floor attenuation factor for a specified number of building floors and PAF represents the partition attenuation factor for a specific obstruction encountered by a ray drawn between the transmitter and receiver in 3-D. Notice that partitions here represent walls in the building, but are given this definition to include those that are not necessarily walls; for example large office buildings with soft partitions. L_{1m} is the path loss intercept.

This technique of drawing a single ray between the transmitter and receiver is called primary ray tracing. Summing the cumulative partition losses along the primary ray has been shown to yield good results and accuracy, according to the authors. Therefore, if a good estimate of n exists, either selected from previously published results or from measurements on the same floor, then the path loss on a different floor can be predicted by adding an appropriate value of FAF (either selected from previous results or from measurements) and then summing the partition losses selected from wall loss databases or tuned with measurements.

There is an alternative way of writing the attenuation factor model, in which FAF is replaced by an exponent that already considers the effects of multiple floor separation, n_{MF}, based on measurements through multiple floors:

$$L = L_{1m} + 10n_{MF}\log r + \sum PAF \qquad (5.10)$$

The attenuation factor model is very simple to implement. It is essential to have a very complete database for materials characteristics, especially if predictions are to be made without prior measurements. Otherwise, it is relatively easy to find wall and floor attenuation factors for different walls. Predictions can also be performed very fast.

In terms of limitations of the model, it does not take into account either diffraction or reflections (waveguiding), which could lead to inaccurate predictions. Also, being an empirical model, the accuracy of the predictions is strongly influenced by the areas where measurements are taken.

5.5.8 Indoor Dominant Path Model (DPM)

The dominant path model (DPM) (Wölfle *et al.*, 2005) is an empirical approach for predicting path loss for different environments, one of which is indoors. It is based on the assumption that the majority of energy from the transmitter to the receiver is transported along a so-called dominant path, in contrast to other methods for which they either only assume direct-ray contributions (ITU, 1997; Seidel and Rappaport, 1992) or most of the ray contributions are considered, even those that correspond to rays heavily attenuated and that do not necessarily represent a significant contribution to the overall field strength (Cheung, Sau and Murch, 1998; Lee, Nix and McGeehan, 201; Seidel and Rappaport, 1994). Thus, the DPM focuses on the dominant signal path, neglecting other paths that are highly attenuated and do not contribute to the overall field strength. Often only a single path is taken, although for some special cases more than one is considered.

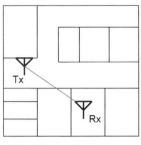

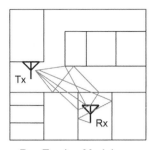

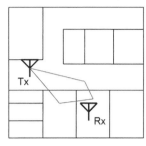

COST 231 Multi-wall Model   Ray Tracing Model   Dominant Path Model

Figure 5.5 Model comparison: multiwall, ray tracing and DPM models.

It also provides a good compromise between the speed/simplicity of a multiwall model and the potential accuracy of ray tracing.

Figure 5.5 shows a comparison of DPM with multiwall and ray tracing models. For multiwall modelling approaches, direct ray contributions are taken into account, neglecting other essential effects such as diffraction and hence making the prediction of path loss rather pessimistic. On the other hand, for optical ray tracing, all possible rays from the transmitter and the receiver are considered, thus increasing the complexity and the computational time to unacceptable levels.

The DPM expression is as follows:

$$L = \frac{\lambda}{4\pi} + 10n \log r + L_e + \sum_{i=1}^{m} f(\phi, i) + \sum_{i=1}^{k} L_i - \Omega \qquad (5.11)$$

The first term in (5.11) represents the frequency-dependent loss (compare with free space loss). The second term is the empirical distance-dependent term for indoor clutter, where the path loss exponent n is taken into consideration. L_e is the path loss intercept, used here as an 'offset' for the calibration of the model. The summation including $f(\phi, i)$ take into account the angle-dependent interactions, such as reflections and diffractions for the m rays, whereas L_k corresponds to accumulated transmission loss of k walls passed. The last term, Ω is the waveguiding gain, which is empirically derived and taken as an extra gain margin due to multiple reflections in corridors (waveguiding effect).

The model combines characteristics of power law, Keenan and Motley and ray-tracing models. Although layout plans of walls of specified material and dimensions are required in preparation of the use of the DPM model, these do not need to be as accurate as for traditional in-building ray-tracing models. This makes the preparation materials process much simpler than that for a full ray-tracing algorithm. Also, no exact location of the diffraction and reflection points is necessary for the determination of the path, which accelerates the processing of the diffracted paths, which is normally the most time-consuming task.

In terms of accuracy, although it neglects some of the fields contributed by secondary paths, the assumption that the most significant path is the one that contributes the most brings a maximum error in the prediction of 3 dB (in a very worst-case scenario).

Finally, key propagation mechanisms are taken into account, such as diffraction, transmission losses (implicitly accounted for in the equation) and reflections (by the use of the empirically derived waveguiding term). Also, execution time according to authors is comparable to empirical models, and much less than ray tracing.

Reported results of the use of DPM show that the model does not perform very well in curved edges and surfaces, since reflections on cylindrical surfaces are not considered. Also, computation speed as reported in AWE Communications GmbH (2007) is larger than for full ray tracing for some scenarios.

On the other hand, the way reflections are modelled through the use of an empirical factor for waveguiding just adds an offset to the overall prediction, which can give reasonable but only limited results in corridors.

Shadowing is not implemented at all in the DPM, which for the case of indoor environments, could represent an important propagation effect to take into consideration, especially in densely cluttered buildings, where the statistical compliance assessments may not be enough for critical areas where coverage is essential.

5.5.9   COST-231 Multiwall Model

A better approach to empirical in-building models is to use wall and floor losses for individual partitions, as suggested in the COST-231 multi-wall model (COST-231, 1999). This model of propagation within buildings was created in a European research collaborative project known as COST-231 (Collaboration in Science and Technology, project 231). It incorporates a linear component of loss, proportional to the number of walls penetrated, plus a more complex term that depends on the number of floors penetrated, producing a loss that increases more slowly as additional floors after the first are added:

$$L = L_{FSL} + L_c + \sum_{i=1}^{W} L_{wi} n_{wi} + L_f n_f^{\left((n_f+2)/(n_f+1) - b \right)} \tag{5.12}$$

where L_{FSL} is the free-space loss for the straight-line (direct) path between the transmitter and receiver; n_{wi} is the number of walls crossed by the direct path of type i; W is the number of wall types; L_{wi} is the penetration loss for a wall of type i; n_f is the number of floors crossed by the path; b and L_c are empirically derived constants that can be tuned with measurements; and L_f is the loss per floor.

The scenario for the multiwall model is depicted in Figure 5.6. Note that the model only accounts for the direct path between the transmitter and the receiver, and wall losses are added as extra losses to the minimum free-space loss factor. If multiple-floor penetration takes place, then the floor loss factor L_{fi} is considered (not the case for Figure 5.6).

The floor loss, that is the last term in (5.12) for the COST-231 multiwall model, is shown in Figure 5.7. Note that the additional loss per floor decreases with the increasing number of floors, due to diffracted waves coming from inside the building and being reflected in contiguous buildings, coming back into the building through windows. In other words, it is clear that two regimes are present; for small spacing between the transmitter and receiver within the building, the signal drops rapidly as the multiple-floor losses on a direct path between the transmitter and the receiver accumulate.

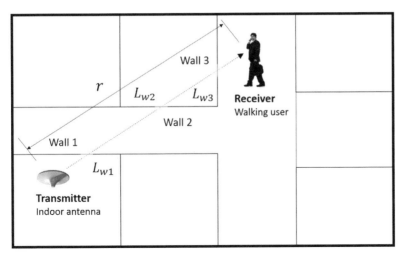

Figure 5.6 COST-231 multiwall model scenario.

Eventually the diffracted paths outside the building, discussed previously, dominate, and these diminish far less quickly with distance. When a reflecting adjacent building is present, the diffraction losses associated with this path are less and this provides a significant increase in the field strength for large separations.

Some recommended values are $L_w = 1.9\,\text{dB}$ (900 MHz), 3.4 dB (1800 MHz) for light walls, 6.9 dB (1800 MHz) for heavy walls, $L_f = 14.8\,\text{dB}$ (900 MHz), 18.3 dB (1800 MHz) and $b = 0.46$. The L_c term can be useful in fitting to measured data, but is generally close to zero.

Note that the losses found from fitting are not necessarily simply the penetration losses of individual walls and floors, but are results of all the mechanisms taking place in the environment interacting together.

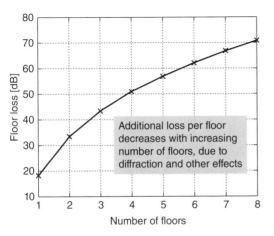

Figure 5.7 Floor loss versus number of floors for the multiwall model.

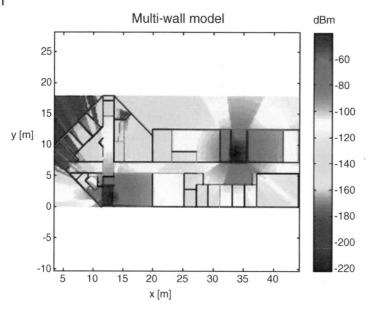

Figure 5.8 Signal strength prediction using the COST-231 multiwall model.

An example of the use of the COST-231 multiwall model is shown in Figure 5.8, for two antennas located in a building for which wall losses have been modelled according to (5.12).

Some of the main issues associated with the COST-231 multiwall model, which can be seen as limitations on the use of it, can be summarized as below:

• Individual material properties can be selected for each wall, which leads to better accuracy than Keenan and Motley model.
• No attempt to account for reflections or diffraction is made; only wall and floor penetration is considered, which will tend to overestimate the path loss.
• Large database of penetration losses for different building materials is needed, and for various frequencies, or retuning for each frequency and building type.
• No account for challenging environments like corridors, atriums, etc., is made in the model.

These issues make the COST-231 multiwall model a simple and easy-to-use option for path loss predictions inside buildings, which can be tuned with measurements and provide reasonable accuracy, as long as the limitations of the model are well understood.

5.6 Physical Models

A *physical model* accounts directly for the important propagation mechanisms, given a description of the geometry and electrical properties of the environment. In a physical model, propagation mechanisms such as reflection, refraction, diffraction and others are taken into account when computing the signal strength at any point. We will examine various physical model approaches in this section, which have been widely used for indoor wireless communication systems.

5.6.1 Introduction to Ray Tracing

One of the most popular physical model methods to predict path loss used for indoor scenarios is known as ray tracing. *Ray tracing* is a sophisticated approach to deterministic physical prediction of in-building propagation. This can be used for site-specific predictions, provided that sufficient detail of the building geometry and materials is available. Building materials often need to be modelled in detail as multiple-layer structures with detailed internal construction to achieve high modelling accuracy. Fine detail of building geometry must also be obtained to account for the wave interactions with walls, floors and the edges of doors and windows. In fact, ray tracing takes into account building characteristics, such as walls and inclined and cylindrical surfaces, as shown in Figure 5.9(a), and predicts how radio waves will propagate.

In Figure 5.9(b) it is shown how ray tracing has been applied to a building with a large corridor and many rooms on both sides. The areas that have increased signal power due to reflection are clearly shown there. For this simulation, details of specific furniture inside the building were not available and therefore the rooms are assumed to be 'empty', as can be clearly observed. A directional antenna was placed on the West corridor.

To perform ray tracing, all possible ray paths between the source and field points should be calculated, which are consistent with Snell's laws of reflection and refraction and including diffracted rays (Saunders and Aragón-Zavala, 2007). The following steps are followed:

- Ray launching sends out test rays at a number of discrete angles from the transmitter.
- The rays interact with objects present in the environment as they propagate.
- Launch new reflected, refracted and diffracted rays whenever a ray hits a surface or edge.
- Use reflection, transmission and diffraction coefficients to find the ray power after each 'bounce'.
- Add the power in all rays in a given area.

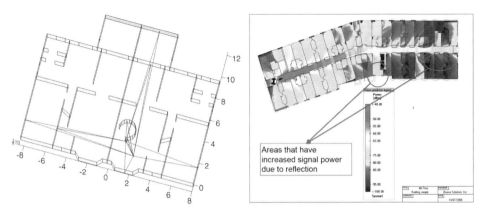

Figure 5.9 Examples of physical model (ray tracing) in a building: (a) reflections caused by walls, horizontal, cylindrical and other surfaces often modelled; (b) signal prediction using ray tracing.

Ray tracing correctly accounts for all the important propagation mechanisms (reflection, refraction, diffraction) on the assumption that there is good knowledge of the building geometry and materials. It can be more accurate provided that the environment information is available and correct.

Despite all of the above-mentioned benefits, ray tracing tends to be computationally slow when there is an increase in resolution and/or complexity of the modelling environment, hence making it impractical for some in-building design scenarios.

The following assumptions are often made when performing ray tracing predictions:

- Each wall is modelled as a plane surface.
- Each surface has known, simple electromagnetic properties (penetration, reflection, diffraction).
- However, walls are rarely consistent materials and windows are complicated!

Various in-building models have incorporated ray-tracing algorithms and will be presented here.

5.6.2 Honcharenko–Bertoni Model

A theoretical in-building ray-tracing model was developed by Honcharenko and Bertoni (Honcharenko *et al.*, 1992). They state that radio propagation is governed by two principal mechanisms: attenuation due to walls and diffraction from obstacles near the floor and in the plenum, with additional diffraction around the corners. The signal diffracts from a cluttered region in the plenum of a building and a cluttered region near the floor.

There is a clear space of height W_c metres between the two cluttered regions, as seen in Figure 5.10. The diffraction factor for wave propagation in a building with a clear space W_c results in excess loss compared with a free-space path. Rays are drawn between the transmitter and the receiver, with transmission and reflection coefficients attached to rays that, respectively, pass through and are reflected by walls.

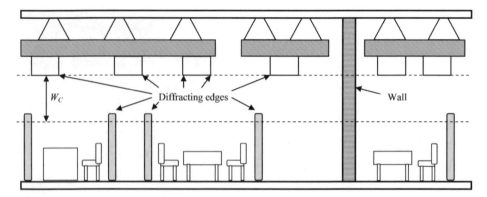

Figure 5.10 Geometry for the Honcharenko–Bertoni ray-tracing model.

Path loss within buildings, according to this model, takes the form of the sum of i individual ray intensities:

$$L = 10 \log \left[\sum_i \frac{L_e \lambda^2}{4\pi r_i^2} \left(\prod_n |\Gamma_n|^2 \prod_m |T_m|^2 \right) \right] \tag{5.13}$$

where r_i is the distance connecting the ith ray between the transmitter and the receiver. The ray undergoes n reflections Γ_n from walls and m transmissions T_m through walls. In addition, the ith ray is subjected to a distance-dependent loss L_e in excess of free-space loss due to diffraction from clutter near the floor and in the plenum. The diffraction loss factor is unity for small distances r_i and approaches an inverse 9.5 power with distance behaviour for large distances in an office building, according to Honcharenko *et al.* (1992). The breakpoint between the extreme behaviours of L_e is dependent on the clear space parameter W_c; for example 30 m at 900 MHz in an office environment where W_c ranges between 1.5 m and 2 m.

Diffraction around corners needs to be taken into account for more accurate results. Therefore, each corner at which the ray is diffracted is taken as an absorbing screen and the field reaching the receiving site via diffracted paths is given as

$$|E_i|^2 = \frac{Z_0 P_e}{4\pi} \frac{\prod_m D^2(\alpha_m)}{\sum_n L_n \prod_n L_n} \tag{5.14}$$

where L_n is the length of the ray path between diffracting sites. The summation term in the denominator accounts for vertical spreading of the ray and the product term accounts for the spreading in the horizontal plane. The diffraction coefficient $D(\alpha_m)$ for a ray bending through an angle α_m at an absorbing screen is given by

$$D(\alpha_m) = \frac{1}{2\pi k} \left[\frac{1}{2\pi + \alpha_m} - \frac{1}{\alpha_m} \right] \tag{5.15}$$

This model can give more accurate results provided that building details are specified, as well as whether constitutive parameters of materials are known for the frequencies of interest. It takes diffraction into account.

The main drawback for ray tracing is the computational time and complexity the predictions may take. This is acceptable for research work, but for a propagation prediction tool could be inappropriate.

5.6.3 Ray-Tracing Site-Specific Model

Site-specific models, also called deterministic models, are based on the theory of electromagnetic wave propagation. Unlike statistical models, site-specific propagation models do not rely on propagation measurements, but on knowledge of greater detail of the environment, and they provide accurate predictions of the signal propagation.

A geometrical-optics based model to predict propagation within buildings of this kind is presented in Seidel and Rappaport (1994). This ray-tracing algorithm predicts multipath impulse responses based on building blueprints. The authors use an

AutoCAD format to import building details from floorplans, leaving only objects whose size is much larger than a wavelength (large objects) in the database. It is claimed by the authors that the model can predict path loss with an overall standard deviation of less than 5 dB, and the power delay profile statistics can also be obtained with high accuracy.

The ray-tracing site-specific model uses geometrical optics to trace the propagation of direct, reflected and transmitted fields. Singly diffracted fields are also computed. This prediction technique uses 'brute-force ray tracing', for which a bundle of transmitted rays is considered, which may or may not reach the receiver. The transmitter and receiver are modelled as point sources in the building, and all possible angles of arrival and departure are considered. Antenna patterns are incorporated to include the effects of antenna beamwidth in both azimuth and elevation. A detailed description of how the rays are traced is included in Seidel and Rappaport (1994).

The complex field amplitude of the ith ray at the receiver is given by

$$E_i = E_0 f_{ti} f_{ri} L_i(r) \prod_j \Gamma(\theta_{ji}) \prod_k T(\theta_{ki}) e^{-jkr} \tag{5.16}$$

where f_{ti} and f_{ri} are the field amplitude radiation patterns of the transmitter and receiver antennas, respectively, $L_i(r)$ is the path loss distance dependence of the ith multipath component, r is the path length in metres, $\Gamma(\theta_{ji})$ and $T(\theta_{ki})$ are the reflection and transmission coefficients, E_i is the field strength of the ith multipath component in V/m, E_0 is the reference field strength, in V/m and e^{-jkr} represents the propagation phase factor due to path length ($k = 2\pi/\lambda$).

This model takes into account all relevant propagation mechanisms affecting indoor radio propagation and, as reported by the authors, this is certainly observed in the error statistics shown, particularly in terms of the standard deviation of the error. It also makes use of available building information from AutoCAD drawings, a standard practice nowadays, which is also implemented in many in-building prediction software tools. Finally, the use of effective building properties to characterize walls and other partitions, as stated by the authors, often simplify the implementation and use of the model.

Brute-force ray tracing is extremely computationally demanding, which anticipates longer computational times than the current dominant-path model (DPM) algorithm, which only takes into account a dominant path based on artificial intelligence techniques for deciding this. Variations in the way dominant rays are chosen could be a solution, but may require additional work.

5.6.4 Lee Ray-Tracing Model

A deterministic spatiotemporal propagation model has been proposed in Lee, Nix and McGeehan (2001) based on ray launching techniques. *Ray launching* sends out test rays at a number of discrete angles from the transmitter. The rays interact with objects present in the environment as they propagate. The propagation of a ray is therefore terminated when its power falls below a predefined threshold. The model considers reflection, transmission and diffraction effects via UTD (uniform theory of diffraction) principles. Transmission and diffraction are considered in this model.

The complex electric field, E_i, associated with the ith ray path is determined by

$$E_i = E_0 f_{ti} f_{ri} L_{FSL}(r) \left[\prod_j \overline{R}_j \prod_k \overline{T}_k \prod_l \overline{D}_l A_l(S_l, S_l') \right] e^{-jkr} \tag{5.17}$$

where E_0 represents the reference field, f_{ti} and f_{ri} the transmitting and receiving antenna field radiation patterns, L_{FSL} is the free-space loss, R_j is the reflection coefficient for the jth reflection, T_k the transmission coefficient for the kth transmission, D_l and A_l are the diffraction coefficient and the spreading attenuation for the lth diffraction and e^{-jkr} is the propagation phase factor, where r is the unfolded ray path length and k is the wavenumber. This model includes diffraction effects and is generalized to multiple interactions, which are added to the reflections and transmissions for all possible rays accounted for in GTD (geometrical theory of diffraction).

This model has been used with full three-dimensional data for both power delay and power azimuth profiles, departure and arrival angles, and coverage predictions, as detailed in Lee, Nix and McGeehan (2001). It is more accurate than statistical or empirical models as it takes into account electromagnetic theory effects (diffraction, transmission, reflection). However, this accuracy strongly depends on building geometry details and material characteristics. As a full ray-tracing algorithm, it is complex and takes a fairly large amount of computer resources, which will take the prediction time to unacceptable levels.

5.6.5 Multichannel Coupling (MCC) Prediction

As stated in Section 5.6.1, one of the most popular deterministic methods for indoor propagation modelling using geometric optics is ray tracing, where the propagation of a ray is followed from the transmitter to a receiver along every path that exactly fulfils the required angle condition for transmission and reflection. The main effort in this method is the search for valid paths. Increasing the number of calculated reflections increases the complexity drastically. The computationally expensive search for paths has to be redone for each combination of transmitter and receiver location.

Another popular method is known as *ray launching*, where the rays are launched in multiple directions from the transmitter. At their first interaction with an object they are split, for example, into the reflected and transmitted rays. These rays are propagated to the next obstacle. There they are split again, etc. The computational effort for this is also very demanding, with the very big disadvantage that the geometrical information of the building has to be processed for each new transmitter position.

Although ray tracing has proven to be a very accurate method, its computational time has made it difficult to be incorporated in a planning tool, which aims to perform some sort of optimization process, for which ideal locations for transmitters need to be found in order to make a more efficient use of the radio resources.

A new method was proposed by Dersch, Liebendörfer and Zehnder (2000), called multichannel coupling (MCC), which, by performing a preprocessing of the building geometry, allows a rapid propagation prediction thereafter, thus allowing the replacement of transmitters fairly quickly and optimization to be performed much more efficiently.

For this method, the first task is the identification of features of the realistic environment relevant for the propagation of electromagnetic waves. The propagation environment is described as a set of two-dimensional rectangular surface elements, arbitrarily oriented in a 3-D space. Each element is assumed to have homogeneous thickness and material properties. These elements A_i usually represent walls, doors, windows, iron bookshelves, conducting blackboards and other objects with a strong influence on the propagation.

Each pair of elements (A_i, A_j) that can directly exchange electromagnetic energy by at least one line-of-sight (LOS) path defines a geometric propagation channel. The channel $\alpha(A_i \rightarrow A_j)$ is the set of all LOS paths conducting from element A_i to element A_j. This is distinct from channel $\beta(A_j \rightarrow A_i)$, which contains the paths with inverse direction. The model for the physical environment is then an assembly of elements interconnected by channels. Each element has transmission, reflection and scattering properties and each channel has a geometrical shape and some propagation characteristics.

The following step in the MCC is the definition of the signal flux that will be propagated through the channels. Different partial contributions of the signal can be superimposed linearly. If a signal that emerges from an element A_i within a channel α is followed (Figure 5.11), it falls at the end of channel α on to an element A_j, where it is split into two different parts. One part is absorbed at A_j. Another part is transmitted and leaves element A_j within another channel β towards a third element A_k. Yet another part perhaps is reflected at A_j and falls into a channel γ in front of A_j. These splittings are described in Dersch, Liebendörfer and Zehnder (2000) by coupling coefficients between the channels:

$$P_\beta = c_{\alpha\beta}P_\alpha \tag{5.18}$$

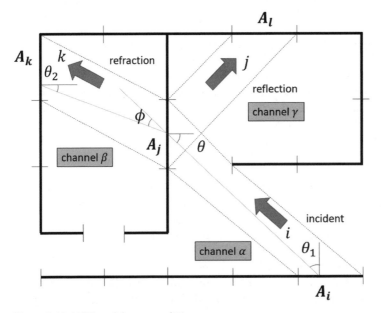

Figure 5.11 MCC model power splitting.

where P_α is the power leaving the channel α at element A_j and P_β represents the power leaving the channel β at element A_k. The coefficient $c_{\alpha\beta}$ couples these two signal rates, containing information of how much energy coming from element A_i reaches element A_k by transmission, reflection or scattering at element A_j. The calculation of such coupling coefficients depends only on the local geometry of the involved elements. It can be performed for all valid channel pairs in a pre-processing step, independently from the presence of transmitters and receivers. In the case of partial changes in the geometry, one may update affected regions only.

Regarding a plane wave incident on element A_j, the angle between the wave vector and the normal $\hat{n}_j$ of element A_j is denoted as θ^j, as seen in Figure 5.11, and the angle ϕ denotes the angle between the emitted wave vector and the angle of ideal reflection or transmission. Therefore, the coupling coefficients are given as follows:

$$c_{\alpha\beta} = \frac{\int_{A_j}\int_{A_i}\Delta_2 f\left(\theta_1^j\right)\frac{\Delta_1 \cos\theta_1^i \cos\theta_1^j}{r_1^2}dA_i dA_j}{\int_{A_j}\int_{A_i}\frac{\Delta_1 \cos\theta_1^i \cos\theta_1^j}{r_1^2}dA_i dA_j} \tag{5.19}$$

$$\Delta_1 = \begin{cases} 1 & \text{if } i \neq j \text{ and ray 1 are not obstructed} \\ 0 & \text{otherwise} \end{cases} \tag{5.20}$$

$$\Delta_2 = \begin{cases} 1 & \text{if ray 2 is not obstructed and hits } k \\ 0 & \text{otherwise} \end{cases} \tag{5.21}$$

From geometric optics and material properties of the element, the fraction $f(\theta^j)$ is assumed to be known and represents the intensity of the plane wave that is emitted towards an element A_k. Once the power that arrives at A_j from A_i by channel α is known, it is possible to calculate the fraction of that power that arrives at element A_k via channel β.

To couple transmitters to the channels previously defined, the coupling of transmitter i to channel β conducting from element A_j to A_k if element A_i is made infinitesimally small is given by

$$c_{i\beta}T_i = \int_{A_k} \sqrt{\frac{P_i^t(\Omega)}{4\pi}}\Delta_1\Delta_2 f\left(\theta_1^j\right)\frac{\cos\left(\theta_2^k\right)}{(r_1+r_2)^2}dA_k \tag{5.22}$$

where $P_i(\Omega)$ is the radiated power of the transmitter into the solid angle Ω.

Now it is time to couple receivers to channels. For this, the power P_α that flows in channel α feeds all receivers that lie in this channel. The power P_α is divided into small cylinders β with cross-section $(\pi d^2)4$ running from element A_i to element A_j. Each of them carries a fraction ω^β of the power P^α. The energy flux Φ^β in cylinder β of a channel α is given by:

$$\Phi_\alpha^\beta = \frac{\Delta^\beta \omega^\beta P_\alpha}{\frac{\pi d^2}{4}\sum_\beta \Delta^\beta \omega^\beta} \tag{5.23}$$

If cylinder β is obstructed, $\Delta^\beta = 0$; otherwise $\Delta^\beta = 1$. A receiver has a given effective area $A(\theta)$ relative to an isotropic antenna in the direction θ of the cylinder. If the receiver is hit by the latter, it is assumed to receive the power $\Phi_\beta^\alpha A(\theta)\lambda^2/4\pi$. The total received power for this receiver then becomes

$$P_{rec} = P_{LOS} + \sum_{\alpha\beta} \Phi_\alpha^\beta A(\theta)\lambda^2/4\pi \qquad (5.24)$$

where P_{LOS} accounts for the line-of-sight contribution that is not included in the channel network.

Finally, the global propagation of electromagnetic energy in the environment considers a combination of the various couplings that have been presented. First, if $i = 1\cdots n$ transmitters are in the system, only one transmission or reflection is considered. From the coupling of transmitters to channels equation derived earlier, the channel power is calculated as follows:

$$P_\alpha^{(1)} = (cT)_\alpha = \sum_i c_{i\alpha}T_i \qquad (5.25)$$

The second transmission or reflection can be included by writing

$$P_\beta^{(2)} = P_\beta^{(1)} + \sum_\alpha c_{\alpha\beta}P_\alpha^{(1)} \qquad (5.26)$$

In general, the power $P_\beta^{(\infty)}$ leaving element A_j in channel β is the transmitted or reflected sum of the incident power at element A_j from the transmitters $i = 1\cdots n$ and the channels $P_\alpha^{(\infty)}$, and is given by

$$P_\beta^{(\infty)} = \sum_i c_{i\beta}T_i + \sum_\alpha c_{\alpha\beta}P_\alpha^{(\infty)} \qquad (5.27)$$

In summary, there is no need to perform ray-tracing computations every time a transmitter location is changed when using this model; it is only necessary to perform building preprocessing knowing basic geometrical characteristics, as well as material constitutive parameters (conductivity, permittivity and permeability) to perform the predictions. This speeds up the algorithm considerably, hence making it suitable for optimization of indoor base stations within a building.

Complex building geometries may result in time and computer demanding preprocessing computations, which could take a considerable amount of time. Also, highly changeable environments such as some airport facilities, for which the building layout and characteristics are quite changeable, may not be a good option for the use of this algorithm for optimization purposes. Finally, the algorithm does not implicitly take into account diffraction effects, although this could be incorporated later once the preprocessing has been performed.

5.6.6 Angular Z-Buffer Algorithm for Efficient Ray Tracing

A deterministic, fully three-dimensional, based on geometrical optics (GO) and the uniform theory of diffraction (UTD) ray-tracing technique to predict the propagation channel parameters in indoor scenarios is presented in Cátedra *et al.* (2000). A very efficient ray-tracing algorithm called angular Z-buffer (AZB) is presented, which allows a reduction in the time necessary to obtain the multipath propagation. The key concept of AZB consists in reducing the number of rigorous tests that have to be made by reducing the number of plane facets (conforming the geometrical description of the environment) that each ray has to treat.

The propagation model is based on GO and the UTD, including the effects related to the transmission, a very important effect in these scenarios. Effects such as direct field, single- and double-reflected fields, single- and double-diffracted fields and combinations of multiple effects are considered. The electric field at the observation point O is given for each as follows:

- Direct field:

$$\vec{E}(O) = \sqrt{\frac{\eta P_r G}{4\pi}} \vec{E}_0(\theta, \phi) \frac{e^{-jk_0 r}}{r} \tag{5.28}$$

- Reflected field:

$$\vec{E}_r(O) = \left[R_\parallel(\theta_i) E_{\theta f}(\pi - \theta_i, \phi_i)\hat{\theta}_i + R_\perp(\theta_i) E_{\phi f}(\pi - \theta_i, \phi_i)\hat{\phi}_i \right] \frac{e^{-jk_0 r_i}}{r_i} \tag{5.29}$$

- Refracted field:

$$\vec{E}_t(O) = \left[T_\parallel(\theta_i) E_\theta(\theta, \phi)\hat{\theta} + T_\perp(\theta_i) E_\phi(\theta, \phi)\hat{\phi} \right] \frac{e^{-jk_0 r}}{r} \tag{5.30}$$

- Diffracted field:

$$D_{\parallel,\perp} = D_1 + D_2 + R_{\parallel,\perp}(D_3 + D_4) \tag{5.31}$$

where

k_0	free-space wavenumber
r	distance between transmitter and observation point
η	free-space wave impedance
P_r	power radiated by the transmitter
G	gain of transmitter antenna
$\vec{E}_0(\theta, \phi)$	normalized radiation pattern of the transmitter antenna
θ, ϕ	spherical coordinates of the observation point referred to the antenna
r_i	distance between the image and the observation point
θ_i, ϕ_i	spherical coordinates of the observation point referred to the image coordinate system
θ_i	incident angle at the facet
d	facet thickness
$R_{\parallel,\perp}$	reflection coefficients for parallel and perpendicular components
$T_{\parallel,\perp}$	transmission coefficients for parallel and perpendicular components
$D_1 \cdots D_4$	components of diffraction coefficients as specified by the UTD

The AZB algorithm consists of dividing the space seen from the source in angular regions and storing the facets of the model in the regions where they belong. In this way, for each ray only the facets stored in its region need be analysed. Also, the facets are ordered in each region according to their distances from the source, since the closer facets have more possibilities for hiding rays.

Details of the application of the AZB to each of the propagation effects (direct ray, reflected ray, transmitted ray and diffracted ray) are presented in Cátedra *et al.* (2000), as well as results of the use of the AZB for predictions on a third floor of a building in Bilbao, Spain. The reduction in processing time is shown to be as much as 93.1%, going from a simulation of 1 h 32 min 42 s without AZB to only 5 min 56 s, having a standard deviation of the prediction error of 7.29 dB.

5.6.7 Intelligent Ray-Tracing (IRT) Model

The mobile radio channel in indoor scenarios is characterized by multipath propagation. Dominant propagation phenomena inside buildings include:

- The shadowing of walls
- Waveguiding effects on corridors due to multiple reflections
- Diffractions around vertical wedges.

In order to accelerate the time-consuming path determination the intelligent ray-tracing (IRT) model (Wölfle, Gschwendtner and Landstorfer, 1997) is based on a preprocessing of the building data, thus combining high accuracy with a short computation time.

Each penetration of a wall, each reflection at the surface of a wall and each diffraction at a wedge is an interaction. The intelligent ray-tracing method considers all propagation paths that fulfil the following criteria:

- Up to 6 reflections (at surfaces of walls/objects)
- Up to 6 penetrations (of walls/objects)
- Up to 2 diffractions (at wedges)
- Up to a total number of 6 interactions (all combinations of reflections, penetrations and diffractions).

IRT is based on the following assumptions:

- Only a few rays to deliver the main part of energy.
- The visibility relation between walls and edges are independent of the position of the transmitter antenna (base station).
- Often adjacent receiver pixels are reached by similar rays.

Based on these considerations, a preprocessing of the building database is made once. In this preprocessing the obstacles in the building database are subdivided into small tiles, as shown in Figure 5.12. The visibility relations between these tiles are determined

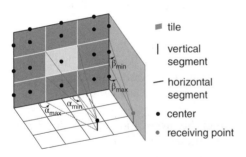

tile

| vertical segment

— horizontal segment

• center

• receiving point

Figure 5.12 Subdivision of the wall into tiles for IRT preprocessing.

and stored. During the prediction these data can be read and have not to be determined again. This accelerates the computation time significantly.

The IRT performs a much more rapid computation of field strength than full ray-tracing models, with very similar accuracy. This is due to the preprocessing performed to determine visibility areas according to the geometry of the building. Key propagation mechanisms are taken into consideration, such as diffraction, reflections and transmission. However, the preprocessing operation takes a considerable amount of time prior to performing predictions, which may be a disadvantage, especially for optimization purposes. Also, as reported in AWE Communications GmbH (2007), although predictions are accurate close to the transmitter, they are much less accurate as the distance between the transmitter and the receiver increases.

5.6.8 Hybrid Parabolic Equation–Integral Equation Indoor Model

The model presented in Theofilogiannakos, Xenos and Yioultsis (2007) is a 3-D hybrid technique for the simulation of wave propagation in complex indoor environments. This method employs a wide-angle parabolic equation (PE) to estimate the electromagnetic field within walls and obstacles, whereas a correction based on the Green function approach provides improved field approximations. This method is applied in 3-D using simple validation models, whereas an approximate 2-D analysis is attempted for a much more realistic office space. The authors state that this hybrid technique is more systematic and automated, compared to ray-tracing methods, having moderate computational requirements.

A first 2-D approach for the development of a PE-based method for indoor communications has been presented in Theofilogiannakos, Xenos and Yioultsis (2007) and applied in simple models of scattering from infinite cylinders. The method combines a 2-D wide-angle PE multistep scheme, based on a Padé approximation of the forward wave equation and a fast Green function formulation. At a first step, the PE is used to calculate an estimate of the electromagnetic field in the interior of the elements comprising the building structure (walls, doors, windows and various objects).

The PE effectively decomposes the 2-D problem in a series of 1-D problems, successively solved by a marching-in-space approach. Still, there are quite important propagation rays that are not taken into account, such as all-backward reflected, diffracted or scattered waves. Hence, a second step is employed to introduce a proper correction of the field by a Green's function approach, using the induced equivalent currents within the walls and other structures, computed by the PE technique, as secondary sources. The resulting correction is, thus, fast and does not involve a matrix system solution.

This model thus presents a rigorous generalization in 3-D of the model stated in Theofilogiannakos, Xenos and Yioultsis (2007), where the PE decomposes the domain in a series of 2-D problems on successive transverse planes. The wide-angle PE, solved by the finite element method (FEM), is combined with the field correction to form the hybrid parabolic equation–integral equation method (PE-IEM). The full 3-D solver is much more cumbersome, compared to the 2-D one, although it can be successfully applied to solve problems within domains that span several wavelengths and are only marginally treatable by other full-wave techniques like FEM in 3-D or the finite-difference time domain (FDTD) method. Application to larger-scale problems, involving realistic 3-D indoor environments, spanning tens or hundreds of wavelengths is definitely feasible, at the cost of increased run times, but can be highly automated provided that a 3-D model of the indoor environment under consideration can be extracted by CAD software. According to Theofilogiannakos, Xenos and Yioultsis (2007), before employing such a tedious task, a wise idea would be to attempt a 2-D analysis first, which would possibly provide some idea of the capabilities of the proposed analysis. Indeed, the hybrid PE-IEM technique has been applied to a 2-D model of an office building floor (excluding the floor and ceiling) and compared to measurements. The computational errors were assessed and a clear improvement comes out as a result of the IEM correction, as opposed to the PE alone.

This IE-PE hybrid method seems to give reasonably accurate results for 2-D environments, showing improvement when compared to PE (parabolic equation) methods and when FDTD (finite difference time domain) or FEM (finite element methods) are not applicable.

There is no clear evidence reported in Theofilogiannakos, Xenos and Yioultsis (2007) or anywhere in the literature that suggests that this method can be applied successfully to complex 3-D environments. In fact, the authors state that: 'Its application in realistic 3D settings, where methods like FEM or FDTD are totally inapplicable, is the next step to be considered. It is deemed feasible by means of appropriate speedup techniques and can potentially offer an alternative to existing ray-tracing techniques.' Therefore, the amount of computational resources and time it takes has not made it even comparable in processing speed to ray-tracing techniques. It might be a promising new algorithm but, so far, it is not there yet!

5.7 Hybrid Models

Indoor propagation models that combine measurements to tune some of its parameters and take into account physical principles that are explicitly included in the modelling approach are known as *hybrid* models. Nowadays, this approach is having increasing popularity, since sufficient accuracy can be obtained with less computational effort than purely physical models. On the other hand, on the basis that carefully calibrated radio measurements are conducted for model tuning, these models can be applied reasonably easily to any building, as is the case for empirical models.

5.7.1 Reduced-Complexity UTD Model

Given the complexity of models such as those given in Section 5.6.4, it is attractive to seek approaches that retain the physical principles while reducing the associated complexity. The model proposed in Cheung, Sau and Murch (1998), validated for use at around 900 MHz, incorporates much of the propagation phenomena suggested by electromagnetic theory, such as UTD, but still retains the straightforwardness of the empirical approach. Computation time is not as heavy as in pure ray-tracing models. Likewise, the empirical factors required in the model can be closely related to theoretical derivations, so that model tuning or curve fitting to measured data may not be required.

Three propagation mechanisms have been incorporated in this model. The first factor is a dual-slope model for the main path, similar to that for microcell propagation. Around 900 MHz, a breakpoint distance of about 10 metres is suggested, denoted as r_b. The second factor included is an angular dependence of attenuation factors. Less power is transmitted through walls when the incidence is oblique, as compared with normal incidence. Therefore, the wall attenuation factor L_{wi} (and likewise the floor attenuation factor L_{fi}) is made to depend on the angle of incidence. A simplified diffraction calculation is also included.

The resulting model is given by

$$L_{\angle}(r) = 10 \log\left(\frac{r}{r_0}\right)^{n_1} u(r_b - r) + 10\left[\log\left(\frac{r_b}{r_0}\right)^{n_1} + \log\left(\frac{r}{r_b}\right)^{n_2}\right] u(r - r_b)$$
$$+ \sum_{i=1}^{W} \frac{n_{wi}L_{wi}}{\cos\theta_i} + \sum_{j=1}^{F} \frac{n_{fi}L_{fi}}{\cos\theta_j}$$

(5.32)

where θ_i and θ_j represent the angles between the ith wall, jth floor and the straight line path joining the transmitter with the receiver and $u(\cdot)$ is the unit step function, defined as

$$u(t) = \begin{cases} 0, & t < 0 \\ 1, & t \geq 0 \end{cases}$$

(5.33)

To keep the diffraction model simple, Cheung, Sau and Murch (1998) utilized only one level of diffraction from corners, including the door and window frames in the building. To perform this, the field was calculated at each corner using the equation defined for $L_{\angle}(r)$ above and the resulting diffracted field was determined using a diffraction coefficient. Thus, the total field at the receiver is computed as the summation of the field from the transmitter and all the corners.

Diffraction coefficients for perfect electrical conductors under UTD (Kouyoumjian and Pathak, 1974) are used in the model and are denoted as $D(r, \phi, r', \phi')$, where (r, ϕ) are the coordinates of the corner relative to the transmitter and (r', ϕ') are the coordinates of the receiver relative to the corner. Hence:

$$L = -10 \log\left[\sum_{m=1}^{M} l_{\angle}(r_m)l_{\angle}(r'_m) \times \left|D\left(r_m, \phi_m, r'_m, \phi'_m\right)\right|^2\right] - 10 \log[l_{\angle}(r)]$$

(5.34)

where M is the number of corners in the building database, m refers to the mth corner and $l_{\angle}(.)$ is a dimensionless quantity given by

$$l_{\angle}(.) = 10^{-L_{\angle}(.)/10}$$

(5.35)

In Cheung, Sau and Murch (1998), at 900 MHz the parameters $n_2 = 2.5$, $r_b = 10$ m, $L_w = 10$ dB (concrete block walls) and $L_w = 5$ dB (hollowed plaster board walls) are obtained from measurements, while n_1 is taken as 1. Good prediction accuracy is claimed. This model combines the best of ray tracing with empirical approaches, which makes it very robust and accurate, especially for long distances where other purely empirical models do not perform very well, as they minimize diffraction effects. However, as was the case with other models, it does not take reflections into account,

Indoor Wireless Communications

which results in minimization of waveguiding, an effect that is more evident in corridors.

5.7.2 Measurement-Based Prediction

Empirical models are usually limited to providing a rather general description of propagation, while higher accuracy for a specific site usually requires detailed physical models. Site-general models are not usually sufficient for an efficient system design in a particular building, while physical models are often too complex to implement in practice, as has been discussed in the previous sections. Thus, hybrid models have been developed to overcome these disadvantages, which aim to combine the best of both approaches.

As an intermediate approach, suitable for high-confidence designs, site-specific measurements may be used to determine the details of the propagation mechanisms and material parameters for a particular building without suffering the high cost of precise entry of wall and floor materials and geometries. Such a *measurement-based prediction* approach is described in detail in Aragón-Zavala *et al.* (2006). A flowchart of the algorithm is depicted in Figure 5.13. Site information such as antenna coordinates and radiation patterns is required for path loss extraction. These parameters along with the signal strength measurements are processed to yield measurements of path loss

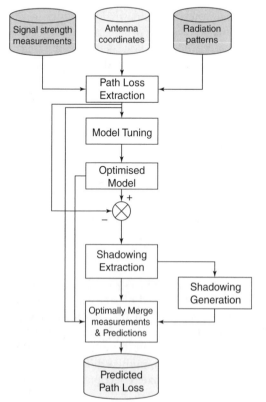

Figure 5.13 Measurement-based prediction (MbP) algorithm.

against location. This loss is made up of two components. The first is a distance-dependent r component, resulting from the bulk characteristics of the propagation medium, known as median path loss L_{50}:

$$L_{50} = k_1 + 10n \log r + k_2 r \tag{5.36}$$

Constants k_1 and k_2 are tuned with the field strength measurements to obtain an optimized model, which often represents a best-fit model for the collected data. The path loss exponent n is determined from this optimization (tuning) process. The second component for the path loss depends on the characteristics of the nearby propagation environment (local clutter). Subtracting the distance-dependent part of Equation (5.36) from the total loss L yields the latter component, known as *shadowing* or *slow fading* L_s:

$$L_s = L - L_{50} \tag{5.37}$$

The derived shadowing parameters are used to generate a set of shadowing predictions for the whole area of interest, mapped on to each prediction point. The result is added to the distance-dependent path loss predictions produced by the propagation model for the area. The result has then very similar characteristics to measurements that would have been produced not only at the measurement route but over the whole prediction area, without the need for detailed building geometry, as with ray-tracing algorithms.

The use of appropriate spatial statistics enables the measurement data to be applied across the whole building, well beyond the measurement route. Reuse of these data, together with empirical models for the wall and floor loss factors, also allows the system design to be optimized for antenna locations and types without remeasuring, even for different frequency bands than the original measurements. See Figure 5.14 for an example, where the directional transmit antennas are marked as black triangles.

One of the most relevant characteristics of this model is that it predicts path loss in areas where other indoor models fail to do so, such as those revealed by slow fading. For example, if there is a floor of a building that has specific signal variations due to clutter, such as walls, furniture, etc., then MbP takes these into account and in fact tunes these variations with the measurements that are used as part of the modelling process. It is true that it requires measurements more often perhaps than other similar models, but the authors believe that this is the case anyway since radio measurements are employed for model tuning and validation – thus accurate radio testing is of extreme importance, as will be discussed in Chapter 7.

5.8 Outdoor-to-Indoor Models

A technique often employed to provide coverage inside buildings is to allow the signal from surrounding macrocells to penetrate the building through external walls, windows and doors; that is macrocell penetration. However, sometimes when an indoor cell is to be deployed in a venue, existing coverage from surrounding macrocells should be determined, to adjust, for example, handover levels in cellular systems. Even though

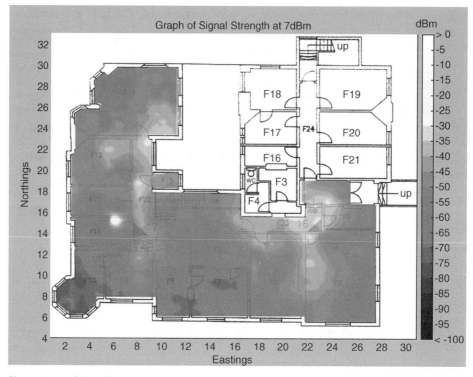

Figure 5.14 MbP prediction example.

measurements are used for this purpose, modelling is also required when measurements are not available in all the building relevant areas.

Outdoor-to-indoor models are utilized to characterize signal propagation inside buildings coming from an external base station. This section highlights the most important ones.

5.8.1 COST-231 Line-of-Sight Model

In cases where a line-of-sight path exists between a building face and the external antenna, the following semiempirical model has been suggested (COST-231, 1999), with the geometry defined in Figure 5.15.

Here r_e is the straight path length between the external antenna and a reference point on the building wall; since the model will often be applied at short ranges, it is important to account for the true path length in three dimensions, rather than the path length along the ground. The loss predicted by the model varies significantly as the angle of incidence, $\theta = \cos^{-1}\left(r_p/r_e\right)$ is varied.

The total path loss is then predicted as follows:

$$L = L_{FSL} + L_e + L_g(1 - \cos\theta)^2 + \max(L_1, L_2) \tag{5.38}$$

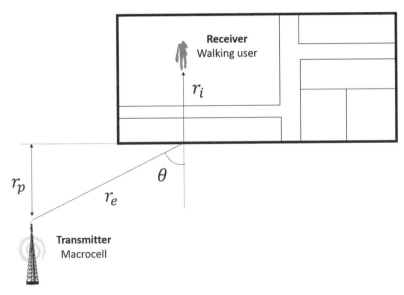

Figure 5.15 COST-231 LOS model geometry.

where L_{FSL} is the free-space loss for the total path length $(r_i + r_e)$, L_e is the path loss through the external wall at normal incidence $(\theta = 0°)$, L_g is the additional external wall loss incurred at grazing incidence $(\theta = 90°)$ and

$$L_1 = n_w L_i \tag{5.39}$$

$$L_2 = \alpha(r_i - 2)(1 - \cos\theta)^2 \tag{5.40}$$

where n_w is the number of walls crossed by the internal path r_i, L_i is the loss per internal wall and α is the specific attenuation (dB/m), which applies for unobstructed internal paths. All distances are in metres.

The model is valid at distances up to 500 m and the parameter values in Table 5.5 are recommended for use in the 900–1800 MHz frequency range. They are in good agreement with measurements from real buildings and implicitly include the effects of typical furniture arrangements.

Table 5.5 COST-231 LOS model parameters.

Parameter	Material	Approximate value
L_e or L_i [dB/m]	Wooden walls	4
	Concrete with non-metallized windows	7
	Concrete without windows	10–20
L_g [dB]	Unspecified	20
α [dB/m]	Unspecified	0.6

The COST-231 model for outdoor-to-indoor penetration losses is normally taken as a reference. It is very simple to implement and fast predictions can be made even without measurement data – taking the recommended parameters given in the paper. Despite its limitations, it can produce reasonable estimations of signal strength if limited information is available. On the other hand, the frequencies upon which the model is valid are restricted to only GSM cellular in Europe (900/1800 MHz). Also, the assumption that the direct ray to the point of interest is the dominant contribution to the overall penetration loss is not entirely accurate, as windows and doors through diffraction can have a significant impact on predictions. Reflections from adjacent buildings are not taken into account and the indoor loss is assumed to depend on only a specific attenuation constant, which could be valid for very short distances close to the better wall – the one closest to the point of interest.

5.8.2 COST-231 Non-Line-of-Sight Model

This is a variation of the LOS discussed in the previous section, which is also documented in COST-231 (1999). The model relates the loss inside a building from an external transmitter to the loss measured outside the building, on the side nearest to the wall of interest, at 2 m above ground level. The loss is given by

$$L = L_{out} + L_e + L_{ge} + \max(L_1, L_3) - G_{fh} \tag{5.41}$$

where $L_3 = ar_i$ and r_i, L_e, α and L_1 are as defined in the COST-231 line-of-sight model (Section 5.8.1). The floor height gain G_{fh} is given by

$$G_{fh} = \begin{cases} nG_n \\ hG_h \end{cases} \tag{5.42}$$

where h is the floor height above the outdoor reference height (m) and n is the floor number, as defined in the figure shown in Section 5.8.1. Shadowing is predicted to be lognormal with location variability of 4–6 dB. Other values are as shown in Table 5.6.

Both the line-of-sight and non-line-of-sight models of COST-231 rely on the dominant contribution penetrating through a single external wall. A more accurate estimation may be obtained by summing the power from components through all of the walls.

The model takes into account other propagation effects observed in measurements, such as floor height gain and dependence on the angle of incidence. This resembles much better the propagation effects in an outdoor-to-indoor scenario. It is also very simple to implement and fast predictions can be made. Transmission losses of the dominant wall can be obtained either from the literature or from measurements; if the

Table 5.6 COST-231 NLOS model parameters.

Parameter	Approximate value
L_{ge}[dB] at 900 MHz	4
L_{ge}[dB] at 1800 MHz	6
G_n[dB per floor] at 900/1800 MHz	1.5–2.0 for normal buildings
	4–7 for floor heights above 4 m

latter case is addressed, these measurements include many other effects, such as reflections and diffraction for wall partitions and openings, which otherwise will not be able to be accounted for using only reported transmission losses.

5.8.3 Broadband Wireless Access (BWA) Penetration Model

Based on the COST-231 outdoor-to-indoor propagation model described in Section 5.8.1, this model is intended to be used for broadband wireless access (BWA) services, as reported in Oestges and Paulraj (2004). It is based on measurements for various buildings in Seattle and San Francisco, USA, considering the impacts of wall material, angle of incidence and receiver antenna height. The COST-231 model, as taken from the authors, computes the total path loss in dB as

$$L = L_{out} + L_{ex} = L_{out} + L_e + L_g(1 - \cos \phi)^2 + n_w L_i \qquad (5.43)$$

where L_{out} is the path loss from the base station to the external wall at the same height as the receiver, L_{ex} is the outdoor-to-indoor excess path loss, L_e is the path loss through the external wall at normal incidence ($\phi = 0°$), L_g is the additional external wall loss incurred at grazing incidence $\phi = \pi/2$, L_i is the path loss per internal wall and n_w is the number of crossed internal walls. The external wall attenuation L_e must take into account the entire considered wall; for example the presence of windows has a significant impact on the performance of the model. The model relies on the attenuation of a dominant component coming through a single wall.

For the case of high-scattering NLOS environments, it may not be possible to isolate any single path, and thus Oestges and Paulraj (2004) suggest a new expression for the excess loss L_{ex}:

$$L_{ex} = L_e + L'_{ge} + n_w L_i \qquad (5.44)$$

where $L_{ge} = L_g(1 - \cos \phi)^2$ has been transformed into an average parameter L'_{ge}. This results from the fact that several dominant waves arrive at the best wall from different angles. It somehow accounts for an average attenuation rather than a worst case for normal incidence. This model is valid for frequencies up to 2.5 GHz.

Another importance addition to the COST-231 model is that the present formulation includes the impact of the floor height in the outdoor path loss L_{out}, using validated models for outdoor-to-outdoor path loss.

Model parameters are reported in the paper as shown in Table 5.7, for a frequency of 2.5 GHz for three different wall materials. Parameters for other frequencies can be obtained by tuning the model with measurements.

Table 5.7 BWA model parameters.

Parameter	Wood	Stucco	Mixed (plastic/metal/glass)
L_e	6.6	6.7	5.2
L_i	2.4	3.5	–
L_g	10	13.3	10
L'_{ge}	5.7	6.4	6.4

This model is relatively simple, taking as reference the COST-231 NLOS model, with slight variations in the way one of the parameters should be taken into account for excess loss. It should be noticed that the model is intended to be used in high-scattering NLOS scenarios. Bear in mind that the model has been tested for fixed broadband wireless access scenarios, which differ from those observed in mobile communications. However, the fundamentals of the model seem appropriate even for a mobile receiver inside the building. BWA is seen as a worst-case scenario when compared to mobile communications, according to the authors. Also, if parameters need to be obtained for other frequencies, various types of measurements are required, as specified in the paper, adding complexity to the model.

5.8.4 Ichitsubo–Okamoto Outdoor-to-Indoor Model (800 MHz–8 GHz)

This model, applicable from 800 MHz to 8 GHz (Okamoto, Kitao and Ichitsubo, 2010), also takes as a reference the NLOS COST-231 model (COST-231, 1999) and, using similar model parameters, a new model is proposed:

$$L = L_{out} + \alpha r_i - hG_h + \alpha_f \log f + \alpha_{LOS} LOS + W \tag{5.45}$$

The new parameters introduced here are: LOS is the condition of the line-of-sight between the transmitter (outdoor) and the place at the window on the measured floor (LOS: LOS = 1, NLOS: LOS = 0) and W is the difference between L_{out}, the propagation loss between the external transmitter and the receiver, located in the road around the building, and L_{in}, the propagation loss between the transmitter and the receiver inside the building:

$$W = L_{out} - L_{in} \tag{5.46}$$

The constant W is taken at the window of the main wall, α_f is the frequency coefficient, α is the penetration distance coefficient (distance attenuation), G_h is the floor height gain and α_{LOS} is the LOS coefficient.

Therefore, the prediction formula for the penetration loss is proposed as follows for NLOS (LOS = 0):

$$L = L_{out} + 0.6r_i - 0.6h + 10 \tag{5.47}$$

The applicable range for the distance from the nearest window is 0–20 m, the applicable frequency is 0.8–8 GHz and the applicable area is a room or hallway with a window. The applicable range of the floor height is 1.5–30 m. It is interesting to note that the authors report a small value of $\alpha_f = -1.1$, obtained within the frequency range over which the model is applicable.

The results obtained by the authors make the model extremely simple to implement, and the fact that the penetration loss is frequency-independent over that frequency range (validated by extensive measurements they performed, as reported in the paper) simplifies even further its implementation. For more accurate results, some of the originally proposed parameters can be tuned with measurements, if desired, and the process is not too complex.

Another advantage is that predictions can be quickly made, even without measurements, for a rough estimate of the levels inside the building. However, the fact that the authors averaged many of the recommended model parameters imposes a restriction on the accuracy of the model and a careful analysis should be made if better accuracy is desired.

5.8.5   Taga–Miura Model Using Identification of Path Passing Through Wall Openings

In the COST-231 model (COST-231, 1999), the penetration point is that on the LOS wall closest to the mobile station regardless of wall structure. The radio waves transmitted by the base station penetrate the wall at this point and propagate inside the building. For the model proposed in Miura, Oda and Taga (2002), not only is this direct path taken into account but also those radio waves propagating through the structural openings along the walls. These wall openings are only doors and windows. Introducing the angle dependency of the losses with the paths that penetrate the indoor area improves the model accuracy, according to the authors (see Figure 5.16).

For the proposed model, the outdoor-to-indoor propagation loss is divided in three parts: the outdoor propagation loss L_{out}, the building penetration loss L_{pn} and the indoor losses L_{in}:

$$L = L_{out} + L_{pn} + L_{in} \tag{5.48}$$

All of the above losses are in dB. The geometry of the model can be seen in Figure 5.16. If there are several wall openings, the received level is the sum of path levels of paths through all such wall openings. For example, for path 1 in the figure, the penetration loss is given by

$$L_{pn,1} = L_e + L_g(1 - \cos \theta_1) + L_{gi} \sin \phi_1 \tag{5.49}$$

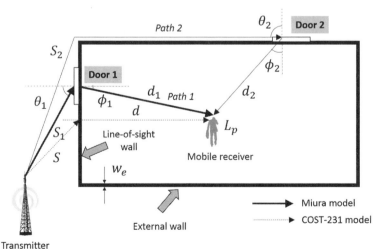

Figure 5.16 Geometry for the Taga–Miura outdoor-to-indoor model.

The value of L_{gi} recommended by the authors according to measurements is 20 dB. For the indoor loss, for simplicity, a constant loss per unit distance is proposed, as follows (for path 1):

$$L_{in,1} = \alpha r_1 \tag{5.50}$$

The constant α is obtained from measurements and is known as the attenuation coefficient. The model was tested at 8.45 GHz using CW measurements, and the following parameters were obtained: $L_g = 20$ dB, $L_e = 17.2$ dB, $\alpha = 0.348$ dB/m and $L_{gi} = 20$ dB.

This model takes into account wall openings, which enhances outdoor-to-indoor predictions (in principle) when compared to the COST-231 model, keeping its simplicity and involving only a few more additions.

There are errors and an offset reported by the authors of 20 dB, who explain that these losses are mainly due to furniture and specific clutter, hence causing radio waves not to propagate directly from the opening to the receiver. It also suffers from most of the disadvantages already mentioned for the COST-231 model approach.

5.9 Models for Propagation in Radiating Cables

Radiating cables are elements that are used in indoor designs to provide an alternative for uniform coverage in tunnels and corridors when antennas may not be suitable, as explained in Chapter 6. Thus, propagation models have been developed in the literature, which considers the use of radiating cables, as explained in this section.

5.9.1 Zhang Model

A radiating cable model for indoor environments is proposed in Zhang (2001). This model is an empirical one, configured for frequencies of around 2 GHz. It takes into consideration both coupling and longitudinal losses of the cable. The equation that shows the overall link loss, combining the cable losses and propagation losses, is

$$L = \alpha z + L_c + L_v + L_b + 10n \log r \tag{5.51}$$

where z is the longitudinal distance along the cable to the point nearest the receiver (m), α is the attenuation per unit length of the cable (dB/m), L_c is the coupling loss referenced to 1 m radial distance from the cable (dB), L_v is the variability in coupling loss (dB), L_b is the loss factor due to blockage (dB), r is the shortest distance between the cable and the receiver (m) and n is the path loss exponent. These parameters are better observed in Figure 5.17.

The particular situation analysed by the author was an academic building with reinforced concrete floors, 3.83 m ceiling height with a suspended false ceiling at 2.83 m, internal brick walls and external glass walls, with rooms interconnected via a corridor, along which ran the leaky feeder. In this case the model parameters were $L_c = 70.9$ dB, $L_v = 3.4$ dB, $L_c = 9.9$ dB, $\alpha = 0.12$ dB/m and $n = 0.6$. Note the very small value of the path loss exponent, indicating how the cable produces a very consistent level of coverage, although the particular parameters would depend strongly on the cable type

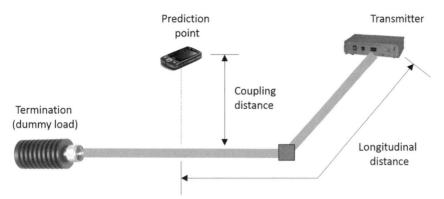

Figure 5.17 Zhang radiating cable model geometry.

and the environment. Coupling loss, longitudinal loss and variability in coupling loss are obtained from the cable's manufacturer.

Since it is an empirical model, it is very easy to implement, predictions are fast and its accuracy is strongly dependent on the quality of the measurements used to tune the model. In its present form, it only can predict path loss accurately in the close vicinity of the radiating cable; however, a more accurate indoor model needs to be incorporated for longer distances, especially as the signal goes through more walls and partitions, as the one described in Section 5.9.3. Also, at the termination of the radiating cable, it is not clearly specified in the paper how distances should be taken into account to avoid discontinuities.

Due to the fact that it depends on manufacturer's cable parameters, a very complete database of various radiating cable types is required for accurate predictions. Otherwise, a methodology for calculating some of these parameters needs to be incorporated.

5.9.2 Carter Model

K. Carter proposed a radio propagation model that considers the radiating cable as a line source and waves are spread in a cylindrical surface (Carter, 1998). A radiating cable straight section is taken into account and terminated with an antenna. In the near field and considering a mono-pole antenna in the receiver, the radio propagation is modelled in linear scale as

$$p_r = p_t \frac{3\lambda^2}{8\pi^2 azdL} \tag{5.52}$$

where p_r is the received power, p_t is the transmit power, λ is the wavelength, az is the longitudinal attenuation, d is the radial distance between the cable axis and the receiver in meters and L is the radiating cable length in metres.

Fading characteristics are not modelled by Carter. On the other hand, Rayleigh fading is compensated for in the customary fade margin applied to the link budget. Wall, clutter and floor losses are modelled separately as empirical correction factors.

One of the restrictions of this model is that only a straight length of radiating cable can be modelled. The author states that changes to cable bending may be overshadowed by diffraction and reflections from the clutter, something that definitely is arguable.

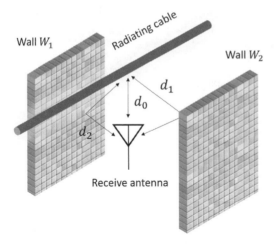

Figure 5.18 Multipath generated by reflected rays and direct ray.

Nevertheless, this model provides a good approximation to estimate signal strength from a leaky feeder, with reported standard deviations of the prediction error of around 5 dB. The author claims that the main source of errors is due to wall losses.

5.9.3 Seseña–Aragón–Castañón Model

An alternative approach to model radio propagation along radiating cables is proposed in Seseña-Osorio *et al.* (2013). Based on the Zhang model described in Section 5.9.1, the modelling of some key propagation mechanisms that are present in a practical indoor environment have been incorporated into this model: reflections, refraction losses, radiating cable paths and cable termination. In particular, first reflected rays and transmission losses are calculated with empirical coefficients, which are not dependent on the wall constitutive parameters or the incident angle.

Figure 5.18 illustrates a radiating cable installed along a corridor. Assuming that the radiating cable generates rays that are perpendicular to the cable axis, there are three paths along which the signal travels. Two paths are generated due to the first reflection of the signal in walls. The distances of reflected signals in walls W_1 and W_2 are d_1 and d_2, respectively. Meanwhile, there is a direct ray that travels from the radiating cable to the receiver, where its distance is d_0. Thus, the received power is composed by the addition of three paths (in watts) and is determined as

$$p_{r,\text{total}} = p_r(d_0) + p_r(d_1)R_1 + p_r(d_2)R_2 \tag{5.53}$$

Note that in equation (5.53) the power in watts is expressed in lowercase and R_1 and R_2 are empirical coefficients.

Figure 5.19 illustrates the scenario when there is a wall between the receiver and the radiating cable. Considering d_1 as the distance between the cable and the receiver, the received power is given by

$$p_{r,\text{total}} = p_r(d_1)T_1 \tag{5.54}$$

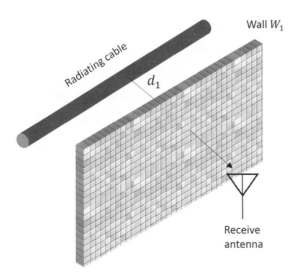

Figure 5.19 Transmission loss generated by a wall.

The received power $p_r(.)$ can be calculated from the Zhang model as follows:

$$p_r = \frac{p_t}{\alpha z l_c l_b d^n} \tag{5.55}$$

where p_r is the received power in watts, p_t is the transmitted power also in watts, αz is the longitudinal attenuation, l_c is the coupling loss, unitless, l_b is a loss factor, unitless, d is the radial distance between the cable axis and the receiver in metres and n is the path loss exponent. The transmission coefficient T_1 is derived empirically from measurements.

Unlike the Carter and Zhang models that only assume straight radiating cable sections, this model takes into account situations where there are cable bends, as depicted in Figure 5.20. The received power for Figure 5.20(a) needs to be calculated as

$$p_{r,\text{total}} = p_r(d_1)T_1 + p_r(d_2)T_2 \tag{5.56}$$

On the other hand, for Figure 5.20(b),

$$p_{r,\text{total}} = p_r(d_1)T_1 + p_r(d_2)K \tag{5.57}$$

The authors in Seseña-Osorio *et al.* (2013) performed an assessment of the proposed model taking the Zhang, Carter and Friis models to compute $p_r(.)$, yielding standard deviations of the errors of around 3.8 dB using the Zhang model. This model is valid for frequencies between 900 MHz and 2.5 GHz.

5.10 Wideband Channel Characteristics

The first part of this chapter was dedicated to narrowband channel models, aimed at predicting the median path loss to determine coverage in an indoor network. Radio

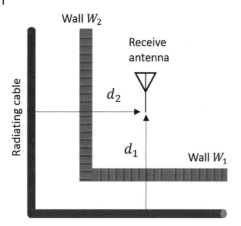

Wall W_2

Receive antenna

d_2

d_1

Wall W_1

Radiating cable

Figure 5.20 Radiating cable scenarios that take into account cable bends.

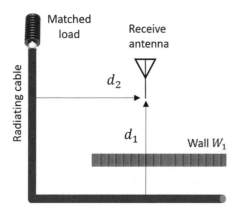

Matched load

Receive antenna

d_2

d_1

Wall W_1

Radiating cable

planning often is interested in estimating coverage levels for antennas deployed in a building. However, another important part of indoor radio planning is concerned with the maximum achievable data rate at a given frequency, and this is greatly determined by the echoes that are produced by multiple versions of the same signal reaching the receiver at different times. This could lead to *intersymbol interference* (ISI), which limits the symbol duration in a digital communication system, directly related to the data rate. The situation is simple to understand: the higher the data rate, the shorter the symbol duration, which for large delays in the various replicas of the original signal will cause a replica (or echo) arriving at the receiver when another symbol is to be transmitted – this is ISI.

The *power delay profile* (PDP) for the channel is one of the most important parameters characterized in wideband channel models. This represents the duration between replicas of the same signal (taps) that are received, and therefore are strongly related to the maximum achievable data rate. Indoor cells have approximately RMS delay spreads in the range 0.01 to 0.05 μs. For more details on the specification of PDP, refer to Saunders and Aragón-Zavala (2007).

In general and since scatterers are to each other inside buildings, the delays that are experienced by the signal taps are not too long compared to those seen outdoors and therefore higher data rates are possible inside buildings. In fact, for LTE systems, using

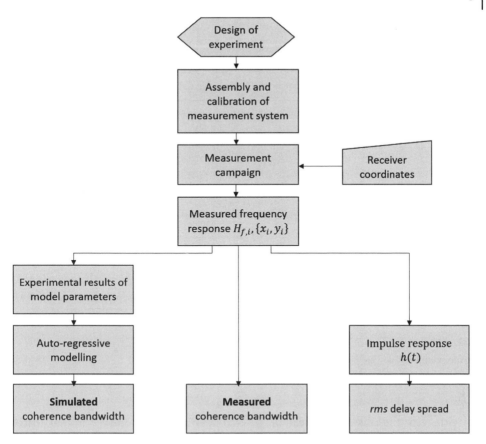

Figure 5.21 Methodology employed to obtain the frequency response of a radiating cable, as explained in Seseña-Osorio *et al.* (2015). https://jwcn.eurasipjournals.springeropen.com/articles/10.1186/s13638-015-0245-1.CC BY 4.0.

MIMO actually greatly improves data rates by taking all those replicas and adding them coherently using OFDM. Nevertheless, the modelling and characteristics need to be known in advance for efficient radio planning.

A good example in the modelling of the wideband characteristics of an indoor channel is given in Seseña-Osorio *et al.* (2015), where the frequency response for a radiating cable in the band 1.3 GHz to 1.8 GHz is obtained. The methodology used is shown in Figure 5.21. First, the experiment was designed by selecting the components, the configuration of the devices, and the sites or regions available for the development of frequency response measurements. Second, the assembly of the measurement system was made as well as its respective calibration. Then, the measurements were carried out and their positions were recorded. Once the frequency responses were obtained, the coherence bandwidth and the impulse response were calculated by using the auto-correlation function and the inverse discrete Fourier transform, respectively. The *rms* delay spread was then calculated by using the impulse response. The model parameters were obtained from measurements and the autoregressive model was applied in each room. Finally, simulated frequency responses and coherence bandwidth were calculated.

Table 5.8 Indoor office wideband channel parameters. Taken from ETSI (1997).

Median channel $\tau_{rms} = 35$ ns		Bad channel $\tau_{rms} = 100$ ns	
Relative delay τ[ns]	Relative mean power [dB]	Relative delay τ[ns]	Relative mean power [dB]
0	0	0	0
50	−3	100	−3.6
110	−10	200	−7.2
170	−18	300	−10.8
290	−26	500	−18
310	−32	700	−25.2

Table 5.9 RMS delay spread (ns) for indoor–indoor scenarios.

Environment	Case A	Case B	Case C
Indoor residential	20	70	150
Indoor office	35	100	460
Indoor commercial	55	150	500

As regards the rms delay spread of the channel, values encountered in most cases are very much lower than those found in either micro- or macrocells. However, the variability around the median value is large, although there is a strong correlation with the path loss and there are occasionally cases where the delay spread is very much larger than the median. In order to provide reasonably realistic simulations both situations must be considered. Table 5.8 gives suitable channels for an indoor office scenario, intended for evaluation purposes at around 2 GHz (ETSI, 1997). Values for the RMS delay spread for indoor-to-indoor environments are also shown in Table 5.9; case A represents low but frequently occurring values, case B represents median values and case C gives extreme values that occur only rarely. The very high cases can occur particularly if there are strong reflections from buildings situated a long way from the building under test.

5.11 Noise Considerations

Noise is an unwanted random-like signal that affects any electronic system. In particular and since received signal levels are very small, noise can considerably affect wireless communication systems and degrade its performance.

Since this book is focused on indoor wireless systems, special attention will be given to those noise sources and calculations that are relevant and present inside buildings. For a deeper analysis and study of noise, refer to Saunders and Aragón-Zavala (2007).

5.11.1 Noise Sources

Sources of noise vary, but in general we can consider the following:

- Any object that has a temperature above $0°$ K will generate *thermal noise*, which results from the excitation of charges (electrons) due to increases in temperature producing a randomly varying potential. This type of noise is also called *white noise* and is present in all frequencies; its power p_N in watts can be computed as follows:

$$p_N = kTB \tag{5.58}$$

where $k = 1.38 \times 10^{-23}\,\mathrm{JK}^{-1}$ is the Boltzmann's constant, T is the device's temperature in Kelvin degrees and B is the bandwidth in hertz. Note that thermal noise is normally taken as the reference or minimum noise in a communication system. In practice, for in-building systems, this noise floor is considered when calculations are to be made to establish whether a system is uplink or downlink limited, and all other noise sources need to be added to this noise floor.

- Receiving aerials 'pick up' *sky noise*, especially those where the elevation angle is small. This type of noise is also wideband, similar to white noise. Indoor systems rarely experience this type of noise, which is more relevant in satellite links, for parabolic dish receiving aerials.

- *Shot noise* occurs in active devices used in amplifiers and receivers, and is generated by the random fluctuations in the flow of electrons crossing the emitter-base or collector-base junctions within a transistor, or any other type of active semiconductor device.

5.11.2 Noise Parameters

It is often common practice to define parameters that establish the quality of the link, such as the *signal-to-noise ratio* (SNR), which establishes the signal level above noise for a system to deliver the expected quality-of-service. In digital systems, the *bit error rate* (BER) is used, which also can be associated with SNR depending on the type of service that is being delivered.

Every time the signal is amplified, repeaters and receivers also add noise to the signal. Therefore, this signal degradation needs to be quantified using SNR as follows:

$$\mathrm{SNR} = \frac{p_S}{p_N} \tag{5.59}$$

where p_S is the signal power in watts and p_N is the noise power also in watts. Note that the *input* SNR is always going to be greater than the output SNR, as additional noise is added to a system, even though both signal and noise power are amplified equally.

To quantify the amount of noise a device adds to the thermal noise, the *noise figure* (NF) is often employed, defined as follows:

$$\mathrm{NF} = 10 \log\left(\frac{SNR_i}{SNR_o}\right) = 10 \log\left(\frac{p_{N,\mathrm{out}}}{kTB}\right) \tag{5.60}$$

where SNR_i is the input SNR, SNR_o is the output SNR, NF is the noise figure in dB, $p_{N,\text{out}}$ is the device output noise power and kTB is the thermal noise. Note that the noise figure represents the amount of noise in dB the device adds above thermal noise and therefore special care should be taken when selecting devices for an in-building system with reasonable low noise figures, particularly at the receiver front end (Saunders and Aragón-Zavala, 2007). Recall that the first component in a receiver contributes the most to the overall SNR.

Another parameter that is usually employed to characterize noise is known as *noise temperature* T_e, which can be related to the noise figure as follows:

$$T_e = \left(10^{NF/10} - 1\right)T \tag{5.61}$$

where T is the reference temperature in Kelvin. Thus the noise temperature is given for a noise source which, when placed at the input of a network, yields the same output noise as if the network were noiseless. Noise temperature is more frequently used in satellite communication systems and is rarely employed for indoor wireless networks.

The *noise floor* in a system $P_{N,floor}$ in dBm is the noise power at a given noise figure for a given bandwidth and is a parameter often employed in receivers, amplifiers or any active component. Thus:

$$P_{N,floor} = 10\log(kTB) + \text{NF} + G_{dB} \tag{5.62}$$

Note that G_{dB} is the gain of the device in dB. Therefore, the *receiver sensitivity* $P_{r,dBm}$ in dBm can be calculated knowing the noise floor and the service SNR requirement in decibels SNR_{dB} as follows:

$$P_{r,dBm} = P_{N,floor} - SNR_{dB} \tag{5.63}$$

Finally, bear in mind that the noise figure is specified to active equipment by manufacturers, but for passive elements (cables, attenuators, splitters), the noise figure is given by the *insertion loss* of the element. Therefore, be aware that lossy cables installed prior to received antennas will have a detrimental effect on the system noise figure and should be avoided.

5.11.3 Considerations for Indoor Wireless Systems

Indoor radio planners and designers need to understand that, for example, passive loss in a distributed antenna system (DAS), which will be further explored in Chapter 9, will increase the total system noise figure.

Noise control is therefore essential for indoor networks, especially in cascaded systems having various active and passive devices that will degrade the system SNR. Gains need to be adjusted according to noise figures and losses of all system components so that the system noise figure can be optimized. In general, low noise figure components should be employed in the receiver front end and use high-gain amplifiers there to further maximize the overall SNR. DAS systems are especially sensitive to noise degradation if noise is not properly accounted for. For UMTS systems noise can definitely be a limiting factor and considerably reduces the data rate and cell size.

For a thorough analysis of noise effects within in-building systems and in particular to DAS, refer to Tolstrup (2011).

5.12   In-Building Planning Tools

Channel modelling is a task that can more efficiently be performed with the use of specialized software. Most of the commercially available packages are either used for outdoor environments and mainly for cellular or are free versions with simple calculations for Wi-Fi networks. As indoor planning has been taking popularity over the last few years, more radio planning tools have incorporated the functionality of performing indoor planning but at different levels. Some of the most popular indoor planning tools used nowadays in the market are briefly presented here.

5.12.1   iBwave Design

This software, developed by the Canadian company iBwave Solutions Inc. (2016), is one of the very few in the market that entirely specializes in in-building radio design and planning, including many key aspects such as:

- Radio propagation modelling, which can be selectable by the user depending on propagation conditions. Model parameters can also be tuned with measurements and optimized for maximized performance.
- Capacity calculations, including challenging venues such as stadiums, airports and shopping centres.
- Automatic link budget calculations including a bill of materials and cost, with a customizable database of thousands of radio parts to be used in the designs.
- Coax, CAT-5 or optical fibre cabling backhaul modelling and routing.
- Automated Access Point placement and optimal antenna placement.
- Network validation and error checking.
- Includes various indoor wireless technologies such as cellular and Wi-Fi.
- Has a mobile version to be used in tablets and other portable devices.

iBwave offers two licensing options:

- iBwave Design Lite, which was created specifically for small to medium size projects. It includes dynamically generated RF calculations, automatic bills of materials, on-screen error validation, 3-D prediction capabilities and KPI compliance checks.
- iBwave Design Enterprise, created to design large and complex in-building wireless networks. It includes multitechnology, multibuilding support, advanced 3-D modelling, coverage and capacity simulations, automatic link budget calculations, error checking and a customizable database of 22 953 parts and growing, as well as propagation and optimization modules to tune propagation model parameters.

The author believes this software is one of the most accurate, reliable and complete suites that can be found commercially, as it integrates all aspects of indoor radio design and planning to deliver full solutions to customers. In addition to this, iBwave offers specialized training and certification programs for those radio engineers and planners willing to develop a career in this area.

5.12.2 WiMap-4G

This tool was manufactured by the German company Brown iposs GmbH (2016) and was developed particularly for planning the air interface of wireless broadband access networks, for example based on the WLAN or WiMAX standard. It does outdoor and indoor radio planning. By the integrated free-space-propagation and COST-231 Walfish–Ikegami model and an interface to a high-performance ray launcher, WiMAP-4G supports a broad variety of applications: from a coarse overview to a detailed analysis of the field strength and from the search for suitable measurement points to the planning of telecommunication networks. They offer a couple of options for licensing:

- Community (free of charge), which includes basic propagation models, project creation and handling, basic antenna patterns and basic visualization features.
- Professional (purchaseable), having additional import/export features, additional advanced propagation models, high definition visualization and an unlimited number of sites and points.

5.12.3 Mentum CellPanner

Mentum CellPlanner, from Infovista (2016), formerly known as TEMS CellPlanner, is an advanced radio network design solution. Developed in close relationship with Ericsson, it provides planning and optimization capabilities including limited in-building systems. Mentum CellPlanner supports planning and optimization activities for LTE, WCDMA (including HSPA) and GSM (including GPRS, EDGE and Evolved EDGE). Mentum CellPlanner enables the import of native TEMS investigation data.

5.12.4 Atrium

Atrium, from Consistel (2016), includes the Atrium Network Design module, which is the RF planning tool of the Atrium modules for 2G, 3G, 3.5G, 4G, WiMAX, Wi-Fi, etc. It automates the planning and design of distributed antenna system in-building wireless networks for whatever technology you require. Given desired indoor coverage areas and KPIs, this planning tool uses a patented automatic antenna placement algorithm, to ensure that coverage and quality objectives are met. Built-in intelligent algorithms automatically generate coverage maps, network schematics, cabling diagrams, the bill of materials (BoM) and link budgets. The following features are available:

- Coverage predictions using your choice of propagation models and automatic generation of coverage maps.
- Automatic tuning of propagation models for enhanced coverage prediction accuracy.
- Built-in and extendable building material database enabling the accurate modelling of various indoor transmission losses.
- Seamless import of walk test log files.

5.12.5 WinProp

WinProp, from AWE Communications GmbH (2016), is the main product of AWE Communications. The development of WinProp started in the mid 1990s and runs on standard PCs under MS Windows (all 32 bit and 64 bit versions are supported). Together with several partners AWE Communications was able to develop a software package ideally suited for propagation modelling in different scenarios (rural, urban, indoor) and for network planning of different air interfaces (2G/2.5G, 3G, WLAN, WiMAX, DVB-H/ DVB-SH). In addition to the above described WinProp product, AWE Communications offers several modules of WinProp (e.g. propagation engines, data conversion modules, etc.) as plug-ins for other radio network planning tools.

5.12.6 CellTrace

CellTrace, from CelPlan Technologies Inc. (2016), is a suite of software tools that addresses radio frequency (RF) planning, design, modelling, analysis and optimization of wireless communication systems within buildings, tunnels, stadiums and in urban environments. It supports a wide variety of air interfaces, including LTE, WiMAX (802.16x), Wi-Fi, CDMA, UMTS, GSM and TETRA. It also supports MIMO technology, distributed antenna systems (DAS), leaky feeder cables and a number of other transmission modes. CellTrace comes in two versions to address the specific needs of the end-user:

- CellTrace I supports the indoor deployments of all wireless technologies, including tunnel environments.
- CellTrace D supports only Wi-Fi network planning in indoor scenarios.

    The following add-on modules are also available for CellTrace:

- CellComp includes all components for indoor network planning.
- CellAntenna offers a convenient facility to generate and edit antenna patterns.

5.12.7 EDX Signal Pro

EDX Signal Pro, from EDX Wireless Inc. (2016), is the principal building block of EDX's comprehensive line of wireless network engineering tools. It is appropriate for any system, including broadband wireless WiMAX, LTE, Wi-Fi, cellular and other mobile radio systems. It offers all of the study types you need to design a basic wireless network, including area studies, link/point-to-point studies and route studies. With the Network Design Add-On Module, you have a carrier-class design tool, complete with automatic system layout, automatic traffic loading and automatic frequency planning.

It includes the release of version 2.3 of EDX® Signal-IQ™, a comprehensive software system for the design of in-building wireless networks, including distributed antenna systems (DAS), picocell, femtocell and 802.11 access point (AP) deployments. With the RF design placed on the floor plan, Signal-IQ allows the user to simulate and display the coverage of the indoor wireless network via a wide range of available RF study options.

Finally, EDX Signal Pro includes an RF design optimization tool that calculates EiRP and predicted coverage in real time as the antenna is moved around the floor plan of the building. This feature allows the user to rapidly try what-if scenarios and optimize the coverage of each antenna in a complex DAS system. Numerous add-ons are available, including SignalProof, Signal MX, Microcell/Indoor and Building Editor.

5.12.8 iBuildNet DAS

iBuildNet DAS, from RanPlan Wireless Ltd (2016), is an advanced solution for designing and optimizing DAS networks and supports multisystem technologies to deliver a seamless network environment. It uses advanced 3-D modelling, fast and accurate 3-D ray tracing and powerful data analysis to optimize antenna location type, power and channel assignment.

iBuildNet DAS can also be used with iBuildNet Tablet Planner, RanPlan's automated on-site tool that allows installers to quote, plan, deploy and optimize a network to deliver maximum coverage and QoE in short time, at low cost. The iBuildNet DAS includes features such as: 3-D building modelling, DAS layout and configuration, an RF propagation engine with a selection of propagation models based on ray tracing, wireless performance prediction and automatic optimization.

5.12.9 Wireless InSite

Wireless InSite, from Remcom Inc. (2016), is a suite of ray-tracing models and high-fidelity EM solvers for the analysis of site-specific radio propagation and wireless communication systems. The software provides predictions of electromagnetic propagation and communication channel characteristics in complex urban, indoor, rural and mixed path environments. This software is to be used for both indoor and outdoor design.

Wireless InSite includes models for urban, indoor and rough terrain. There are several modelling options for indoor as follows:

- High-fidelity ray-tracing models
  - X3D (multithreaded and GPU accelerated). The X3D Ray model is Remcom's first ray-based model to use GPU acceleration to reduce calculation time. X3D is a 3-D propagation model, so there are no restrictions on geometry shape or transmitter/receiver height. The model includes effects from reflections, diffractions, transmissions, and atmospheric absorption. Applications include urban, indoor, and indoor–outdoor propagation scenarios.
  - Full 3D. Remcom's Full 3D model is a general propagation model that can be applied to urban, indoor and indoor–outdoor propagation. The model can handle arbitrary geometry and transmitter/receivers at any height. Users have the option of performing the ray tracing with the shooting and bouncing ray (SBR) method, or Eigenray, which is based on the method of images. The model includes effects from reflections, diffractions and transmissions.
- Empirical propagation models
  - Wall Count. The Wall Count model is the Wireless InSite Real Time method for indoor calculation. Direct rays are constructed between the transmitter and the receiver, and every wall intersected in the indoor geometry is counted. Each intersection adds an additional 3 dB of loss to free-space path loss.

5.13 Conclusion

Mathematical models for narrowband and wideband channels are a very useful tool for any radio planner, as they assist in the estimation of radio coverage as well as determination of time-domain and frequency-domain characteristics of the channel. However, and since the in-building radio channel is complex due to the large variability of materials, geometries and characteristics, indoor channel models are still the subject of research and development efforts worldwide. It is therefore wise to keep an eye on recent publications and literature available in the form of research and white papers, both from academia and industry. The range of frequencies and applications is growing as newer services are offered and therefore keeping this up to date is a priority.

The modelling of building penetration loss (BPL) has not shown consistent results to date, with a large variety of studies following different methodologies and therefore arriving at contradictory conclusions in some cases – for some authors BPL increases with frequency, whereas for others it decreases. Nevertheless, some progress has been made and dependencies with clutter and depth have also been investigated.

When coming to the choice of an in-building planning tool, there are various aspects to take into consideration, such as: cost, customer support, flexibility, operating system and computing requirements, computation speed, accuracy, range of propagation models to choose, compatibility with other applications and measurement data, etc. I am presenting a survey of some of the most popular suites but the final decision has to be made by the engineer responsible for designing the indoor network.

Finally, one of the key elements in the success of channel models is the use of accurate measurements, both to obtain them and to validate their predictions. Note that the equipment used in these measurements should be properly validated and calibrated for accurate results. This includes the miscellaneous RF components associated with the measurements. Radio measurements for in-building planning is the subject of Chapter 7.

References

Akerberg, D. (1988) Properties of a TDMA picocellular office communication system, in Proceedings of the IEEE Global Telecommunications Conference Globecom '88, Hollywood, USA, pp. 1343–1349.

Aragón-Zavala, A., Belloul, B., Nikolopoulos, V. and Saunders, S.R. (2006) Accuracy evaluation analysis for indoor measurement-based radio-wave-propagation predictions, *IEE Proceedings on Microwaves, Antennas and Propagation*, **153** (1), 67–74.

AWE Communications GmbH (2007) *Propagation Models and Scenarios: Indoor*, Report, http://www.awe-communications.com.

AWE Communications GmbH (2016) URL: http://www.awe-communications.com.

Brown iposs GmbH (2016) URL: http://www.brown-iposs.com.

Carter, K. (1998) Prediction propagation loss from leaky coaxial cable terminated with an indoor antenna, in Proceedings of 8th Virginia Tech/MPRG Symposium on Wireless Communications, pp. 71–82.

Cátedra, M., Aldana, F.S., Gutiérrez, O., González, I. and Pérez, J. (2000) Propagation model based on ray tracing for the design of personal communication systems in indoor environments, *IEEE Transactions on Vehicular Technology*, **49** (6), 2105–2112.

CelPlan Technologies Inc. (2016) URL: http://www.celplan.com.

Cheung, K.W., Sau, J.H.M. and Murch, R.D. (1998) A new empirical model for indoor propagation prediction, *IEEE Transactions on Vehicular Technology*, **47** (3), 996–1001.

Consistel (2016) URL: http://www.consistel.com.

COST-231 (1999) Final report, COST Action 231: Digital mobile radio towards future generation systems, *European Commission/COST Telecommunications*, Brussels, Belgium.

Dersch, U., Liebendörfer, M. and Zehnder, M. (2000) Multi-channel coupling: interactive prediction of 3D indoor propagation, in Proceedings of the International Zurich Seminar on Broadband Communications, Zurich, Switzerland, pp. 215–221.

EDX Wireless Inc. (2016) URL: http://edx.com.

ETSI (1997) European Telecommunication Standards Institute, *Selection Procedures for the Choice of Radio Transmission Technologies of the Universal Mobile Telecommunications System (UMTS)*, DTR/SMG-50402.

Hata, M. (1980) Empirical formula for propagation loss in land mobile radio services, *IEEE Transactions on Vehicular Technology*, **29**, 317–325.

Honcharenko, W., Bertoni, H.L., Dailing, J.L., Qian, J. and Yee, H.D. (1992) Mechanisms governing UHF propagation on single floors in modern office buildings, *IEEE Transactions on Vehicular Technology*, **41** (4), 496–504.

iBWave Solutions Inc. (2016) URL: http://www.ibwave.com.

Infovista S.A.S. (2016) URL: http://www.infovista.com.

ITU (1997) International Telecommunication Union, ITU-R Recommendation P.1238: Propagation data and prediction models for the planning of indoor radiocommunication systems and radio local area networks in the frequency range 900 MHz to 100 GHz, Geneva.

Keenan, J.M. and Motley, A.J. (1990) Radio coverage in buildings, *BT Technology Journal*, **8** (1), 19–24.

Kouyoumjian, R.G. and Pathak, P.H. (1974) A uniform geometrical theory of diffraction for an edge in a perfectly conducting surface, *Proceedings IEEE*, **62** (11), 1148–1461.

Lee, B.S., Nix, A.R. and McGeehan, J.P. (2001) Indoor space–time propagation modelling using a ray launching technique, in Proceedings of the International Conference on Antennas and Propagation, pp. 279–324.

Miura, Y., Oda, Y. and Taga, T. (2002) Outdoor-to-indoor propagation modelling with the identification of path passing through wall openings, in Proceedings of the IEEE Personal Indoor and Mobile Radio Conference, PIMRC '02, pp. 130–134.

Oestges, C. and Paulraj, A.J. (2004) Propagation into buildings for broad-band wireless access, *IEEE Transactions on Vehicular Technology*, **53** (2), 521–526.

Okamoto, H., Kitao, K. and Ichitsubo, S. (2010) Outdoor-to-indoor propagation loss prediction in 800-MHz to 8-GHz band for an urban area, *IEEE Transactions on Vehicular Technology*, **58** (3), 1059–1067.

Okumura, Y., Ohmori, E., Kawano, T. and Fukuda, K. (1968) Field strength and its variability in VHF and UHF land mobile radio service, *Reviews of Electronics and Communications Laboratory*, **16**, 825–873.

RanPlan Wireless Ltd (2016) URL: http://www.ranplan.co.uk.

Remcom Inc. (2016) URL: http://www.remcom.com.

Saunders, S. and Aragón-Zavala, A. (2007) *Antennas and Propagation for Wireless Communication Systems*, 2nd edition, John Wiley & Sons, Ltd, Chichester. ISBN 0-470-84879-1.

Seidel, S.Y. and Rappaport, T.S. (1992) 914 MHz path loss prediction models for indoor wireless communications in multi-floored buildings, *IEEE Transactions on Antennas and Propagation*, **40** (2), 207–217.

Seidel, S.Y. and Rappaport, T.S. (1994) Site-specific propagation prediction for wireless in-building personal communication system design, *IEEE Transactions on Vehicular Technology*, **43** (4), 879–891.

Seseña-Osorio, J.A., Aragón-Zavala, A., Zaldívar-Huerta, I.E. and Castañón, G. (2013) Indoor propagation modelling for radiating cable systems in the frequency range of 900–2500 MHz, *Progress in Electromagnetics Research B*, **47**, 241–262.

Seseña-Osorio, J.A., Zaldívar-Huerta, I.E., Aragón-Zavala, A. and Castañón, G. (2015) Analysis and experimental evaluation of the frequency response of an indoor radiating cable in the UHF band, *EURASIP Journal on Wireless Communications and Networking*, Springer, Vol. **2015** (1), pp. 1–12. DOI: 10.1186/s13638-015-0245-1.

Siwiak, K., Bertoni, H.L. and Yano, S.M. (2003) Relation between multipath and wave propagation attenuation, *IEEE Electronics Letters*, **39** (1), 142–143.

Theofilogiannakos, G.K., Xenos, T.D. and Yioultsis, T.V. (2007) An efficient hybrid parabolic equation–integral equation method for the analysis of wave propagation in highly complex indoor communication environments, *Wireless Personal Communications*, **43** (2), 495–510.

Tolstrup, M. (2011) *Indoor Radio Planning: A Practical Guide for GSM, DCS, UMTS, HSPA and LTE*, 2nd edition, John Wiley & Sons, Ltd, Chichester. ISBN 0-470-71070-8.

Tuan, S.C., Chen, J.C., Chou, H.T. and Chou, H.H. (2003) Optimization of propagation models for the radio performance evaluation of wireless local area networks, in Antennas and Propagation Society International Symposium, Vol. 2, pp. 146–149.

Wölfle, G., Gschwendtner, B.E. and Landstorfer, F.M. (1997) Intelligent ray tracing – a new approach for field strength prediction in microcells, in 47th IEEE Vehicular Technology Conference, Vol. 2, pp. 790–794.

Wölfle, G., Wahl, R., Wertz, P., Wildbolz, P. and Landstorfer, F.M. (2005) Deterministic propagation model for the planning of hybrid urban and indoor scenarios, in 16th International Symposium on Personal, Indoor and Mobile Radio Communications, PIMRC '05 pp. 659–663.

Zhang, Y.P. (2001) Indoor radiated-mode leaky feeder propagation at 2.0 GHz, *IEEE Transactions on Vehicular Technology*, **50** (2), 536–545.

6

Antennas

So far, we have discussed the way radio waves propagate through various media and how the specific constitutive parameters of materials have an effect on this propagation. Propagation effects such as reflection, refraction, scattering and diffraction are present in many practical situations, hence making the task of estimating the received signal strength in a wireless link a challenge. Furthermore, we have defined the concept of path loss and described various propagation models that aim at predicting this path loss more accurately, so that system designers can make use of these tools in designing efficient radio systems. When the link budget concept was examined, we mentioned that we needed 'antennas' at transmitter and receiver ends in order to send signals wirelessly, but did not attempt to go any further in our analysis – we simply assumed that antennas produce a gain while transmitting or receiving that should be accounted for. It is time now to discuss in detail what antennas are, to gain a better understanding and to be able to select such devices for specific applications for indoor wireless projects. Note that the aim of this chapter is not to provide a thorough analysis of antennas and electromagnetic effects, as in Hayt and Buck (2005), Kraus and Fleisch (1999) and Sadiku (2007), but rather to understand their basic characteristics, specifications and performance parameters needed to select the right antennas for given indoor requirements.

6.1 The Basics of Antenna Theory

Most fundamentally, an antenna is a way of converting the guided waves present in a waveguide, feeder cable or transmission line into radiating waves travelling in free space, or vice versa, as seen in Figure 6.1. The fields trapped in the transmission line travel in one dimension towards the antenna, which converts them into radiating waves, carrying power away from the transmitter in three dimensions into free space.

The art of antenna design is to ensure that this process takes place as efficiently as possible, with the antenna radiating as much power from the transmitter into useful directions, particularly the direction of the intended receiver, as practically as can be achieved.

I think most of us have experienced the typical case of using a paper clip or a clothes hook as an antenna and it works! To be more specific, it couples some of the radiation, but not necessarily in an efficient way. Now think about a satellite link for a geostationary orbit (GEO), where the signal has to travel 36 000 km from the satellite to the ground station; if the receive antenna is not properly designed, the extremely tiny received signal

Indoor Wireless Communications: From Theory to Implementation, First Edition. Alejandro Aragón-Zavala.
© 2017 John Wiley & Sons Ltd. Published 2017 by John Wiley & Sons Ltd.

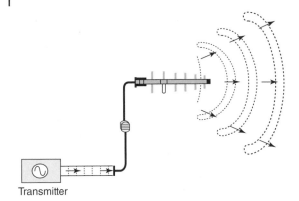

Figure 6.1 Antenna seen as a transition region between guided and unguided waves.

will be superseded by noise and not properly coupled – thus the importance of designing efficient antennas!

6.1.1 Conditions for Radiation

As a direct consequence of Maxwell's equations (Maxwell, 1861, 1865), a group of charges in uniform motion (or stationary charges) do not produce radiation. However, radiation does occur if any of the following conditions is met (Figure 6.2):

- The velocity of the charges in a given wire is changing in time.
- The charges are reaching the end of the wire and reversing direction, producing radiation.

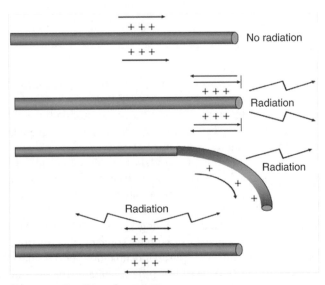

Figure 6.2 Conditions for radiation.

- The speed of the charges remains constant, but their direction is changing, thereby creating radiation.
- The charges are oscillating in periodic motion, causing a continuous stream of radiation.

Antennas can therefore be seen as devices that cause charges to be accelerated in ways that produce radiation with desired characteristics. Similarly, rapid changes of direction in structures that are designed to guide waves may produce undesired radiation, as is the case when a printed circuit track carrying high-frequency currents changes direction over a short distance.

6.1.2 Antenna Regions

The space surrounding an antenna is usually subdivided into regions, as seen in Figure 6.3. These regions are so designated to identify the field structure in each. Although no abrupt changes in the field configurations are noted as the boundaries are crossed, there are distinct differences amongst them. The boundaries separating these regions are not unique, although various criteria have been established and are commonly used to identify these regions.

The *reactive near-field region* is defined as that portion of the near-field region immediately surrounding the antenna wherein the reactive field predominates.

The *radiating near-field* or *Fresnel region*, close to an antenna, is that for which the field patterns change very rapidly with distance and include both radiating energy and reactive energy; although the radiating field predominates, the field oscillates towards and away from the antenna, appearing as a reactance that only stores but does not dissipate energy.

Further away, in the *far-field* or *Fraunhofer region*, the reactive fields are negligible and only the radiating energy is present, resulting in a variation of power with direction that is independent of distance. Within the far-field region, the wavefronts appear very closely as spherical waves, so that only the power radiated in a particular direction is of importance, rather than the particular shape of the antenna. Measurements of the power radiated from an antenna have either to be made well within the far field, or else special account has to be taken of the reactive fields.

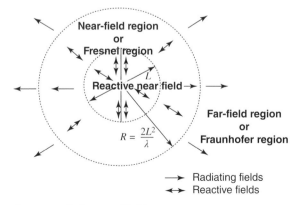

Figure 6.3 Definition of field regions.

Figure 6.4 Antennas mounted close to metallic structures affecting radiation characteristics.

In siting an antenna, it is particularly important to keep other objects out of the near field, since they will couple with the currents in the antenna and change them, which in turn may greatly alter the designed radiation and impedance characteristics. An example of this is shown in Figure 6.4, where an antenna has been mounted on a hoist basket but too close to the metal structure of the basket, thus distorting the antenna radiation characteristics. This can be particularly relevant in buildings with exposed metallic structures such as stadiums or sports arenas, where antennas may need to be mounted close to these structures and therefore extra care should be taken.

6.2 Antenna Parameters

Antennas are often characterized by various parameters that define specific characteristics of antenna performance. When selecting antennas, the evaluation of such parameters is very important and, thus, a good understanding of each is essential. Antenna parameters define the performance and characteristics of an antenna, as well as make them distinct from others. These parameters are often used when designing an indoor wireless system and are also given by antenna manufacturers to assist system designers in the correct selection of the antenna type. As we will see later on, the choice of antenna parameters is not only influenced by their technical performance but also by its mechanical characteristics and appearance – there are specific indoor environments in which antennas need to be 'disguised'.

6.2.1 Radiation Pattern

The radiation pattern of an antenna is a plot of the far-field radiation from the antenna. More specifically, it is a plot of the power radiated from an antenna per unit solid angle or its radiation intensity. *Radiation intensity* in a given direction is the power radiated from an antenna per unit solid angle (steradian) and is specified in the far-field of the antenna as follows:

$$U = r^2 S_{\text{rad}} \tag{6.1}$$

where U is the radiation intensity in watts, r is the radial distance from the antenna to the point of interest in metres and S_{rad} is the radiated power density in W/m^2.

Radiation patterns are usually plotted by normalizing the radiation intensity by its maximum value and plotting the result. An example of a radiation pattern in 3-D is given in Figure 6.5, showing its main characteristics.

Antenna manufacturers often specify radiation patterns using only two cuts: azimuth (horizontal) and elevation (vertical), as seen in Figure 6.6. Therefore, for propagation work, it is necessary to interpolate the values given for these cuts to work out an approximate value of gain for given intermediate angles.

Radiation patterns are often given in polar coordinates (Figure 6.7), having angle and, instead of specifying absolute gain, a relative attenuation with respect to maximum gain (boresight) is given. For example, for 30° in azimuth, the radiation pattern plot may indicate an attenuation of 3.8 dB and, if the maximum gain of the antenna is 7 dBi, then the absolute gain at this angle would be $7 - 3.8 = 3.2$ dBi.

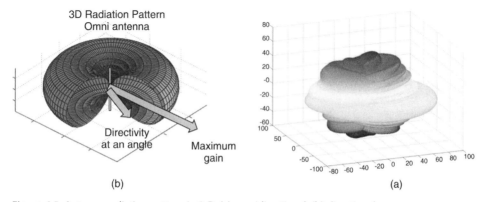

Figure 6.5 Antenna radiation pattern in 3-D: (a) omnidirectional; (b) directional.

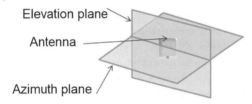

Elevation plane

Antenna

Azimuth plane

Figure 6.6 Antenna cuts – horizontal (azimuth plane) and vertical (elevation plane).

For a radiation pattern, various lobes can be identified, as shown in Figure 6.8, as follows:

- Main lobe, which includes the direction of maximum radiation.
- Side lobe, expressing the amplitude of the largest side lobe, usually expressed in decibels relative to the peak of the main lobe.
- Back lobe, with radiation diametrically opposite the main lobe.

The nulls indicate the angles at which no radiation occurs, either for transmit or receive, according to the reciprocity theorem, as also shown in Figure 6.8. In fact, it is through the nulls that adaptive (intelligent) antennas steer their main radiation pattern towards the desired user and places interferers in the null directions.

The shape of the radiation pattern changes across frequency bands; thus, manufacturers specify radiation pattern plots at given frequencies. For accurate design work, the radiation pattern having the closest frequency to that in which the system will operate must be used.

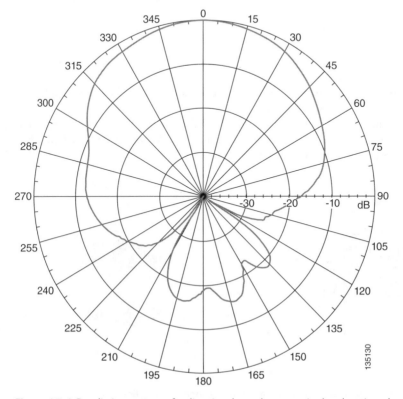

Figure 6.7 2-D radiation pattern of a directional panel antenna in the elevation plane.

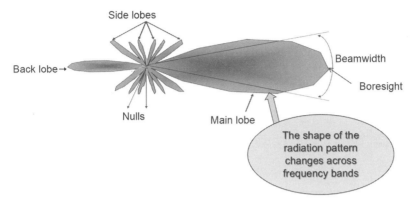

Figure 6.8 Radiation pattern lobes.

The direction of maximum radiation is normally known as the *boresight*. The beamwidth is the angular separation between two identical points on opposite sides of the pattern maximum.

In summary, various parts of a radiation pattern are referred to as lobes. These can be further classified as follows:

- *Radiation lobe*: portion of the radiation pattern bounded by regions of relatively weak radiation intensity. *Major lobe (main beam)*: the radiation lobe containing the direction of maximum radiation.
- *Minor lobe*: any lobe except a major lobe; this includes the side lobes as well as the back lobe. For indoor radio systems, radiation has to be directed towards a desired area and minimized in other directions, in both horizontal and vertical directions.
- *Side lobe*: a radiation lobe in any direction other than the intended lobe. These side lobes are quite relevant for interference control, especially in sectors.
- *Back lobe*: radiation lobe whose axis makes an angle of approximately 180° with respect to the beam of an antenna. This back lobe is particularly important for interference when installing a directional antenna, or for through-floor penetration if an omnidirectional antenna is mounted on a ceiling.

On the other hand, there are additional parameters that need to be specified for antenna radiation patterns, as follows:

- *Null*: angle of minimum radiation or reception; for some applications, nulls are placed at specific directions to cancel interferers.
- *Boresight*: direction of maximum radiation; the maximum antenna gain is often specified at this direction.

Let us illustrate some of these properties with a practical example. The Andrew DB992HG28N-B directional antenna is presented in Figure 6.9, with a maximum gain of 16 dBi – we will see later how the antenna gain is specified, but for now let us take it as it is. Horizontal and vertical beamwidths are shown, as well as main and side lobes. Radiation patterns are often specified in polar coordinates by manufacturers, using two orthogonal cuts: horizontal and vertical.

16-dBi Flat Panel Antenna
Andrew, DB992HG28N-B

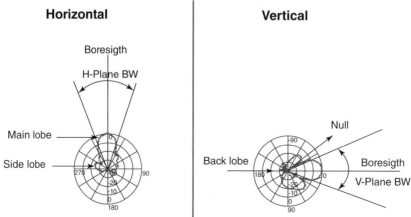

Horizontal

Boresigth

H-Plane BW

Main lobe

Side lobe

270

90

180

Vertical

Null

-90

Back lobe

180

Boresigth

V-Plane BW

90

Figure 6.9 Antenna radiation pattern example.

6.2.2 Directivity

In order to define the concept of directivity, first it is important to introduce an isotropic radiator, which will be used as a reference of directivity and gain. An *isotropic radiator* is a hypothetical antenna that radiates power equally in all directions, as seen in Figure 6.10. This cannot be achieved in practice, but acts as a useful point of comparison.

The power density S_{rad} depends on radiated power P_{rad} over a sphere of radius r and is given by

$$S_{rad} = \frac{P_{rad}}{4\pi r^2} \tag{6.2}$$

Thus, according to Equation (6.2), the radiation intensity of an isotropic source is

$$U_0 = \frac{P_{rad}}{4\pi} \tag{6.3}$$

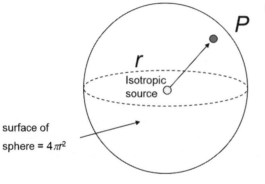

Figure 6.10 Isotropic antenna.

P

r

Isotropic
source

surface of
sphere = $4\pi r^2$

If Equation (6.3) is plotted as a radiation pattern, a sphere is produced, the radiation intensity of which is constant in all directions.

Once we have understood the concept of an isotropic source, it is time now to define directivity. The *directivity* of an antenna is defined as the ratio of the radiation intensity in a given direction from the antenna to the radiation intensity averaged over all directions. Thus, the directivity d is a function of direction, as follows:

$$d(\theta, \phi) = \frac{\text{Radiation intensity of antenna in direction } (\theta, \phi)}{\text{Mean radiation intensity in all directions}}$$

$$= \frac{\text{Radiation intensity of antenna in direction } (\theta, \phi)}{\text{Radiation intensity of isotropic antenna radiating the same total power}}$$

$$= \frac{U}{U_0} \tag{6.4}$$

Notice that the isotropic source is considered a reference in the denominator since it radiated equally in all directions. Sometimes directivity is specified without referring to a direction. In this case the term 'directivity' implies the maximum value. It is also common to express the directivity in decibels. The use of the isotropic antenna as a reference is then emphasized by giving the directivity units of dBi:

$$D = 10 \log d \tag{6.5}$$

Notice that we use d to express directivity in linear scale and D to indicate it in decibels. We will use this convention throughout this book when the same parameter can be expressed in linear and logarithmic scales. The directivity of an antenna increases as its beamwidth is made smaller, since the energy radiated is concentrated into a smaller solid angle. For large antennas, with a single major lobe, the half-power beamwidth of the antenna in the azimuth and elevation directions may be related to its directivity by the following approximate formula:

$$d \approx \frac{41,253}{\theta_{HP}^\circ \phi_{HP}^\circ} \tag{6.6}$$

Where θ_{HP}° and ϕ_{HP}° are the vertical and horizontal half-power beamwidths in degrees.

To illustrate directivity, suppose we need to find the maximum directivity of an antenna with the following radiation power density:

$$\mathbf{S}_{rad} = \frac{s_0 \sin \theta}{r^2} \mathbf{a}_r \text{ W/m}^2 \tag{6.7}$$

We need to write an expression for the directivity as a function of the vertical and horizontal directional angles. Therefore, to determine the maximum directivity d_{max}, Equation (6.4) is used as follows:

$$D_{max} = \frac{U_{max}}{U_0} = \frac{4\pi U_{max}}{P_{rad}} \tag{6.8}$$

Recall that U_0 is the radiation intensity of an isotropic antenna. Now, for our antenna and by using Equation (6.1), radiation intensity U is given by

$$U = r^2 \frac{s_0 \sin \theta}{r^2} = s_0 \sin \theta \tag{6.9}$$

The maximum radiation intensity U_{max} occurs at $\theta = 90°$ and is $U_{max} = s_0$. To compute the maximum directivity, we must now calculate the radiated power P_{rad} over the whole surface, given S_{rad}. Thus, taking $\mathbf{ds} = r^2 \sin \theta \, d\theta d\phi \mathbf{a}_r$ yields

$$P_{rad} = \oint \mathbf{S}_{rad} \cdot \mathbf{ds} = \int_0^{2\pi} \int_0^\pi \frac{s_0 \sin \theta}{r^2} \mathbf{a}_r \cdot r^2 \sin \theta d\theta d\phi \mathbf{a}_r$$

$$= s_0 \pi \int_0^{2\pi} \left(\frac{1}{2} - \frac{1}{2} \cos 2\theta \right) d\theta = \pi^2 s_0 \tag{6.10}$$

The maximum directivity can now be calculated:

$$d_{max} = \frac{4\pi s_0}{\pi^2 s_0} = \frac{4}{\pi} \tag{6.11}$$

Finally, an expression for the directivity can be obtained as follows:

$$d = \frac{4}{\pi} \sin \theta \tag{6.12}$$

6.2.3 Radiation Resistance and Efficiency

The equivalent circuit of a transmitter and its associated antenna is shown in Figure 6.11. The resistive part of the antenna impedance is split into two parts, a radiation resistance R_r and a loss resistance R_l. The power dissipated in the radiation resistance is the power actually radiated by the antenna and the loss resistance is power lost within the antenna itself. This may be due to losses in either the conducting or dielectric parts of the antenna.

Since only the radiated power normally serves a useful purpose, it is useful to define the radiation efficiency e of the antenna as

$$e = \frac{\text{Power radiated}}{\text{Power accepted by antenna}} = \frac{R_r}{R_r + R_l} \tag{6.13}$$

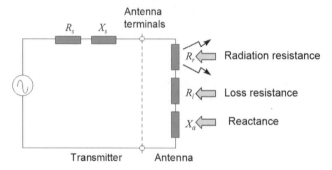

Figure 6.11 Equivalent circuit of a transmitting antenna.

An antenna with high radiation efficiency therefore has high associated radiation resistance compared with the losses. The antenna is said to be resonant if its input reactance $X_a = 0$. Thus, antennas that are designed to resonate at certain frequencies have their optimum behaviour at those frequency bands; that is an antenna designed to be used for 802.11 g Wi-Fi cannot be used to operate in the LTE 800 MHz band.

6.2.4 Power Gain

The power gain g, or simply the gain, of an antenna is the ratio of its radiation intensity to that of an isotropic antenna radiating the same total power as accepted by the real antenna. When antenna manufacturers specify simply the gain of an antenna they are usually referring to the maximum value of g. From the definition of efficiency e, the directivity and the power gain are then related by:

$$g(\theta, \phi) = ed(\theta, \phi) \tag{6.14}$$

Gain may be expressed in dBi to emphasize the use of the isotropic antenna as reference:

$$G = 10 \log g \tag{6.15}$$

It is also common to express power gain in dBd if a dipole antenna gain is used as a reference:

$$G[\text{dBd}] = G[\text{dBi}] - 2.14 \tag{6.16}$$

Although the gain is, in principle, a function of both azimuth and elevation together, it is common for manufacturers to specify patterns in terms of the gain in only two orthogonal planes, usually called cuts. In such cases the gain in any other direction may be estimated by assuming that the pattern is separable into the product of functions g_θ and g_ϕ which are functions of only θ and ϕ, respectively. Thus:

$$g(\theta, \phi) = g_\theta(\theta)g_\phi(\phi) \tag{6.17}$$

To understand the concept of power gain in an antenna better and why dBi units are used, an example is shown in Figure 6.12. This example shows how a dipole antenna gain

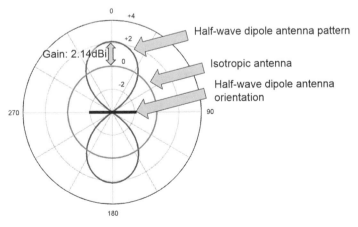

Figure 6.12 Power gain of an antenna with respect to an isotropic source.

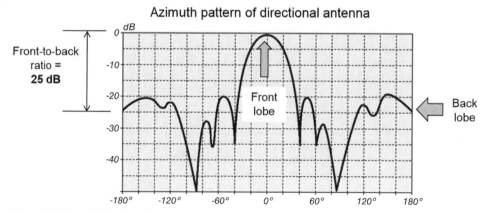

Figure 6.13 Front-to-back ratio.

is represented having as reference an isotropic radiator. In this case, at boresight, the dipole exhibits 2.14 dB more gain than the isotropic source; therefore, the dipole is said to have a gain of 2.14 dBi. If dBd units were used, then the dipole gain would be 0 dBd (since the dipole is the reference).

A very important parameter related to antenna power gain between the front and back lobes now needs to be described. The front–back ratio is defined as the gain of the direction of maximum radiation, also known as boresight, compared to the gain of the antenna at 180° from the specified azimuth, as seen in Figure 6.13. Manufacturers often express this parameter in decibels.

For directional antennas, the front–back ratio represents how well the antenna isolates any radiation in the back lobe. This is particularly important for interference and leakage control (signal spillage outside of a building) issues. The choice of a high front-to-back ratio ensures that leakage is kept under control for contiguous rooms or spaces, if directional antennas mounted on walls or pillars are used, or controlled in contiguous floors for ceiling-mounted omnidirectional antennas.

In the radiation pattern plot in Figure 6.13, the difference in gain between the 0° (boresight) and 180° (back lobe) is 25 dB. This means that the maximum gain is 25 dB above the back lobe gain.

6.2.5 Bandwidth

The bandwidth of an antenna is defined as the range of frequencies within which the performance of the antenna, with respect to some characteristic, conforms to a specified standard. The bandwidth can be considered to be the range of frequencies, on either side of the centre frequency, where the antenna characteristics (e.g. input impedance, radiation pattern, beamwidth, polarization, side lobe level, gain, beam direction, radiation efficiency) are within an acceptable value for those at the centre frequency.

For broadband antennas, the bandwidth is usually expressed as the ratio of the upper-to-lower frequencies of acceptable operation. For example, a 10:1 bandwidth indicates that the upper frequency is 10 times greater than the lower.

For narrowband antennas, the bandwidth is expressed as a percentage of the frequency difference (upper minus lower) over the centre frequency of the bandwidth. For example, a 5% bandwidth indicates that the frequency difference of acceptable operation is 5% of the centre frequency of the bandwidth.

The bandwidth of an antenna expresses its ability to operate over a wide frequency range. It is often defined as the range over which the power gain is maintained to within 3 dB of its maximum value. The bandwidth is usually given as a percentage of the nominal operating frequency. The radiation pattern of an antenna may change dramatically outside its specified operating bandwidth.

As a practical example of the antenna bandwidth, let us assume that we need an antenna to extend the coverage of our wireless LAN (Wi-Fi) network at home. If we use a Yagi antenna intended to operate in the UHF frequency band, it is very likely that we will get some coupling but not the desired performance, since this Yagi antenna has been designed to operate at a different frequency band.

6.2.6 Reciprocity

Provided the environment is linear (which is a very good assumption for mobile radio propagation), then:

> If a voltage is applied to the terminals of an antenna A and the current measured at the terminals of another antenna B then an equal current will be obtained at the terminals of antenna A if the same voltage is applied to the terminals of antenna B.

Therefore the antenna gain is the same whether used for receiving or transmitting, and all of the gain and pattern characteristics are fully applicable in either the transmit or receive mode. This is known as the reciprocity theorem and is shown graphically in Figure 6.14.

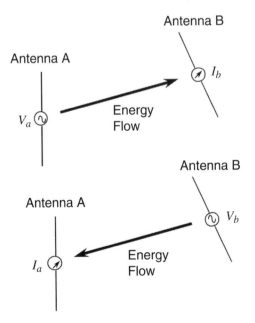

Figure 6.14 Reciprocity theorem.

According to the reciprocity theorem, it does not matter which end transmits when measuring or predicting propagation between two points in an environment.

For example, we can analyse the radiation pattern characteristics of an antenna in both ways: either when it is used to transmit, having a well-defined behaviour, or an identical behaviour when it receives electromagnetic waves, say, in a given direction, for which a specific gain is measured.

6.2.7 Receiving Antenna Aperture

If an antenna is used to receive a wave with a power density of S watts per square metre, it will produce a power in its terminating impedance (usually a receiver input impedance) of P_r watts. The constant of proportionality between P_r and S is A_e, the *effective aperture* of the antenna in square metres:

$$P_r = A_e S \tag{6.18}$$

For some antennas, such as horn or dish antennas, the aperture has an obvious physical interpretation, being almost the same as the physical area of the antenna (Figure 6.15), but the concept is just as valid for all antennas. The effective aperture may often be very much larger than the physical area, especially in the case of wire antennas. Note, however, that the effective aperture will reduce as the efficiency of an antenna decreases.

Antenna directivity d is related to the effective aperture as follows:

$$d = \frac{4\pi}{\lambda^2} A_e \tag{6.19}$$

To get an idea of the effective aperture for a GSM indoor directional antenna, let us calculate the effective aperture for a beam antenna having a half-power beamwidth of 30° and 35° in perpendicular planes intersecting in the beam axis at 900 MHz. The

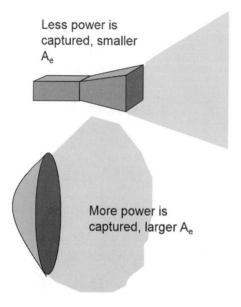

Less power is captured, smaller A_e

More power is captured, larger A_e

Figure 6.15 Receiving antenna aperture example for two antennas (horn and dish).

maximum antenna gain is also to be computed. Minor lobes are small and may be neglected. The antenna has a radiation resistance of $50\,\Omega$ and loss resistance of $8\,\Omega$. Thus, the effective aperture for this beam antenna can be determined from its gain/ directivity, as they are related according to Equation (6.19). The wavelength needs to be computed first:

$$\lambda = \frac{c}{f} = \frac{3 \times 10^8}{900 \times 10^6} = 0.333 \text{ m} \tag{6.20}$$

From Equation (6.6) and since side lobes are very small compared to the main lobe, an approximation of the directivity from the beamwidth can be obtained:

$$d = \frac{41,253}{30 \times 35} = 39.3 \tag{6.21}$$

The maximum effective aperture is

$$A_e = \frac{\lambda^2}{4\pi} d = \frac{(0.333)^2 (39.3)}{4\pi} = 34.75 \text{ cm} \tag{6.22}$$

To calculate the gain, efficiency must be known. Since radiation and loss resistance values are given, the antenna efficiency can be obtained:

$$e = \frac{50}{50 + 8} = 86.2\% \tag{6.23}$$

Thus

$$G = 10 \log(ed) = 10 \log(0.862 \times 39.3) = 10 \log(33.87) = 15.3 \text{ dBi} \tag{6.24}$$

6.2.8 Beamwidth

The beamwidth of an antenna represents the angular separation between two identical points on opposite sides of the radiation pattern. It is expressed as an angle and gives an indication of the directivity of the antenna.

There are two types of beamwidth often used, which are described as follows:

- *Half-power beamwidth (HPBW)*. This is the angle between the two directions in which the radiation intensity is one half value of the maximum (Figure 6.16).
- *First-null beamwidth (FNBW)*. Angular separation between the first nulls of the radiation pattern.

As expected, the more directive an antenna, the higher is the gain, but this happens at the expense of potentially not illuminating some desired areas.

There is a trade-off between the HPBW and the side lobe level: if the HPBW is increased, the side lobe level decreases. This has the advantage of reducing interference issues as the side lobes are reduced, at the expense of reduced gain and directivity.

6.2.9 Cross-Polar Discrimination

The cross-polar discrimination (XPD) of an antenna for a given direction is the difference in dB between the peak co-polarized gain of the antenna and the cross-polarized gain of the antenna in the given direction. It can also be seen as a measure of

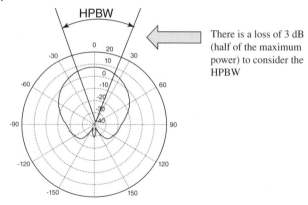

There is a loss of 3 dB (half of the maximum power) to consider the HPBW

Figure 6.16 Half-power beamwidth (HPBW).

the strength of a co-polar transmitted signal that is received cross-polar by an antenna as a ratio to the strength of the co-polar signal that is received.

One of the applications for which XPD is used is *polarization diversity*. Polarization can be used to reduce co-channel interference. Due to XPD, when a horizontally polarized antenna receives a co-channel signal sent from a vertically polarized antenna (and vice versa), the effective signal strength is reduced by several dB.

The XPD can be calculated as follows:

$$\mathrm{XPD} = 20 \log\left(\frac{E_{12}}{E_{11}}\right) \tag{6.25}$$

E_{11} is the co-polarized component and E_{12} is the co-polarized component, assuming vertical polarization, in the scenario depicted in Figure 6.17.

6.2.10 Polarization Matching

The polarization state of an antenna describes the direction of the electric field that the antennas transmit (in the far field), similarly to how polarization is defined for an electromagnetic wave.

Polarization can be linear, circular or anything in between (elliptical), though for most mobile systems linear polarization is preferred. Circular polarization is often used for satellite communications applications, to overcome all depolarizing effects from which the wave suffers when it passes through the atmosphere.

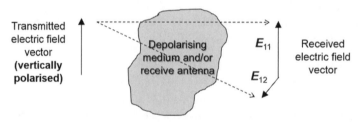

Figure 6.17 Cross-polar discrimination (XPD).

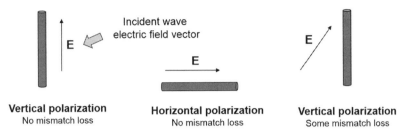

Vertical polarization
No mismatch loss

Horizontal polarization
No mismatch loss

Vertical polarization
Some mismatch loss

Figure 6.18 Polarization mismatch loss.

By applying reciprocity, it is clear that the polarization response of the antenna is the same when used in the receive mode (i.e. the antenna will produce maximum power when the incoming wave is aligned to the antenna polarization).

There is polarization mismatch if the polarization state of the incoming wave and that of the receiving antenna are different, as shown in Figure 6.18. If the incident wave's electric field vector is given by

$$\mathbf{E_i} = E_i \boldsymbol{\rho}_w \tag{6.26}$$

and the receiving antenna electric field polarization vector is

$$\mathbf{E_a} = E_a \boldsymbol{\rho}_a \tag{6.27}$$

then the polarization loss factor (PLF) is

$$\text{PLF} = |\boldsymbol{\rho}_w \cdot \boldsymbol{\rho}_a|^2 \tag{6.28}$$

6.3 Antenna Types

6.3.1 Linear Wire

The most basic antenna is that made of a single wire, which is fed by an alternating current whose wavelength is comparable to the size of the wire. In this category, there are various types of wire antennas, which are described in this section.

If a length of a two-wire transmission line is fed from a source at one end and left open-circuit at the other, then a wave is reflected from the far end of the line. This returns along the line, interfering with the forward wave. The resulting interference produces a standing wave pattern on the line, with peaks and troughs at fixed points on the line (upper half of Figure 6.19). The current is zero at the open-circuit end and varies sinusoidally, with zeros of current spaced half a wavelength apart. The current flows in opposite directions in the two wires, so the radiation from the two elements is almost exactly cancelled, yielding no far-field radiation.

If a short section of length $L/2$ at the end of the transmission line is bent outwards, it forms a dipole perpendicular to the original line and of length L (lower half of Figure 6.19). The currents on the bent section are now in the same direction and radiation occurs. Although this radiation does change the current distribution slightly,

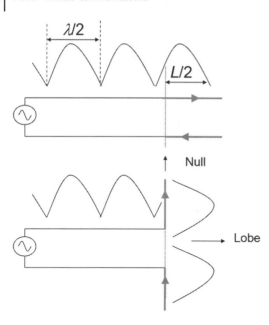

Figure 6.19 Half-wave dipole.

the general shape of the current distribution remains the same and a sinusoidal approximation may be used to analyse the resulting radiation pattern. This type of antenna is known as a *dipole.*

Some qualitative results may be deduced:

- Since the dipole is rotationally symmetric around its axis, it must be omnidirectional, whatever the current distribution.
- In a plane through the transmission line and perpendicular to the plane of the page in Figure 6.19, the distance from the arms of the dipole to all points is equal. Hence the radiation contributions from all parts of the dipole will add in phase and a lobe will always be produced.
- The current always points directly towards or away from all points on the axis of the dipole, so no radiation is produced and a null appears at all such points.

The radiation characteristics of a dipole antenna resemble a doughnut-shaped pattern, as shown in Figure 6.5. A dipole is an omnidirectional antenna in the azimuth plane, whereas its radiation characteristics vary in the elevation plane.

If the length of the dipole is changed, so its radiation pattern changes, since the current distribution changes and the fields are added differently to that of exactly a half-wave length (Saunders and Aragón-Zavala, 2007).

For a dipole at resonance, its impedance becomes purely resistive. A half-wave dipole, the most popular one, has an impedance of 73 Ω and a gain of 2.15 dBi, which is taken as a reference for expressing antenna gains in dBd.

6.3.2 Loop

A loop antenna is a radio antenna consisting of a loop (or loops) of wire, tubing or other electrical conductor with its ends connected to a balanced transmission line. Within this

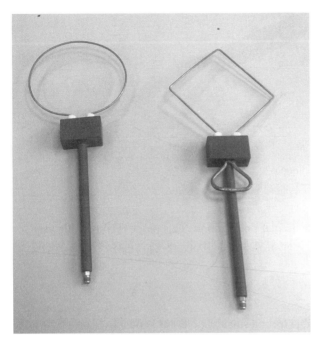

Figure 6.20 Loop antennas.

physical description there are two very distinct antenna designs: the small loop (or magnetic loop) with a size much smaller than a wavelength and the resonant loop antenna with a circumference approximately equal to the wavelength. Figure 6.20 shows an example of loop antennas.

Small loops have a poor efficiency and are mainly used as receiving antennas at low frequencies. Except for car radios, almost every AM broadcast receiver sold has such an antenna built inside it or directly attached to it. These antennas are also used for radio direction finding. A technically small loop, also known as a magnetic loop, should have a circumference of one tenth of a wavelength or less. This is necessary to ensure a constant current distribution round the loop. These antennas are more typically used in those applications for which signal-to-noise ratio (SNR) performance is more important than efficiency.

Self-resonant loop antennas are larger. They are typically used at higher frequencies, especially VHF and UHF, where their size is manageable. They can be viewed as a form of folded dipole and have somewhat similar characteristics. The radiation efficiency is also high and similar to that of a dipole.

The helical antenna (Figure 6.21) can be considered a vertical array of loops, at least for the case when the diameter of the helix is small compared to a wavelength. The result is normal mode radiation with a higher gain than a single loop, providing an omni-directional antenna with compact size and reasonable efficiency, but rather narrow bandwidth. It is commonly used for hand-portable mobile applications where it is more desirable to reduce the length of the antenna below that of a quarter-wave monopole.

In the case where the diameter is around one wavelength or greater, the mode of radiation changes completely to the axial mode, where the operation of the antenna is

Multi-band (GSM/PCS/3G)   Single-band (GSM)

Figure 6.21 Portable helical (helix) antennas.

similar to that of a Yagi but with circular polarization. This mode is commonly used for satellite communications, particularly at lower frequencies where a dish would be impractically large.

6.3.3 Antenna Arrays

The maximum directivity available from a single dipole was shown in Figure 6.20 to be limited to that which can be achieved with a dipole a little over one wavelength. In some applications this may not be sufficient. One approach to improving on this is to combine arrays of dipoles, or of other antenna elements, where the amplitude and phase with which each element is fed may be different. The fields produced by the elements then combine with different phases in the far field, and the radiation pattern is changed. This also allows the radiation pattern to be tailored according to the particular application or varied to allow beam scanning without any physical antenna motion. If the amplitude and phase weights are controlled electronically, then the beam can be scanned very rapidly to track changes in the communication channel.

Arrays may be linear or planar, as explained in Saunders and Aragón-Zavala (2007). A linear array allows beam steering in one dimension, permitting directivity to be obtained in a single plane; hence an omnidirectional pattern can be synthesized. A planar array has two dimensions of control, permitting a narrow pencil beam to be produced.

In general, antenna arrays can be made with the combination of any antenna types, but the most popular ones are those made with dipoles. We will see later on in this chapter some examples of practical antenna arrays.

6.3.4 Travelling Wave and Broadband

A travelling-wave antenna (Figure 6.22) is a class of antenna that uses a travelling wave on a guiding structure as the main radiating mechanism. Travelling-wave antennas fall into two general categories: slow-wave antennas and fast-wave antennas, and both are usually referred to as *leaky-wave antennas*.

In slow-wave antennas, the guided wave is a slow wave, meaning a wave that propagates with a phase velocity that is less than the speed of light in free space. Such a wave does not fundamentally radiate by nature, and radiation occurs only at discontinuities (typically the feed and the termination regions). The propagation

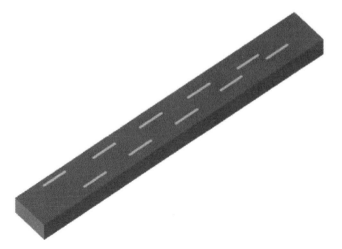

Figure 6.22 Travelling-wave antenna.

wavenumber of the travelling wave is therefore a real number (ignoring conductor or other losses). Because the wave radiates only at the discontinuities, the radiation pattern physically arises from two equivalent sources, one at the beginning and one at the end of the structure. This makes it difficult to obtain highly directive single-beam radiation patterns. However, moderately directive patterns having a main beam near end-fire can be achieved, although with a significant sidelobe level. For these antennas there is an optimum length depending on the desired location of the main beam. Examples include wires in free space or over a ground plane, helixes, dielectric slabs or rods, and corrugated conductors. An independent control of the beam angle and of the beam width is not possible.

By contrast, the wave on a leaky wave antenna (LWA) may be a fast wave, with a phase velocity greater than the speed of light. This type of wave radiates continuously along its length and hence the propagation wavenumber kz is complex, consisting of both a phase and an attenuation constant. Highly directive beams at an arbitrary specified angle can be achieved with this type of antenna, with a low sidelobe level. The phase constant of the wave controls the beam angle (and this can be varied, changing the frequency), while the attenuation constant controls the beamwidth. The aperture distribution can also be easily tapered to control the sidelobe level or beam shape. Leaky-wave antennas can be divided into two important categories, uniform and periodic, depending on the type of guiding structure.

Broadband antennas are those designed to operate over a very wide bandwidth. This can be achieved either by using a slotted-wave antenna or by integrating many microstrip (patch) antennas to have a multiband performance.

6.3.5 Microstrip

Patch antennas are based upon printed circuit technology to create flat radiating structures on top of dielectric, ground-plane-backed substrates. The appeal of such structures is in allowing compact antennas with a low manufacturing cost and high reliability. It is in practice difficult to achieve this at the same time as acceptably high bandwidth and efficiency. Nevertheless, improvements in the properties of the dielectric materials and in design techniques have led to enormous growth in their popularity and there are now a large number of commercial applications. Many shapes of patch are possible, with varying applications, but the most popular are rectangular (pictured), circular and thin strips (i.e. printed dipoles).

In the rectangular patch, the length is typically up to half of the free-space wavelength. The incident wave fed into the feed line sets up a strong resonance within the patch, leading to a specific distribution of fields in the region of the dielectric immediately beneath the patch, in which the electric fields are approximately perpendicular to the patch surface and the magnetic fields are parallel to it. The fields around the edges of the patch create the radiation, with contributions from the edges adding as if they constituted a four-element array. The resultant radiation pattern can thus be varied over a wide range by altering the length and width.

A major application of patch antennas is in arrays, where all of the elements, plus the feed and matching networks, can be created in a single printed structure. An example of a patch antenna is shown in Figure 6.23.

6.3.6 Yagi-Uda

Another array-based approach to enhancing the directivity of dipole antennas is to use parasitic elements. Parasitic elements are mounted close to the driven dipole and are not connected directly to the source. Instead, the radiation field of the driven element induces currents in the parasitics, causing them to radiate in turn. If the length and position of the parasitic elements are chosen appropriately, then the radiation from the parasitics and the driven element add constructively in one direction, producing an increase in directivity. The classic form of such an antenna is the Yagi-Uda, or simply Yagi antenna, illustrated in Figure 6.24 and widely used as a television reception antenna. For in-building applications, this type of antenna is very widely used when

Figure 6.23 Patch antenna.

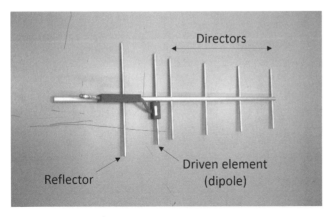

Figure 6.24 Yagi-Uda antenna.

two buildings are to be interconnected in a 'bridging' application, especially for Wi-Fi. Also, these types of antennas are used when repeaters are employed, and are installed on rooftops to pick up the signal from a 'donor' macrocell. The use of repeaters will be expanded in Chapter 9.

Typically, the driven element is made a little shorter than $\lambda 2$ in order to permit a good match to $50\,\Omega$. Elements in the radiation direction, called directors, are made a little shorter than the driver element, and an element very close to $\lambda 2$ is placed behind and called the reflector. Increasing the number of directors increases the gain, although the improvement diminishes according to how far the director is from the driven element. A four-director Yagi can have a gain of up to around $12\,\mathrm{dBi}$.

6.3.7 Aperture Antennas

By definition, an aperture antenna is that in which the beamwidth is determined by the dimensions of a horn, lens or reflector. In other words, those antennas for which their directivity is proportional to their aperture are classified as aperture antennas. Since some of these antennas are very important, sections in this chapter have been dedicated to describe them in more detail. We have included this term here, 'aperture antennas', since it is used in the literature to refer to those types of antennas with similar characteristics to those described earlier.

6.3.8 Horn

The horn antenna is a natural evolution of the idea that any antenna represents a region of transition between guided and propagating waves (Figure 6.1). Horn antennas are highly suitable for frequencies (typically several gigahertz and above) where waveguides are the standard feed method for they consist essentially of a waveguide whose end walls are flared outwards to form a megaphone-like structure, as shown in Figure 6.25. In the case illustrated, the aperture is maintained as a rectangle, but circular and elliptical versions are also possible. The dimensions of the aperture are chosen to select an appropriate resonant mode, giving rise to a controlled field distribution over the aperture.

Figure 6.25 Horn antennas.

The best patterns (narrow main lobe, low side lobes) are produced by making the length of the horn large compared to the aperture width, but this must be chosen as a compromise with the overall volume occupied. A common application of horn antennas is to be used as the feed element for parabolic dish antennas in satellite systems.

6.3.9 Monopole

Reflector antennas rely on the application of image theory, which may be described as follows. If an antenna carrying a current is placed adjacent to a perfectly conducting plane, that is the ground plane, then the combined system has the same fields above the plane as if an image of the antenna were present below the plane. The image carries a current of equal magnitude to the real antenna but in the opposite direction, and is located an equal distance from the plane as the real antenna but on the other side. This statement is a consequence of Snell's law of reflection, given the Fresnel reflection coefficients for a perfect conductor. Examples of reflector antennas include monopoles and parabolic dishes.

The *monopole antenna* results from applying image theory to the dipole. If a conducting plane is placed below a single element of length $L/2$ carrying a current, then the combined system acts essentially identically to a dipole of length L except that the radiation only takes place in the space above the plane, so the directivity is doubled and the radiation resistance is halved (Figure 6.26). The quarter-wave monopole thus approximates the half-wave dipole and is a very useful configuration in practice for mobile antennas, where the conducting plane is the car body or handset case.

Monopoles are very popular for portable antennas, although in recent years they have been replaced by sleeve dipoles.

6.3.10 Parabolic Reflector (Dish)

The parabolic (dish) antenna extends the reflector antenna concept to curved reflectors (Figure 6.27). In this case, the number of images is effectively infinite and the locations of

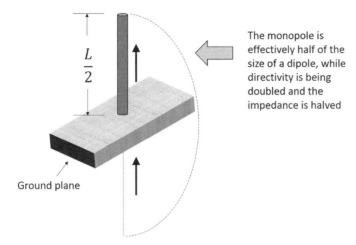

The monopole is effectively half of the size of a dipole, while directivity is being doubled and the impedance is halved

Ground plane

Figure 6.26 Monopole antenna.

the images are such as to produce a parallel beam from the reflector, provided that the driven element is placed at the focus of the parabola. All the energy is effectively focused on the driven element, thus producing a very directive antenna.

The gain of a parabolic dish can be increased essentially arbitrarily by enlarging the size of the dish; this makes it appropriate for long-range communications applications, such as satellite or space systems. They are also used for terrestrial applications where the distance is considerable and very narrow beams are needed; for example microwave links, since diffraction losses may affect the link performance. Also, for Wi-Fi systems, parabolic dishes are the preferred choice when connecting two buildings.

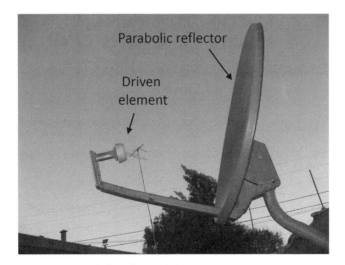

Parabolic reflector

Driven element

Figure 6.27 Parabolic reflector dish.

In general, the gain of a parabolic dish antenna is given by

$$g = e\left(\frac{\pi \mathrm{Dia}}{\lambda}\right)^2 \tag{6.29}$$

Remember that e is the antenna efficiency (typically around 55%–65%) and Dia is the diameter of the dish. The gain here is specified in linear scale. On the other hand, the half-power beamwidth θ_{HP}° of the parabolic dish (in degrees) can be approximated as

$$\theta_{HP}^{\circ} \approx \frac{70 \times \mathrm{Dia}}{\lambda} \tag{6.30}$$

6.3.11 Smart Antennas

The limiting factor on the capacity of a cellular mobile system is interference from co-channel mobiles in neighbouring cells. Adaptive antenna technology can be used to overcome this interference by intelligent combination of the signals at multiple antenna elements at the base station and potentially also at the mobile. In order to perform this joining together efficiently and accurately, a thorough knowledge of the propagation channel will be required for every pair of antennas from the base to the mobile.

In a phased array, a set of antenna elements is arranged in space and the output of each element is multiplied by a complex weight and combined by summing, as shown in Figure 6.28 for a four-element case.

The complete array can be regarded as an antenna in its own right, with a new output y. The radiation patterns of the individual elements are summed with phases and amplitudes depending on both the weights applied and their positions in space; this yields a new combined pattern.

If the weights are allowed to vary in time, the array becomes an adaptive array and can be exploited to improve the performance of a mobile communication system by choosing the weights so as to optimize some measure of the system performance. Typically this would be done by estimating the desired weights using a digital signal

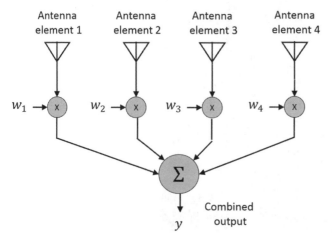

Figure 6.28 Four-element phased receive array.

processor (DSP) and applying them in complex baseband to sampled versions of the signals from each of the elements. The same approach can be used on both transmit and receive due to the reciprocity of the channel and the antenna elements themselves, but there are considerable challenges associated with assessing the downlink channel state with sufficient accuracy to achieve the full potential of adaptive antennas.

The basic aim of a mobile adaptive antenna system is to improve the performance of the system in the presence of both noise and interference; this is achieved by effectively steering the radiation antenna patterns dynamically in order to cancel interferers (direct nulls to them) and maximize wanted user coverage. This concept has been further exploited in what is known as space division multiple access, or SDMA, which offers the potential for greatly increasing system capacity in future mobile systems.

6.4 Antenna Performance Issues

6.4.1   Mean Effective Gain (MEG)

The performance of a practical mobile antenna in its realistic operating environment may be very different from what would be expected from measurements of the gain of the antenna in isolation. This arises because the mobile is usually operated surrounded by scattering objects that spread the signal over a wide range of angles around the mobile. The question arises, given this complexity, as to what value of mobile antenna gain should be adopted when performing link budget calculations. The concept of a mean effective gain that combines the radiation performance of the antenna itself with the propagation characteristics of the surrounding environment was introduced by Taga (1990) to address this and is described in some detail here.

Consider a mobile antenna that receives power from a base station after scattering has occurred through a combination of buildings, trees and other clutter in the environment. The total average power incident on the mobile is composed of both horizontally and vertically polarized components, PH and PV, respectively. All powers are considered as averages, taken after the mobile has moved along a route of several wavelengths. The mean effective gain (MEG) of the antenna, G_e, is then defined as the ratio between the power that the mobile actually receives and the total that is available:

$$G_e = \frac{P_{rec}}{P_H + P_V} \tag{6.31}$$

The received power at the antenna can then be expressed in spherical coordinates taking into account the three-dimensional spread of incident angles as follows:

$$P_{rec} = \int_0^{2\pi} \int_0^{\pi} \left[P_1 G_\theta(\theta, \phi) P_\theta(\theta, \phi) + P_2 G_\phi(\theta, \phi) P_\phi(\theta, \phi) \right] \sin\theta \, d\theta \, d\phi \tag{6.32}$$

where P_1 and P_2 are the mean powers that would be received by ideally θ (elevation) and ϕ (azimuth) polarized isotropic antennas, respectively, G_θ and G_ϕ are the corresponding radiation patterns of the mobile antenna and P_θ and P_ϕ represent the angular distributions of the incoming waves. The following conditions must be satisfied to ensure that the functions are properly defined:

$$\int_0^{2\pi}\int_0^{\pi} \left[G_\theta(\theta,\phi) + G_\phi(\theta,\phi) \right] \sin\theta \, d\theta \, d\phi = 4\pi \tag{6.33}$$

$$\int_0^{2\pi}\int_0^{\pi} P_\theta(\theta,\phi) \sin\theta \, d\theta \, d\phi = 1 \tag{6.34}$$

$$\int_0^{2\pi}\int_0^{\pi} P_\phi(\theta,\phi) \sin\theta \, d\theta \, d\phi = 1 \tag{6.35}$$

The angular distribution of the waves may be modelled by, for example, Gaussian distributions in elevation and uniform in azimuth as follows:

$$P_\theta(\theta,\phi) = A_\theta \exp\left\{ -\left[\theta - \left(\frac{\pi}{2} - m_V \right) \right]^2 \times \frac{1}{2\sigma_V^2} \right\} \tag{6.36}$$

$$P_\phi(\theta,\phi) = A_\phi \exp\left\{ -\left[\phi - \left(\frac{\pi}{2} - m_H \right) \right]^2 \times \frac{1}{2\sigma_H^2} \right\} \tag{6.37}$$

where m_V and m_H are the mean elevation angles of the vertically and horizontally polarized components, respectively, σ_V and σ_H are the corresponding standard deviations and A_θ and A_ϕ are chosen to satisfy (6.34) and (6.35). The precise shape of the angular distribution is far less important than its mean and its standard deviation.

Since the arrival angle has been assumed uniform in azimuth, any variations from omnidirectional in the radiation pattern will have no impact upon the MEG. Although this assumption is likely to be valid in the long term as the mobile user's position changes, there may be short-term cases where this is not so and the power arrives from a dominant direction. This may particularly be the case in a rural setting where a line-of-sight or near-line-of-sight path exists. Figure 6.29 shows the radiation pattern obtained from a pair of dipoles arranged apart and fed with equal phase and amplitude. This could, for example, represent an attempt to reduce radiation into the human head by placing a null in the appropriate direction. This results in a gain pattern of the form:

$$G_\theta = \cos^2\left(\frac{\pi}{4} + \frac{\pi}{4}\cos\phi \right) \tag{6.38}$$

Also shown is the arrival angle distribution, assumed to be Gaussian in azimuth and shown with a standard deviation of 50° and a mean of 0°. Calculation of (6.32) in comparison with the assumption of a uniform arrival angle in the azimuth plane yields the results shown in Figure 6.30. This is performed with the centre of the arriving waves both within the pattern null and directly opposite and is shown as a function of the standard deviation of the spread relative to the mean. Considerable gain reduction is evident in both cases, particularly when the angular spread is small.

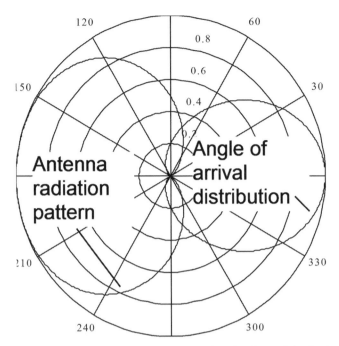

Figure 6.29 Azimuth radiation pattern and angle-of-arrival distribution.

6.4.2 Radiation Pattern Extrapolation

An issue that has become more important for in-building antennas is that of finding gain values everywhere in space. When performing propagation predictions in outdoor environments, using the models described in Chapter 5, distances between base stations and mobiles are large compared to the base station antenna height, and therefore the

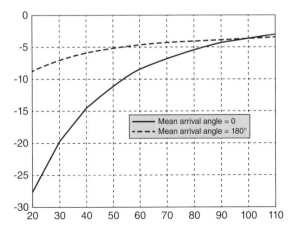

Figure 6.30 Reduction in mean effective gain due to angular spreading.

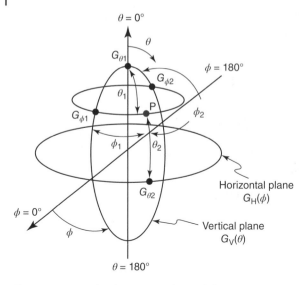

Figure 6.31 Angular distance weight model.

signal would be estimated for radiation angles very near to the base station antenna horizontal plane where the manufacturer typically specifies the antenna radiation pattern. However, for indoor cells, elevation angles typically span the whole range, and hence simple extrapolation methods will lead to unacceptable prediction errors.

Two methods were found in the literature, both published in recent years, which seem to address this extrapolation problem suitably. The first one, published by Gil *et al.* (1999), suggests that the generalized gain in any direction $P(\theta, \phi)$, $G(\theta, \phi)$ is obtained from the previous ones by weighting them with the relative angular distances between the direction of interest and the horizontal (θ_2) and the vertical (θ_1, ϕ_1, ϕ_2) planes, that is the four points on the sphere closest to the point of interest in Figure 6.31. The basic idea of the model is that the weight by which the value of the gain on a given radiation plane is inversely proportional to the angular distance, so that the closer the direction of interest is to the given radiation plane, the higher the weight. Therefore, continuity of the extrapolation is ensured on each plane.

The final formulation for this method, also known as the *angular distance weight model*, is given by

$$G(\theta, \phi) = \frac{\left[\phi_1 \cdot G_{\phi 2} + \phi_2 \cdot G_{\phi 1}\right] \cdot \dfrac{\theta_1 \cdot \theta_2}{(\theta_1 + \theta_2)^2} + \left[\theta_1 \cdot G_{\theta 2} + \theta_2 \cdot G_{\theta 1}\right] \cdot \dfrac{\phi_1 \cdot \phi_2}{(\phi_1 + \phi_2)^2}}{\left[\phi_1 + \phi_2\right] \cdot \dfrac{\theta_1 \cdot \theta_2}{(\theta_1 + \theta_2)^2} + \left[\theta_1 + \theta_2\right] \cdot \dfrac{\phi_1 \cdot \phi_2}{(\phi_1 + \phi_2)^2}}$$

(6.39)

where all the angles and gain values are defined in Figure 6.31. This method is recommended as an alternative for pattern extrapolation for indoor environments, since, as reported in Aragón-Zavala (2003), when compared with anechoic chamber measurements taken at various planes, it shows an improvement in accuracy of around

2.5 dB of standard deviation of error when compared to other methods. However, there are specific elevation and azimuth angle combinations for which some degree of ambiguity exists and the gain is not accounted for properly.

The second method, presented in Vasiliadis, Dimitrou and Sergiadis (2005), combines the two principal normalized patterns using 360° horizontal and 180° vertical gain samples. The azimuth pattern is the planar cut at $\theta = \pi/2$ while the elevation pattern is the cut at $\phi = 0$. The authors define the logarithmic counterparts of these patterns as G_H and G_V is follows:

$$G_H = 10 \log[hor(\phi)] \tag{6.40}$$

$$G_V = 10 \log[vert(\theta)] \tag{6.41}$$

an arbitrary point in space is defined, an approximated antenna gain is calculated weighting the horizontal and vertical samples of gain as follows:

$$\hat{G} = \frac{G_H(\phi) \cdot w_1 + G_V(\theta) \cdot w_2}{\sqrt[k]{w_1^k + w_2^k}} \tag{6.42}$$

where k is defined as the normalization-related parameter and weight functions are given by:

$$w_1(\theta, \phi) = vert(\theta) \cdot [1 - hor(\phi)] \tag{6.43}$$

$$w_2(\theta, \phi) = hor(\phi) \cdot [1 - vert(\theta)] \tag{6.44}$$

The key idea of this method is that the estimation of a radiation pattern sample involves data of the other principal pattern and is weighted as a function of angular distance between the point of interest and the sample, in a cross-weighting manner between the two principal cuts, which according to the authors, is more efficient than the summing algorithm or the angular distance weight model.

If A_1 and A_2 are defined as the overall normalized weights of the two sample radiation cuts then

$$\hat{G}(\theta, \phi) = G_H(\phi) \cdot A_1 + G_V(\theta) \cdot A_2 \tag{6.45}$$

$$A_1 = \frac{w_1}{\sqrt[k]{w_1^k + w_2^k}} \tag{6.46}$$

$$A_2 = \frac{w_2}{\sqrt[k]{w_1^k + w_2^k}} \tag{6.47}$$

The initial weights A_1 and A_2, which define mathematically the degree of participation of each sample-cut into the overall approximation, are normalized by $\sqrt[k]{w_1^k + w_2^k}$ so that the following relationship is met:

$$A_1^k + A_2^k = 1 \tag{6.48}$$

If $A_1 = A_2 = 1$ is considered, then the summing algorithm is obtained, where the estimated gain is simply

$$G_{SA}(\theta, \phi) = G_H(\phi) + G_V(\theta) \tag{6.49}$$

Omnidirectional patterns are reconstructed without any error, according to Vasiliadis, Dimitrou and Sergiadis (2005). For directive patterns, the error is reduced when compared to the summing algorithm and performs slightly better than the angular distance weight model. In order to improve the performance of the suggested algorithm, a hybrid technique is suggested in Vasiliadis, Dimitrou and Sergiadis (2005), which combines the summing algorithm with the original proposed method, yielding

$$G_{hyb}(\theta, \phi) = G_{SA}(\theta, \phi) \cdot w_3 + \hat{G}(\theta, \phi) \cdot (1 - w_3) \tag{6.50}$$

additional weighting function w_3 that actually bridges the two algorithms is suggested as follows:

$$w_3(\theta, \phi) = \sqrt[n]{hor(\phi) \cdot vert(\theta)} \tag{6.51}$$

Prameter n can be optimized to minimize the estimation error. The optimum value of n is thus associated with the directivity of the antenna under test. For patterns with extreme directive characteristics, containing considerably more radiation nulls and sidelobes, a parameter $n \leq 6$ can provide satisfactory performance for all antenna types.

6.4.3 Reliability of Radiation Patterns

Antenna characteristics are very relevant when designing an indoor system, since the performance of such a system depends strongly on some key parameters, especially the radiation pattern. Interference and coverage can be properly controlled if the antenna can be selected according to system requirements. Therefore, the correct antenna selection strongly depends on the accuracy of antenna parameters provided by manufacturers.

A comparison of antenna measurements performed in an anechoic chamber and the radiation patterns provided by antenna manufacturers was made in Aragón-Zavala (2003) to assess the accuracy of these data and the impact of the use of such patterns in signal strength predictions.

Figure 6.32 shows the corresponding measured cuts for a directional antenna. It is noticeable that the maximum absolute gain obtained from the anechoic chamber measurements is less than that reported by the manufacturer at the frequency in which

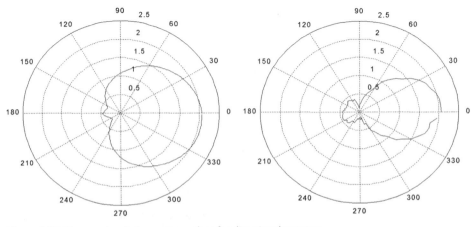

Figure 6.32 Measured radiation pattern plots for directional antenna.

Table 6.1 Measurements versus manufacturer's errors in azimuth and elevation planes.

Directional 5027-000	Azimuth [dB]	Elevation [dB]
Mean	3	−0.1
Standard deviation	1.3	1.5

it was tested, for both cuts. This difference never exceeds 2 dB, and hence is not considered critical for design purposes. It may be caused by imperfections in the antenna calibration prior to the anechoic measurements, as well as differences in the fabrication and performance of different antennas, even if they are of the same type. The shape of the pattern is very similar, although in the elevation cut the back lobe shows predominantly higher gain than that reported by the manufacturer.

The error statistics of the comparison of measured absolute gain values and manufacturer data are presented in Table 6.1. The standard deviation of the error is relatively low, and hence suggests that although there are discrepancies in various single gain values in both planes, this degradation does not impose a major problem.

In summary, even though manufacturers supply radiation pattern data for all their products at different frequencies, special care must be taken when using these patterns for accurate propagation work having the following special considerations:

- The manufacturer measures the antenna radiation patterns in an anechoic chamber, which is an ideal free-space style environment. Some manufacturers do not even attempt to measure the patterns, but rather provide theoretical predictions based on the radiation intensity equations they use. In practice, patterns may be heavily distorted due to multipath, which tends to spread lobes and fill nulls, or the presence of objects near the antenna, distorting the pattern and affecting the VSWR.
- Unless the radiation pattern is used at the exact frequency at which the antenna measurements were performed, there will be some uncertainty as to the exact values of gain at the operating frequency and therefore approximations should be made.
- Individual antenna units, although manufactured having the same product number and characteristics, may exhibit different performances.

Therefore it is much better to think of the antenna radiation pattern as a guide to the areas that need to be illuminated rather than a precise beam for controlling coverage.

6.5 Antenna Measurements

The aim of antenna measurement techniques can be either to ensure that the antenna under test (AUT) meets manufacturer specifications or simply to characterise it. These measurements are often focused on specific antenna parameters, such as gain, radiation pattern, bandwidth, beamwidth, impedance and power reflection.

As described before, the antenna pattern is the response of the antenna to a plane wave incident from a given direction or the relative power density of the wave transmitted by the antenna in a given direction. For a reciprocal antenna, these two patterns are identical (reciprocity theorem). A multitude of antenna pattern measurement

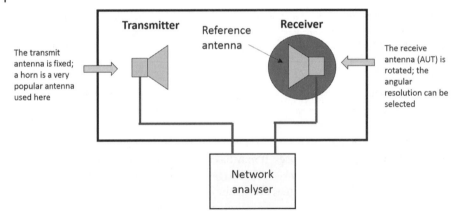

Figure 6.33 Antenna measurement setup in an anechoic chamber.

techniques have been developed and are extensively used. Three methods will be briefly described here.

The first technique developed was the far-field range, where the AUT is placed in the far-field of a range antenna. Due to the size required to create a far-field range for large antennas, near-field techniques were developed that allow the measurement of the field on a surface close to the antenna (typically 3 to 10 times its wavelength). This measurement is then predicted to be the same at infinity. A third common method is the compact range, which uses a reflector to create a field near the AUT that looks approximately like a plane wave.

A typical (although not the only) scenario to perform antenna measurements is an anechoic chamber. The anechoic chamber provides a controlled environment, an all-weather capability, security and electromagnetic interference minimization. Before calibration, it is necessary to set up a reference antenna, which is arranged in the chamber as shown in Figure 6.33. The transmit antenna is stationary at one end of the chamber while the reference antenna is placed on a turntable in the receive mode. The two antennas are connected likewise to the appropriate ports of a network analyser outside the chamber via feeder cables. It must be noted that the antennas need to be at the correct height and aligned properly facing each other. Next, the receive antenna is rotated at different angles, first at a given plane (say the E-plane) and then at the orthogonal plane (the H-plane), to obtain radiation pattern values at distinct angles. More cuts can be achieved if desired, although manufacturers only provide horizontal and vertical cuts.

Figure 6.34 shows an example of an anechoic chamber. Anechoic chambers are normally wrapped with absorber materials to minimize reflection and other unwanted propagation effects, which will alter the results of the radiation patterns.

6.6 MIMO (Multiple-Input Multiple-Output)

The most recent, and arguably most powerful, application of adaptive antennas is the class of systems known as multiple-input multiple-output (MIMO) systems, which use

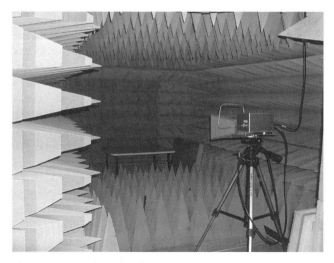

Figure 6.34 Anechoic chamber.

space–time coding (STC) to realize substantial performance and capacity gains. This technology has been formally adopted by various wireless standards, amongst others cellular LTE. The improvement in capacity due to this configuration has been proved to be far beyond what could be expected with SISO (single-input single-output) systems, even with the use of advanced modulation and coding techniques. The fact that MIMO takes the benefit of multipath makes it more suitable for in-building wireless systems.

In MIMO systems, multiple antennas must be available at both the transmitter and the receiver. These systems can then use STC to exploit the rich spatial multipath scattering present in many radio channels. MIMO systems may be seen as the ultimate extension of adaptive antenna technology to include diversity and SDMA as special cases, but being more powerful than either.

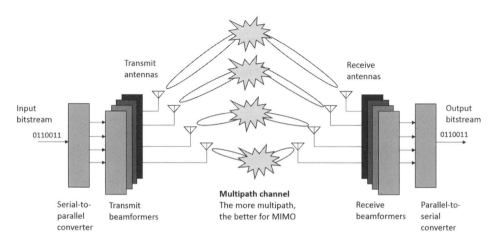

Figure 6.35 MIMO concept.

A simple example is given in Figure 6.35. At the transmitter, an input bit stream is divided into three parallel bit streams, each of which occurs at one-third the rate of the original. Each of these is transmitted via an independent beamformer, which produces three separate radiation patterns as a combination of the three transmit array elements. The radiation patterns are chosen to produce a maximum gain in the directions of three scatterers. Reradiation from each of these scatterers is received by three separate beams formed at the receiver by three further beamforming networks. After demodulation, the three bit streams can be recombined to produce the original input bit stream. Provided the scatterers are sufficiently separated in angle to be resolved by the beams formed, the three bit streams can all occupy the same spectrum without interfering, and the channel capacity has been increased by a factor of three. Clearly the capacity gain from such a system is limited by the amount of scattering in the multipath environment and by the number of antenna elements available at both transmitters. Nevertheless, the potential to increase channel capacity by a potentially unlimited degree offers great potential for high bit-rate systems.

Many wireless technologies have started to incorporate MIMO in their standards; for example, Wi-Fi 802.11n considers the use of MIMO to enhance the data throughput that can be achieved; Wi-MAX also considers in its standard the use of MIMO to increase the capacity of the link to levels that otherwise would not be achievable using only special modulation and coding techniques; cellular 4G systems, such as LTE, also contemplates the use of MIMO to enhance its capacity and minimize interference even further, taking advantage of multipath propagation effects. To know more about MIMO systems, good references can be found in Brown, Kyritsi and de Carvalho (2012) and Costa and Haykin (2010).

In-building distributed antenna systems are now equipped with MIMO to enhance the data performance in the building. In practice, only 2×2 configurations are feasible due to installation restrictions. Alternatively, cross-polarized antennas can also be used to minimize the space restrictions in MIMO and potentially larger configurations could be possible.

The optimal antenna separation for MIMO to provide uncorrelated paths and thus maximize MIMO performance is around 3 to 7 wavelengths. From this, it is evident that the larger the frequency the easier it is to install MIMO within a restricted space; this sets a trade-off between available bandwidth at large frequencies having a reduced antenna separation against larger antenna separation and lower path loss at low frequencies.

In Tolstrup (2011) some recommendations are given for practical indoor MIMO DAS deployments. Some of these are summarized as follows:

- Do not place the antennas in the same cluster too far apart, as this potentially could worsen the 'near-far' effect, where a mobile is close to one antenna.
- Both antennas should be visible to a mobile receiver, to avoid impacting on power control and dynamic range.
- Ideally, both antennas should be placed so that they can service the same areas, for example illuminating the same corridor; since this will result in a much more uniform RF level between the two MIMO paths, taking advantage of the multipath and scattering inside the building (less ISI).
- Ensure there is a good SNR in both MIMO paths.

- The ideal antenna separation would also depend on the delay spread and wideband channel characteristics of the local environment; for example recommended separations could be given for different types of environments with distinct wideband channel characteristics.
- For multiband indoor solutions, the lowest frequency will dictate the antenna separation.
- If passive MIMO DAS is to be used, two parallel DAS installations would be required, thus increasing cost and complexity, using the same cable distance on both links to avoid ISI.

6.7 Examples Of In-Building Antennas

Antennas that can be used for indoor wireless systems come in various shapes, sizes and materials. Although it is not technology-specific, we will use the cellular industry as a good example to show the requirements that indoor antennas must fulfil in a practical deployment.

6.7.1 In-Building Cellular Antenna Requirements

In-building cellular antennas are a special type of antenna, which need to have distinct characteristics to those for outdoor applications. These characteristics are not only related to RF performance but also to physical properties and aesthetics. Some of the most important requirements are:

- Discrete appearance, as people often are not keen on walking in front of a radiating element, even if this radiation is harmless and non-ionizing. Also, aesthetically, they should not represent an 'intrusion' to existing landscape within a building.
- Compliance with radiated power density and specific absorption rate (SAR) levels set by international and local standards.
- Linear polarization is the preferred polarization state for in-building antennas; although, for certain applications, circular polarization may also be used.
- High level of integration between systems and technologies deployed by different operators, so wideband and multiband in-building antennas are increasingly of interest.
- Uniform coverage to illuminate the designated areas efficiently. This is especially important for challenging in-building environments, in which multipath propagation can cause the coverage to be non-uniform.
- Wide vertical beamwidth is of special importance, to illuminate floors above and below. Horizontal beamwidth varies, depending on the application, and hence omnidirectional or directional antennas can be chosen.

6.7.2 Omnidirectional

If coverage is to be provided in a facility with large spaces, such as halls, conference rooms, airport terminals, etc., and antennas can be mounted on the ceiling, omni-directional antennas are the preferred choice. Antenna locations are thus chosen so that they are not mounted close to windows or external walls, and therefore signal spillage can be better controlled.

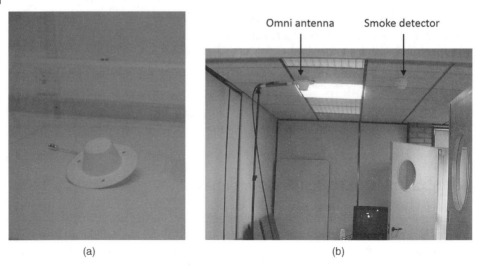

Figure 6.36 Omnidirectional antennas: (a) ceiling-mounted; (b) comparison of an omni antenna with a smoke detector.

An example of a ceiling-mounted antenna for indoor coverage at 900 MHz is shown in Figure 6.36(a). A particular requirement of indoor antennas is a very wide beamwidth, consistent with a discrete appearance, so this particular antenna has been designed to look similar to a smoke detector, as shown in Figure 6.36(b). Linear polarization is currently used almost universally for indoor communications, but there are potential benefits in the use of circular polarization. This has been shown to substantially reduce fade depth and RMS delay spread due to the rejection of odd-order reflections as well as reducing polarization mismatch loss. Similarly, a reduction in antenna beamwidth has been shown to substantially reduce the delay spread in line-of-sight situations, but this effect must be traded against the difficulty of providing a reasonably uniform coverage area.

Increasingly, indoor antennas and the associated feed powers also typically have to be compliant with specific requirements on radiated power density and specific absorption rates. It is also increasingly desirable to achieve a high level of integration between the systems and technologies deployed by different operators so wideband and multiband indoor antennas are increasingly of interest, providing, for example, Wi-Fi, 3G and LTE technologies in a single antenna housing.

For other wireless technologies, such as Wi-Fi, omnidirectional antennas may have different shapes and configurations. The vast majority of wireless local area networks can be found operating at two frequency bands: the 2.4 GHz ISM band (IEEE 802.11b and 802.11g standards) and the 5.4 GHz band (802.11a standard), with maximum data rates from 11 Mbps (802.11b) up through to 54 Mbps (802.11g/a) and up to over 100 Mbps (IEEE 802.11n, operating in either frequency band).

Spatial diversity is often employed in Wi-Fi access points to overcome multipath fading effects and combine the various replicas of the received signal coherently, achieving substantial spatial diversity gain. Indeed, in the 802.11n standard multiple antennas are an

Figure 6.37 External omnidirectional antennas with spatial diversity.

absolute requirement to achieve high data rates. Omnidirectional antennas are preferred for some applications, but this depends on whether uniform coverage is required; that is if the access point and antennas are located in the middle of a room. In the early days of WLAN, these antennas were external, as depicted in Figure 6.37.

Some WLAN access points have integrated antennas, which are often microstrip elements, designed to provide coverage underneath the access point in an 'umbrella' fashion (Figure 6.38). Floor penetration is sometimes difficult to achieve, especially at the relatively low transmit powers used in access points (50–200 mW EIRP depending on the regulatory regime in the country of use).

For TV reception, omnidirectional antennas span from the early days of an analogue TV loop (Figure 6.20) or dipole 'rabbit ears' (Figure 6.39a) to the most recent digital HDTV era and more sophisticated designs (Figure 6.39b). Perhaps the impression we have as users is that digital TV allows a much better reception than analogue, but this

Figure 6.38 Wi-Fi access point with integrated antennas.

Figure 6.39 TV indoor reception antennas: (a) dipole 'rabbit ears'; (b) HDTV panel.

may be a misleading conclusion; with poor SNR in analogue TV, 'snowy' or 'blurred' images could be received, whereas in digital TV, one can go from perfect reception to none at all almost instantly! Therefore, a good understanding of antenna characteristics along with the right choice of antenna can be the difference between good and poor TV service in a household or venue.

6.7.3 Directional

When coverage is to be maximized in a certain direction, the use of directional antennas is recommended. Also, in circumstances where antennas cannot be mounted on ceilings and only wall/pillar mounting is possible, directional antennas should be employed. In-building directional antennas have narrow azimuth and elevation beamwidth, hence allowing better coverage control in confined areas such as corridors or tunnels. Due to this reduced beamwidth, they are more directive and therefore their maximum gain is higher than an omnidirectional one (around 5–7 dBi for directional panels).

Directional antennas are often wall-mounted, suitable for buildings with large walls or pillars. These antennas can be very useful for sectorizing buildings and for containing signals internally where the external wall loss is low and/or the macrocell interference could be high. They also have a large front-to-back ratio to minimize back lobe radiation and provide better leakage control.

Directional antennas are also used as donor macrocell antennas (when using repeaters) or for bridging two buildings. For these cases, a Yagi-Uda is the preferred choice, due to its high directivity and gain (around 17 dBi for UHF).

Popular models of directional antennas are shown in Figure 6.40. Note that most of them are panels whose size depends on the specific frequency and application, but which in most cases also have discrete appearance. If these antennas are wall-mounted, they should not interfere with the wall design and the surrounding environment, making their design rather simple.

In Wi-Fi systems, when coverage enhancement is required, especially for corridors, tunnels or to connect two buildings, directional antennas with a narrow beamwidth are employed. In this case, parabolic reflectors, Yagi-Uda antennas and phased-array panels

Figure 6.40 Examples of indoor directional antennas: (a) sharksfin; (b) flat panel.

are often used. As the number of channels that can be used is very limited (only three non-overlapping channels in the 2.4 GHz band in the many countries where 11 or 12 channels are available), interference management and sectorization (also known as zoning for indoor systems) is also important, and hence stringent directional requirements must be enforced to maximize system performance. Such high-gain antennas will usually increase effective transmit power beyond the regulatory limits, so transmit power from the access point should be reduced pro-rata; however, the gain is still effective in increasing the range at the receiver.

6.7.4 Macrocell

Strictly speaking, macrocell antennas do not belong to the 'in-building' category. However, they impact the design of an in-building radio system when installed in macrocell sites and therefore the author believes they should be taken into consideration.

Panels are a common choice for macrocell directional antennas and are often used close to buildings to guarantee some degree of outdoor-to-indoor penetration. These panels are downtilted in the elevation plane to control coverage and hence minimize co-channel interference with adjacent cells. However, there is a trade-off between how much coverage can be obtained and the level of interference that is allowed; thus the C/I parameter (carrier-to-interference ratio) is often used to delimit the maximum amount of necessary downtilt.

A common way to control building penetration is also by adjusting the panel antenna's orientation and downtilt.

These macrocell directional antennas have a larger horizontal beamwidth than the in-building ones, to delimit sector coverage areas in three-sector or six-sector configurations, depending on the technology.

6.7.5 Multiband

If various wireless technologies are to be deployed in a building and antenna locations are restricted, site sharing may need to be the option for radio planners. The use of multiband antennas is strongly encouraged under these circumstances, but be aware that sometimes it is physically hard to achieve a wide frequency range in a single antenna housing. Some common multiband antenna configurations are:

- Dual- or triple-band cellular, covering, for example, 800/1900/2.1 GHz bands for 2G/ 3G/LTE.
- Cellular/Wi-Fi antennas, which cover the range 800 MHz to 2.5 GHz.
- The combination of various frequency bands that are required to be deployed in a common housing.

It is very difficult to design a single antenna element at a reasonable size for a broad frequency range. There is often a compromise that needs to be made and if multiple bands are needed; the elements or patches (depending on the specific antenna type or technology) would then be fabricated separately and then merged within the same housing. De-coupling distances should be maintained and thus bandwidth versus size (including housing) is a trade-off.

Modern antenna design technologies allow smaller elements that can operate at larger bandwidths. However, for practical multiband antennas, it is still not feasible to construct single elements that could cover the entire band. Nevertheless, antenna sizes permit easy installation and deployment, even within narrow spaces.

6.7.6 Deployment Considerations

The selection of antenna locations is very important in radio planning. Since coverage can be optimized if the correct locations can be guaranteed, efforts are often made to negotiate the installation of antennas in key places within a building. However, there are some issues and considerations that need to be made, which are summarized as follows:

- Sometimes the best antenna locations for coverage are not necessarily the best for installation, due to space or other constraints. Thus, the closest location needs to be pursued.
- Antennas should be 'disguised' when deployed inside buildings – people often feel 'nervous' about having a radio element on top of their heads!
- There may be some areas of spaces where antennas cannot be deployed; for example VIP lounges. Radio planners would then need to be prepared to select alternative locations and still obtain good coverage.
- Omnidirectional antennas are better for large spaces to be deployed in a central location, frequently on the ceiling. Directional antennas preferably are used to provide coverage to corridors or spaces for which the only option is to mount antennas on walls.
- When deploying distributed antenna systems, cable run paths need to be carefully considered and taken into account, to avoid excessive losses due to long cables.
- Avoid installing omnidirectional antennas close to external walls or windows, since uncontrolled leakage would be produced.

6.8 Radiating Cables

A radiating cable (leaky feeder) is a special type of coaxial cable where the screen is slotted to allow radiation along the cable length, exhibiting controlled leakage provided by slots in the outer conductor, as shown in Figure 6.41.

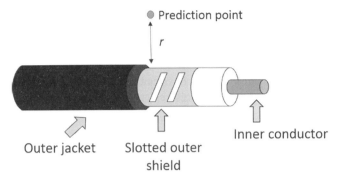

Figure 6.41 Radiating cable.

With careful design, a radiating cable produces virtually uniform coverage, which makes it very suitable for areas where confined coverage is desired; that is a path loss exponent of unity. In open environments, local mean power falls off roughly as $1/r$ (cylindrical waves).

6.8.1 Structure

The structure of a radiating cable is very similar to that of a coaxial feed, as seen in Figure 6.42. The term 'leaky feeder' implies a type of coaxial cable with specific characteristics.

The physical description of a radiating cable can be given starting from the inner parts and moving towards the outer parts. At the centre of the cable there is a copper (Cu) tube, filled with air from the inside. A low-loss foam polyethylene dielectric is used as an insulator and separates the inner from the outer layer of copper foil. The last is considered to be without thickness and with periodic slots that have various lengths, shapes and distances to each other according to the mode that the cable is radiating. Usually the outer conductor is also called the cable shield.

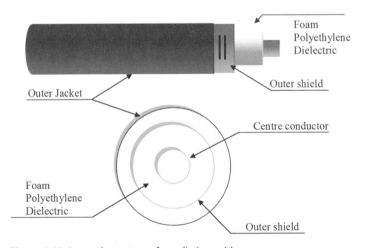

Figure 6.42 Internal structure of a radiating cable.

The cable jacket, which covers the external copper foil, is fabricated from materials that depend on the environment in which the cable is going to be used. For example, it might be fire-retardant, low-smoke and fume, halogen-free or even resistant to corrosive gases.

In order to maximize the effectiveness of the cable, its longitudinal attenuation must be minimized. Materials used in cables are carefully selected according to their performance and cost. As a result, the dielectric is usually foam polyethylene with permittivity around 1.25 and wave velocity of about 88%.

6.8.2 Applications

Radiating cables were originally conceived to provide subterranean radio propagation, for example in railway tunnels and coal mines. They are ideal for providing uniform coverage in such locations, overcoming the complexity and potential uncertainty arising from some propagation considerations when antennas are used. Radiating cables have also been used as an alternative distribution system for in-building scenarios, especially for areas that are difficult to cover with conventional antennas such as long corridors and airport piers.

A very attractive application for which radiating cables are used is for multioperator wireless solutions, due to their large bandwidth. Therefore, many technologies can be accommodated sharing the same transmission medium and hence optimizing the use of resources and minimizing costs. For example, it is common to find radiating cable installations in road tunnels for VHF broadcasting as well as UHF cellular communications.

6.8.3 Propagation Modes

There are two basic types of propagation modes and any other mode can be analysed as a synthesis of those two. The first one is the *coupled mode*, where the electromagnetic energy is concentrated in the close vicinity of the cable; the second one is the *radiating mode*, in which the energy is moving transverse to the cable (see Figure 6.43).

6.8.3.1 Coupled Mode

In this mode, the power flow is parallel to the cable's axis. Cables operating in this mode have lower longitude attenuation and higher coupling attenuation compared to those of

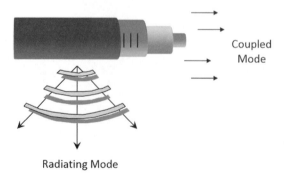

Figure 6.43 Radiating cable propagation modes.

the radiating mode. These kinds of cables are used mainly in tunnels where the length of the cable is large and the aperture of the coverage area is small. Due to its large coupling loss, this mode does not perform well in open areas where the distance between the cable and the receiver is large (typically larger than 15 m). The propagation mechanism of this mode is based on scatters existing in the vicinity of the cable, while the absence of those, theoretically, will lead to no radiation since all the energy will be confined very close to the cable.

6.8.3.2 Radiating Mode
Cables that work in this mode have large longitude attenuation and small coupling loss. These cables are preferred for in-building environments like offices or airports where usually the length of the cable is small (typically a few decades of metres) and the coverage area wide. The propagation mechanism of this mode is based on in-phase addition of all apertures. These modes appear only where there is a very well-defined slot arrangement and over a well-defined radiated mode frequency, although they are not narrowband. There is no tuning effect comparable to this encountered with antennas. The main characteristics of the radiating mode are low scattering, definite waves and high radiation.

6.8.4 Parameters

A radiating cable functions as both a transmission line and as an antenna. The radiating slots are effectively individual antennas and propagation can be analysed as the summation of all of these antennas individually. It is usually more convenient, however, to analyse radiation from radiating cables as if the cable radiated continuously along its length. The amount of radiation is quantified by two main parameters: coupling and insertion (longitudinal) loss.

6.8.4.1 Coupling Loss
The *coupling loss* is defined as the ratio between the power transmitted into the cable and the power received by a half-wavelength dipole antenna located at a fixed reference distance from the cable in free space.

6.8.4.2 Insertion Loss
The *insertion loss* is a measure of the longitudinal attenuation of the cable, usually expressed in decibels per metre at a specific frequency. For a given cable size, the insertion loss increases as the frequency of operation increases.

6.8.4.3 Bandwidth
A very important property of the radiating cables, which makes them useful for telecommunications and wireless applications, is that they are wideband; that is they have a very large bandwidth. To understand this property, it must be considered that only part of the cable is active for the calculation of the radiation in a given point and not the whole cable. This condition implies limitation about the number of slots that are going to be used for the calculations.

The bandwidth of a radiating cable depends strongly on the number of slots that exist in one period.

6.8.5 Practical Considerations

Leaky feeders are an attractive solution for cellular coverage in areas difficult to illuminate with other distribution systems, such as tunnels, airport piers and long corridors. For multioperator solutions, a leaky feeder often offers a viable solution, since its reasonably large bandwidth can accommodate many frequency bands to support multioperator designs.

When installing radiating cables, however, special care must be taken to ensure that the cable is suspended carefully with reasonable separation from metal cabling trays and other cables to avoid major impacts to its radiation characteristics. Compliance with fire and cable installation standards must also be taken into account.

For large tunnels, leaky feeders are also accompanied by bidirectional amplifiers or repeaters, which amplify the signal when longitudinal losses are beyond acceptable limits.

Finally, when selecting or installing a radiating cable, various parameters should be examined carefully, such as losses (coupling and insertion), operating frequency, bandwidth, VSWR and size, which impact physical installation of the cable. Bear in mind that special care should be taken when installing the radiating cable, to avoid mounting it along metal surfaces. If the environment at which the cable is to be installed requires it, fire regulations compliance should be met for the chosen cable. Finally, cable rigidity might be important if cable bends are to be done according to the designated building geometry and areas in which the cable is to be installed.

6.9 Conclusion

Antennas are one of the key elements in a radio link, so this is the case for in-building deployments. In fact, interference and leakage control could somehow be achieved within a contained environment surrounded by walls and partitions, but could also in principle become a challenge since the propagation environment is very complex. In addition, antenna radiation pattern characteristics dictate the way antennas radiate or receive radio waves, and thus careful antenna extrapolation methods need to be considered for accurate prediction work.

Since people spend most of their time indoors, the aesthetics and size of in-building antennas is crucial: the aim is to provide sufficient coverage whilst being unnoticed. On the other hand, antenna location selection comes from practice and experience, where a strong compromise between location feasibility and radio performance needs to be made.

Finally, for some hard-to-reach areas where the use of antennas to provide coverage is not recommended, radiating cables offer a good alternative. Although this technology has been on the market for many decades, it has only been in recent years that installations have become less expensive and affordable for indoor wireless deployments.

References

Aragón-Zavala, A. (2003) In-building cellular radio system design using measurements, PhD dissertation, University of Surrey, UK.

Brown, T., Kyritsi, P. and de Carvalho, E. (2012) *Practical Guide to MIMO Radio Channel: With MATLAB Examples*, 1st edition, John Wiley & Sons, Ltd, Chichester. ISBN 978-1-119-94523-9.

Costa, N. and Haykin, S. (2010) *Multiple-Input Multiple-Output Channel Models: Theory and Practice*, 1st edition, John Wiley & Sons, Ltd, Chichester. ISBN 978-0-470-39983-5.

Gil, F., Claro, A.R., Ferreira, J.M., Pardelinha, C. and Correia, L.M. (1999) A 3-D extrapolation model for base station antennas' radiation patterns, in IEEE Vehicular Technology Conference, Amsterdam, Netherlands, Vol. 3, pp. 1341-1345, September 1999.

Hayt, W.H. and Buck, J.A. (2005) *Engineering Electromagnetics*, 7th edition, McGraw-Hill Higher Education, USA. ISBN 978-007124449-7.

Kraus, J.D. and Fleisch, D. (1999) *Electromagnetics*, 5th edition, McGraw-Hill Higher Education, USA. ISBN 978-007116429-0.

Maxwell, J.C. (1861) On physical lines of force, The London, Edinburgh and Dublin Philosophical Magazine, Fourth Series, pp. 162-195, March 1861.

Maxwell, J.C. (1865) A dynamical theory of the electromagnetic field, Philosophical Transactions of the Royal Society of London pp. 459–512, 1865.

Sadiku, M.N.O. (2007) *Elements of Electromagnetics*, 4th edition, Oxford University Press, Oxford (UK)/New York (USA). ISBN 978-0-19530048-3.

Saunders, S. and Aragón-Zavala, A. (2007) *Antennas and Propagation for Wireless Communication Systems*, 2nd edition, John Wiley & Sons, Ltd, Chichester. ISBN 0-470-84879-1.

Taga, T. (1990) Analysis for mean effective gain of mobile antennas in land mobile radio environments, *IEEE Transactions in Vehicular Technology*, **39** (2), 117–131.

Tolstrup, M. (2011) *Indoor Radio Planning: A Practical Guide for GSM, DCS, UMTS, HSPA and LTE*, 2nd edition, John Wiley & Sons, Ltd, Chichester. ISBN 0-470-71070-8.

Vasiliadis, T.G., Dimitrou, A.G. and Sergiadis, G.D. (2005) A novel technique for the approximation of 3-D antenna radiation patterns, *IEEE Transactions on Antennas and Propagation*, **53** (7), 2212–2219.

7

Radio Measurements

When designing an in-building radio network, predictions using the models presented in Chapter 5 allow us to check if the system requirements for the network are satisfied, especially in terms of radio coverage, and initially an approximation can be made using simulated data. However, in order to refine the model accuracy and to validate whether these predictions resemble reality, measurements are required. These measurements are not only used for validating radio coverage levels but can also help to visualize the wideband performance of the network, which in principle could limit the maximum achievable data rate, depending on propagation conditions. This chapter outlines the main characteristics of narrowband and wideband radio measurements often employed in indoor wireless communications.

Radio measurements are therefore an essential ingredient of the overall in-building design process. It is rather difficult to visualize their importance if we have not analysed in depth the motivation, benefits and need for measurements. By the end of this chapter, you are expected to have a clear understanding of the value of measurements for in-building design projects; learn the different types of in-building measurement systems and their characteristics; appreciate the impact of inaccuracies while performing measurements; and learn various guidelines and recommendations for in-building testing, which will assist you in your design.

The care and accuracy with which these measurements should be performed determine to a great extent the quality of the predictions and will affect the overall design performance. Any radio engineer, researcher or designer should be aware of key elements and observations that should be accounted for when performing indoor radio measurements, and hopefully this chapter will provide this valuable information.

7.1 The Value of Measurements

So far, we have reviewed and understood antennas and propagation concepts to dimension various radio networks, especially for indoor wireless communication systems. Propagation modelling, as presented in Chapter 5, sets up the basis for designing radio systems. These designs, in principle, work well on paper, as demonstrated by various researchers and engineers in the past. However, how can we make sure our system works appropriately for any building?

Indoor Wireless Communications: From Theory to Implementation, First Edition. Alejandro Aragón-Zavala.
© 2017 John Wiley & Sons Ltd. Published 2017 by John Wiley & Sons Ltd.

In planning a mobile communication network or developing mobile equipment, it is essential to characterize the radio channel to gain insight into the dominant propagation mechanisms (Chapter 4) that will define the performance experienced by users. This characterization allows the designer to ensure that the channel behaviour is well known prior to the system deployment, to validate the propagation models used in the design process and to ensure that the equipment used provides robust performance against the full range of fading conditions likely to be encountered. Furthermore, after a system has been deployed, measurements allow field engineers to validate crucial design parameters that show how the system is performing and how it may be optimized. In summary, measurements are a very useful tool for any radio system designer:

- To gain insight into key propagation mechanisms and validate the propagation models used in the design. When we conduct measurements, we can validate that the theory applies well in the real world, and even gain more information of special propagation effects in challenging indoor environments.
- To characterize the radio channel, and in this way the radio system designer can make decisions on the system's deployment and equipment specifications.

In the context of in-building scenarios, measurements can be used to:

- Tune parameters of empirical models
- Create synthetic channel models
- Determine existing coverage levels in a building or perform testing of new candidate antenna locations (design survey) and
- .Validate the design of a recently deployed system and test for compliance with coverage requirements.

Figure 7.1 shows an example of what can be done with radio measurements. In the top plot, measurements for an airport terminal were conducted for an antenna placed right at the centre of the plot, in the GSM900 band (900 MHz). From these measurements, a radio propagation model (lower left, red line) has been tuned with these measurements (lower left, blue lines) – for this model, as seen in Chapter 5, path loss is changing with distance according to a mathematical model for which the coefficients have been tuned with the measurements. An estimation of the percentile of area covered at least with a certain signal strength level (lower right) can also be determined from measurements, and this is regularly a requirement that needs to be satisfied for design acceptance. Practical radio designs are very rarely performed to achieve a minimum signal strength value within 100% of the designated area (coverage area).

7.1.1 Tuning Empirical Path Loss Models

This is usually the starting point for the rollout of any new network, whether in a new frequency band or in a new environment. Detailed measurements are made of the path loss for sites in a range of environments, and models of the propagation path loss are fitted to them for later application within planning tools. For example, Figure 7.2 shows typical measured values of the path loss as individual data points after removing fast fading. The curve in the same figure represents an empirical power law model of the form described in Chapter 5, fitted to minimize the mean and the standard deviation of the error between the model and the measurements. The mean error can

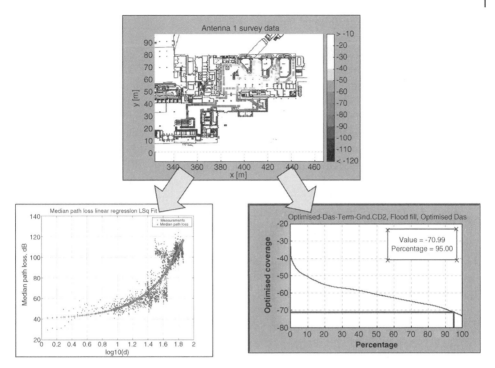

Figure 7.1 The value of radio measurements.

always be reduced to zero by choosing the model offset appropriately, so the usual way of assessing the goodness of such a model is via the standard deviation of the differences between the measurements and predictions. In the case illustrated the error standard deviation is 4.7 dB, with an error mean of only 0.05 dB, which is typical of a well-fitted model. Accurate measurements are therefore crucial in achieving an accurate path loss model.

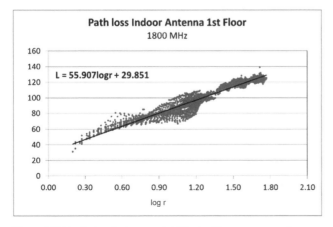

Figure 7.2 Typical path loss model fitted with measurements.

7.1.2 Creating Synthetic Channel Models

When developing and characterizing base or mobile station equipment, it is often necessary to subject the equipment to the full range of signal variations that will be encountered in practical operation. However, it is often desirable to do this in a repeatable fashion, so that different equipment can be directly compared without introducing measurement uncertainties. In such situations a synthetic channel model may be used, implemented in hardware and/or software, which produces entirely realistic channel variations in a laboratory environment. Although the channel is synthetic in such systems, the starting point is always extensive field trials to accurately determine the variation of the statistical parameters of the channel with the environment.

Measurements for this purpose will typically involve the detailed capture of both wideband and narrowband fading statistics, carefully indexed to the locations and environments in which they are recorded. This facilitates detailed subsequent analysis to determine the dependence of channel parameters on the environments in which they are experienced. These parameters are stored and loaded into a channel simulator to replicate given environments.

7.1.3 Validating Indoor Radio Designs

Once an in-building radio system has been deployed in a venue, measurements assist radio planners and engineers to assess the performance of the system against various test parameters, such as traffic load, interference, full power capacity, etc. This is the point at which adjustments can be made to the network and fine parameter tuning is performed. Without a way to verify coverage levels, quality-of-service parameters, handovers, leakage, capacity thresholds, etc., it would be impossible to finalize the design process. Accurate measurements are thus also needed here, which, in some cases and depending on contractual obligations, may need to be conducted periodically.

7.2 Methodology for Indoor Measurements

The indoor radio measurement process constitutes an essential ingredient for any specialized in-building radio engineer, since measurements are a valuable asset that accompany any indoor radio design. Therefore, before going into the specifics of these measurements, an overview of the entire process is needed.

7.2.1 Measurement Campaign Plan

The very first step in any measurement process needs to be a document that states all the details of the measurements that are to be taken. At this stage, indoor radio planners and designers decide which type of measurements need to be done, the venue (or venues) and locations, frequency bands, characteristics of the measurements, the type of equipment, an anticipated schedule, etc. A more in-depth view of a measurement campaign plan (MCP) is given in Section 7.6.1.

7.2.2 Preliminary Site Visit

A visit to the building under consideration may be required to identify areas in which measurements can be performed (walk routes), potential test antenna locations, etc. Photos are often taken and archived here, for their inclusion in the MCP document. Sometimes this preliminary visit allows radio planners to gather some existing coverage levels from surrounding macrocell sites or other wireless technologies, which will help them plan ahead the measurement campaign.

7.2.3 Site Acquisition and Permissions

Once the measurement campaign plan has been finalized, site acquisition is next. Permissions to have access to certain areas of the designated venue need to be granted, as well as security clearance for some high-security buildings, such as airports. The measurement campaign plan document accompanied by an official letter from the company responsible for the measurements is often enough to have access to the site.

7.2.4 Equipment Checklist

In the MCP there is a section that deals with test equipment to be utilized in the measurement campaign. All equipment, especially transmitters, receivers and antennas, must be checked and verified for calibration prior to the measurement date, to guarantee that accurate data are to be gathered.

7.2.5 Measurement Campaign

Measurements are conducted on the specified date for which access to the site has been granted. Data needs to be recorded and backed up and photos taken as part of the measurement report.

7.2.6 Data Postprocessing

Although not necessarily an essential part of the measurement process, sometimes the postprocessing of data may need to be performed during the measurement campaign and therefore needs to be included in the process. Sometimes this is only done to validate the collected data. If this is the case, it may be called a 'fast-track' postprocessing, just to verify that the data collection was correct. The 'full' postprocessing would then need to be performed back at the office.

7.2.7 Postvisit to Site

For network validation, once the in-building radio solution has been installed, measurements may be required, and a site revisit will need to be planned and scheduled. It is better to consider this in the initial site acquisition step, although sometimes this may not be possible and a second round of negotiations would then need to be done.

7.3 Types of Measurement Systems

Before going to the specifics of radio measurement equipment, let us examine briefly the two types of measurement systems available for radio measurements. These are narrowband channel sounding and wideband channel sounding.

7.3.1 Narrowband Measurements

To characterize the narrowband behaviour of the mobile radio propagation channel, continuous wave (CW) measurements are often performed by transmitting an unmodulated single tone carrier. Note that the path loss is essentially a narrowband effect so that CW measurements are adequate for most path loss models, although the UWB (ultra-wideband) system path loss is a possible exception.

A fixed radio transmitter is often employed for such measurements, connected to many test antennas, and the mobile receiver is carried by a pedestrian user or is even on a mobile robot platform. Occasionally the receiver remains fixed and the base station is moved to perform the narrowband measurements, which is particularly useful if it is desired to compare signal paths between multiple base stations and a mobile location.

An example of narrowband channel testing equipment can be seen in Figure 7.3. Note that for the measurements that were undertaken there, a radio scanner along with

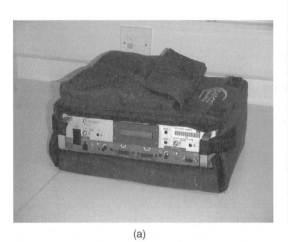

(a)   (b)

Figure 7.3 Typical narrowband indoor testing equipment: (a) radio scanner; (b) antenna under test mounted on a tripod.

Figure 7.4 Portable spectrum analyser and software used for indoor narrowband measurements.

software installed in a laptop was used as the receiver equipment, whereas antennas were deployed on tripods and located at the candidate tested antenna locations.

Portable spectrum analysers as shown in Figure 7.4 can also be employed for specific narrowband measurements, such as the one shown here with a fabricated dipole antenna, for hidden node margin measurements (Aragón-Zavala, Brown and Castañón, 2016). This facilitates the collection of samples at specific frequency bands that may not be available in many commercial radio scanners.

Sometimes portable spectrum analysers may not be available for indoor measurements and alternative solutions are employed. Figure 7.5 shows a ZVL6 spectrum analyser from Rhode and Schwarz that was employed for specific correlation measurements, for which repeatability was important and relevant and therefore a trolley was used to support the ZVL6 (Aragón-Zavala, Jevremovic and Jemmali, 2014). Configurations similar to this are often encountered in research projects.

For indoor radio coverage design, narrowband measurements are employed to determine coverage and leakage levels. The ones that are frequently used in the market are continuous wave (CW), code-scanning and engineering test mobiles.

7.3.1.1 CW Measurements

A generic narrowband channel sounding system is presented in Figure 7.6. Notice that, depending on the environment in which the measurements are to be performed (indoor or outdoor), the navigation and positioning system changes. Annotations have been made to indicate the cases for in-building testing use.

The data acquisition system is normally composed of a laptop computer that has specialized data acquisition software installed, and includes all the interfacing drivers to communicate with the receiver and navigation and positioning systems. Signal envelope or phase measurements can be performed with the above-mentioned system, and the design of the receiver determines whether this amplitude or phase is measured. If phase is of interest, then this phase can be measured relative to a fixed reference, by demodulating the signal in two channels; that is in-phase and quadrature. In some applications the absolute phase might be required and then it is required that the local

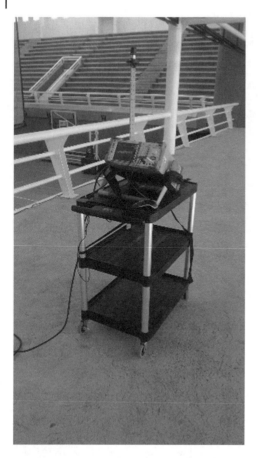

Figure 7.5 Spectrum analyser mounted on a trolley to perform special narrowband correlation measurements.

oscillators of transmitter and receiver are phase-locked, and hence only phase variations due to propagation characteristics are recorded.

When selecting measurement receivers, the dynamic range is an essential parameter to take into account, since this will determine both the maximum signal strength that can be detected as well as the receiver noise floor. For example, for in-building measurement campaigns in which large distances are to be covered, a large dynamic range is desired, especially if the transmitter can only be placed at a distant location from many rooms that are of interest. Furthermore, the maximum signal strength that can be recorded is often of interest here if walk tests are performed very close to the transmitting antennas. In all cases, the full extent of signal fast fading has to be considered within the required dynamic range.

The first type of narrowband measurements we will analyse are CW measurements. They are used to characterize the narrowband behaviour of the mobile radio propagation channel. CW testing is also quite adequate for tuning path loss models, especially if accurate propagation work is required – we are interested in the radio propagation of an unmodulated carrier throughout a building, rather than on specific network-related issues.

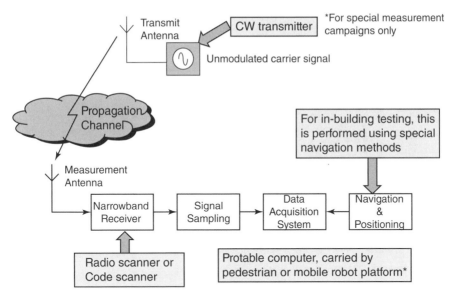

Figure 7.6 CW measurement system for in-building radio testing.

For CW work, the transmitter is situated at a fixed location and an unmodulated single tone carrier is transmitted, as shown in Figure 7.7. The receiver (often a radio scanner) is carried by the person conducting the measurements, as depicted in Figure 7.8. Special care must be taken to hold the receive antenna at an average person's height, to avoid coupling with the body. The dynamic range of the receiver is essential to determine the maximum detected signal strength and the receiver noise floor.

As CW measurements only use unmodulated single tone carriers, they are not suitable for measuring network performance and therefore other narrowband measurement equipment should be used.

7.3.1.2 Code Scanning

Code scanning is used for 3G cellular systems to log all the decoded cells, scrambling codes and CPICH (pilot) levels when performing a measurement campaign. Typically,

Figure 7.7 Transmitter and antenna for a CW measurement.

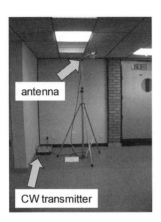

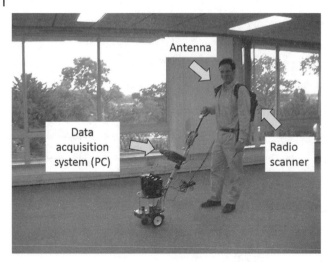

Figure 7.8 Example of a CW measurement for in-building testing.

the user defines the start and end channels and the code scanner will scan and measure all the channels in the specified range. The purpose of this channel scan in 3G systems is to measure potential pilot polluters, find unexpected neighbours and establish the baseline for the noise level of the radio channel.

7.3.1.3 Engineering Test Mobiles

Engineering test mobiles are used when mobile network parameters are to be monitored, such as handover failures, idle/dedicated mode call statistics, blocking issues, dropped call rates, interference effects affecting network performance, network benchmarking, etc. This method is used only with operating mobile networks, and although test mobiles can measure signal strength levels, their accuracy and dynamic range are not sufficient for accurate propagation modelling work. A couple of examples of in-building measurement systems using engineering test mobiles include TEMS (Ascom) and E6474A indoor wireless measurement system (Agilent). Figure 7.9 shows an example of a radio engineer conducting an in-building radio survey using engineering test mobiles.

One of the advantages of engineering test mobiles is that they can resemble user's experience within the network, so they are very useful for validation or benchmarking testing. They also can assist operators in detecting failures and adjusting network parameters, not necessarily related to coverage.

Amongst the limitations of test engineering mobiles is their limited accuracy for signal strength measurements, which make them appear as a not viable alternative for propagation prediction work.

7.3.1.4 Comparative Analysis

A comparative table for the three narrowband measurement systems described earlier is shown in Table 7.1. Note that although test engineering mobiles lack sufficient accuracy for propagation prediction work, CW measurements can perform this task very well, and complement each other with test engineering mobiles as the network performance analysis that cannot be performed using CW testing can be done without any problem

Figure 7.9 Engineering test mobiles.

Table 7.1 Comparative analysis table of indoor narrowband testing methods.

Continuous wave (CW)	Code scanner	Engineering mobiles
Function: Sends un-modulated signal carrier (tone) for accurate propagation work	**Function:** Detects scrambling codes and decodes signals for identifying potential pilot polluters	**Function:** Scans all specified channels for a given technology; used for network performance assessment
✓ Accurate signal strength collection - ideal for propagation modelling and tuning work and for some elements of installed systems verification	✓ Provides signal quality and strength for each code individually	✓ Can monitor network performance from a user perspective
✗ Not suitable for monitoring network performance	✗ Limited dynamic range and sample rate – not suitable for propagation modelling	✗ Poor accuracy in signal strength collection
✗ Unable to determine signal quality		
✗ Need to be careful to avoid measurement of unwanted signals (co-channel, adjacent channel and blocking)		

using test mobiles. Likewise for code scanning, which is unique to decode information relevant to 3G systems, even though it lacks a sufficient dynamic range and sampling rate.

At this stage, it is very important to appreciate that the three systems complement each other and should be selected depending on the type of testing that is needed. There

Figure 7.10 Wideband channel sounder used for MIMO measurements.

are measurement campaigns whose aim and scope is such that two or more of these systems need to be used.

7.3.2 Wideband Measurements

Wideband channel parameters such as delay spread, delay profile, average delay and coherence bandwidth are of special interest when channel characterization is desired and are especially important for system performance. Various techniques may be employed, some of which use the principle of transmitting many narrowband signals, either sequentially or simultaneously. However, due to the limitations of the above-mentioned methods, genuine wideband sounding techniques are required. Examples of such techniques are periodic pulse sounding and pulse compression.

An example of the use of an Elektrobit wideband channel sounder for MIMO measurements is shown in Figure 7.10, where an 18 × 8 MIMO configuration was tested.

There are various wideband measurement procedures that can be carried out either in the time domain or in the frequency domain. For the time domain case, the *impulse response* of the channel is measured. On the other hand, for the frequency domain, the *frequency response* or *transfer function* of the channel is measured.

The impulse response of the channel can be measured either by transmitting a short pulse or by transmitting a wideband spread-spectrum signal. In the short duration pulse case, the received signal is sampled; that is the convolution of the pulse with the channel's impulse response. Thus, the ratio of peak to average power must be high in order to detect the low power multipath. In the wideband spread-spectrum case, the signal is modulated by a pseudo-random sequence and the receiver signal is correlated with the original signal. In this case, the duration of the signal must be short compared to the channel coherence time.

Transmit antenna   Receive antenna

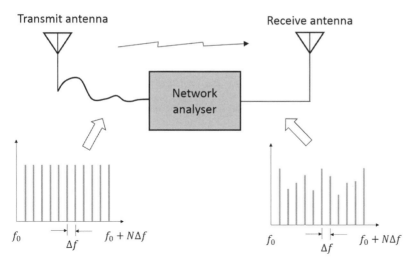

f_0 Δf $f_0 + N\Delta f$ f_0 Δf $f_0 + N\Delta f$

Figure 7.11 Measurement system used to obtain the channel's frequency response.

For the frequency domain approach, the frequency response is directly measured. Figure 7.11 shows an example of a measurement system using a network analyser as the main component. Thus, it is possible to obtain the channel's impulse response by applying the inverse Fourier transform to the obtained frequency response.

The transmit antenna (antenna TX) is plugged to one of the network analyser ports and a set of discrete frequencies $f_k = f_0 + k\Delta f$ is swept, with $0 \leq k \leq N$, having equal increments Δf. At the same time, N complex samples of the channel's frequency response $H(k)$ are measured in the receiver antenna (antenna RX), connected to the other port of the network analyser. Assuming that $f_0 = 0$, the baseband complex frequency response of the channel is (Monk and Wingbier, 1956)

$$H(k) = H(f)|_{f=k\Delta f} = \sum_{i=1}^{L} \beta_i e^{-j2\pi k\Delta f \tau_i} e^{-j\phi_i} \tag{7.1}$$

where β_i, τ_i and ϕ_i represent the magnitude, arrival time and phase, respectively, of the L individual paths. Since the measurement system is band-limited, the system measures windowed frequency characteristics of the channel. Thus, the measured frequency response is given by

$$H_{\text{meas}}(k) = W(k) \sum_{i=1}^{L} \beta_i e^{-j2\pi k\Delta f \tau_i} e^{-j\phi_i} \tag{7.2}$$

where $W(k)$ represents the effects of filtering in the frequency domain.

The baseband impulse response of the channel can therefore be obtained by performing the inverse discrete Fourier transform of the measured samples $H_{\text{meas}}(k)$ as

$$h(\tau) = \frac{1}{N} \sum_{k=0}^{N-1} H_{\text{meas}}(k) e^{j2\pi \tau k \Delta f} \tag{7.3}$$

The frequency-domain measurement system can measure a time span of $T_m = 1/\Delta f$. The time resolution Δt is obtained by using the inverse of the measurement bandwidth, yielding

$$\Delta t = 1 \Big/ \sum\nolimits_{k=1}^{N} f_k \tag{7.4}$$

7.4 Measurement Equipment

7.4.1 Transmit Equipment

Transmitters produce the signal used for the measurements, especially in those cases where antennas need to be deployed at specific locations, and there are issues that need to be monitored to make sure errors will not be produced. An example of a Rohde and Schwarz signal generator often employed for in-building radio testing is depicted in Figure 7.12.

To start with, transmitters should undergo calibration and validation periodically, to verify that manufacturer's specifications are still being met and especially prior to the measurement campaign.

Transmitters should transmit signal levels at the specified power and at a given frequency, maintaining a degree of frequency stability. Since transmitters are active devices, they are subject to change of performance parameters with temperature, humidity, supply voltage variations, etc. Selection of the transmitter to use must account for a compensation of these effects in order to minimize frequency and power drifts.

Another issue with transmitters has to do with their power supply. If power supply is a signal generator, the performance of the transmitter may be jeopardized by the generator's transients, noise and voltage variations, so a careful monitoring of power supply levels is of importance. On the other hand, if the power supply is a battery, the output frequency and power must be checked when operating over large periods of time, preventing battery discharge periods and having if possible available redundant batteries.

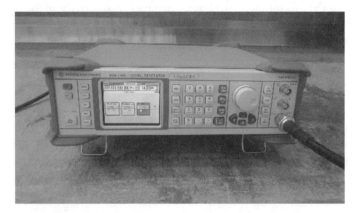

Figure 7.12 Transmitter employed for in-building radio testing.

Figure 7.13 Radiating cable testing supporting equipment.

Additional transmit equipment should be used along with the radio transmitter to support the measurement campaign. The following items are usually required:

- Tripod for supporting the antenna (see Figure 7.3b), preferably made of nylon or wood to avoid disturbing the propagation environment being measured. For radiating cables, suitable support tools should also be used (Figure 7.13).
- Signal source, carefully calibrated and preferably placed at a well-ventilated place, to avoid changes in power and frequency stability due to an increase in temperature.
- Full-charged main and spare battery packs for battery-powered transmitters.
- Coaxial jumper cables and connectors, with the corresponding RF adapters to properly fit any type of connector in the cables to any connector installed in the transmit equipment and antenna.
- In-building transmit antennas, of the type and frequency according to the campaign specifications and characteristics.

7.4.2 Receive Rquipment

Receivers should be selected according to the specific characteristics of the measurement campaign. For example, if coverage is desired for accurate propagation prediction work, then a CW transmitter should be employed using a radio scanner. If 3G code scrambling is to be performed, then code scanners are to be selected. Engineering test mobiles are chosen when network performance parameters are to be monitored, as well as benchmarking amongst operators and when parameters are to be optimized, after the system has been deployed.

One of the key parameters to consider for in-building measurements for receivers is their *dynamic range*, to make sure distant rooms and areas can still be tested without reaching the receiver's noise floor. This is particularly important for large facilities such as airports, retail stores or shopping centres.

Signal reception can be achieved using spectrum analysers, radio scanners or engineering test mobiles, as discussed earlier. In all cases, accurate measurements strongly depend on the periodicity of which calibration is performed on the receivers, to guarantee signals are within specified limits.

Finally, receivers must be calibrated prior to commencing the measurement campaign, to guarantee that any measurement uncertainties can be known and measurement errors can be compensated.

A signal of known power from a well-calibrated signal generator can be injected at the frequency of interest and samples of the recorded signal strength by the scanner will need to be within these injected values. The frequency range is varied as well as the transmitter power, to cover the entire dynamic range of the radio scanner. The transmitter employed should be a carefully calibrated laboratory signal generator and the received signal strength can be compared with a calibrated spectrum analyser, as the known reference.

If channels are to be scanned over a wide frequency range, then it is important to ensure that the receiver is calculated at least at the extremes and centre of the measurement range to determine whether any significant differences exist.

Several issues should be taken into account when preparing receive equipment for a measurement campaign. These are listed as follows:

- Full-charged main and spare battery packs should be taken.
- The chosen measurement receiver should be calibrated and packed according to measurement campaign needs. It might be the case that two or more types of receivers could be required if many types of testing need to be performed in the same measurement campaign; for example propagation measurements and existing coverage for benchmarking.
- Antenna and mounting arrangements are to be considered for the walking person conducting the measurements. Receive antenna height should be clear of head sight if propagation measurements are to be performed. Also, it is usually better to use an antenna that does not require a ground plane, to avoid radiation pattern distortion (e.g. a sleeve dipole).
- A portable computer must be prepared with the collection software and drivers installed. It is important to bear in mind that preferably special laptops/palmtops are to be used here, since come portable computers may be sensitive to motion and their hard disk can be damaged. Ruggedized portable computers are the preferred option for in-building measurements.
- Also, make sure that the floorplan of the building to be measured has been imported into the collection software, and depending on the type of testing, even external areas may be needed.

7.4.3 Miscellaneous Testing Components

Passive elements, such as cables, attenuators, antennas and splitters (Figure 7.14) form an integral part of a measurement system. These elements, although reasonably stable, do suffer degradation with time and therefore periodic validation of their main RF properties is required. Passive elements are frequently the cause of measurement failures in field systems. RF connectors must be tightened to the correct torque, as recommended by the manufacturer. If they are too loose they will produce intermittent errors, which are hard to diagnose and introduce additional loss into the system. If they are too tight

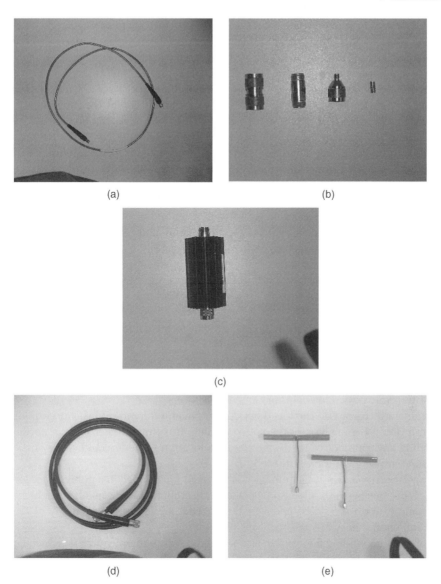

Figure 7.14 Examples of passive elements used in a measurement survey: (a) coaxial cables; (b) connectors; (c) attenuators; (d) RF feeder; (e) antennas.

they will deform, causing a change in the presented electrical impedance and inducing unwanted reflections within the transmission line. Mechanical vibration in the mobile then effectively varies the transmission line impedance dynamically, causing severe and unpredictable variation in the system calibration. Most RF connectors are only rated to several hundred mating cycles, so cables and connectors used for field measurement work should be replaced at regular intervals.

It is important to use cables with an appropriate specification: for example, cables with $50\,\Omega$ and $75\,\Omega$ characteristic impedance look identical but will not operate correctly.

Cables degrade swiftly when used in field measurement systems, producing additional losses and unwanted reflections. This may occur, for example, due to corrosion or to the cable being bent to too small a radius. A measurement with a network analyser can check both the loss and impedance of a cable. A specialized time domain reflectometer is also a useful tool for characterizing and locating any cable faults. This operates by sending short pulses along a cable and measuring the time and amplitude of the reflections.

Antennas are elements that should not be neglected from this validation process, as for very accurate propagation measurements, a few decibels in antenna gain or decimal places in VSWR can be critical. Therefore, ideally, a radiation pattern characterization at least in azimuth and elevation at the operating frequency would be desired, as well as a VSWR measurement over its operating frequency range to evaluate how much power will not be radiated from the antenna. This is essential for antennas that have been designed specifically for the measurement campaign plan but is optional if the antennas are purchased from known and recognized antenna manufacturers. Ideally the antenna radiation pattern would be characterized in situ, but typically it is sufficient to use a standard test route with known characteristics to establish an appropriate value for the mean effective gain (MEG). The key point is to regularly monitor the antenna characteristics over time and ensure that the same setup is used through a given measurement campaign.

Finally, some accessories are highly recommended to take along with transmit and receive equipment, which are described as follows:

- Power meter, to confirm power measurements. This is very important, especially for propagation prediction measurements, as the link budget needs to take an accurate figure of transmit power for path loss calculations.
- Digital camera to record all the locations of the transmit antenna. This is useful, for when performing the analysis of the collected data, it is much easier to refer to the surrounding environment of the tested antenna location to understand some of the propagation effects taking place there.
- Temporary security barriers, if equipment is to be left unattended in a public place for any length of time. Signs should also be placed indicating that RF transmit equipment is active and caution should be taken by the general public not to touch any of the devices and equipment.
- Paper copy of the building floor plans, as it is often easier to identify on paper specific areas of the building, prior to loading the digital map in the collection portable computer.
- Clipboard, to hold any relevant documents (permissions, floorplans, etc.).
- Measuring tape or electronic laser measure, as the measurement of distance may be required.
- Tape measure and compass.

7.4.4 Buyer's Guide

An example of some equipment vendors for in-building measurement equipment include:

- Andrew (transmitters, receivers, antennas, radiating cables)
- Anritsu (transmitters, receivers, power metres)

- Rohde & Schwarz (transmitters, receivers)
- Agilent (transmitters, receivers)
- PCTEL (receivers)
- Motorola (receivers)
- Sagem (receivers)
- Tektronix (receivers)
- Kathrein (antennas)
- RFS (radiating cables).

These are amongst the most popular and known, but by no means the only ones. Also, bear in mind a few recommendations when buying test equipment:

- *Price*: need to assess cost versus benefit. Prices may change for equipment of similar characteristics between two vendors.
- *Technical specifications meeting measurement requirements*: buy the equipment that you are to use for many measurement campaigns according to the type of technology, the type of testing, etc. If it is a 'one-off' test, it might be much better to rent the equipment for a number of days only.
- *Upgrades*: as wireless technology is progressing fairly quickly, the test equipment needs to be upgradeable; otherwise it tends to become obsolete in a short period of time.
- *Technical support*: when things go wrong, it is much better to have a vendor who is really proving high-quality customer support than someone not interested in helping you out! Be careful, as this is something every vendor offers, but not all provide with the highest quality. It is useful to get recommendations from other people who have bought similar equipment.

7.5 Types of Indoor Measurement Surveys

The types of measurement surveys can be classified differently, according to the type of environment or the characteristics of the measurement campaign. For example, when designing radio systems, drive testing is used for getting samples from various locations around a large transmitting cell, such as a macrocell-type base station. This is applicable for cellular, Wi-MAX or satellite-based terrestrial coverage. However, if the measurements are to be conducted inside buildings, for instance, for Wi-Fi or cellular, in-building measurements are to be performed. Since our end-of-chapter propagation experiment is related to conducting measurements inside several buildings in your university campus, our focus on this section will be on highlighting special techniques and guidelines to conduct radio measurements inside buildings.

7.5.1 Design Survey

When designing a new portion of a mobile network for any cell type (macro, micro or picocell), it is common to conduct site-specific surveys to ensure that any design assumptions made are valid. For example, while propagation models may be used to determine an appropriate general location for a new macrocell, a survey may be conducted to ensure that a specific proposed base station location provides coverage

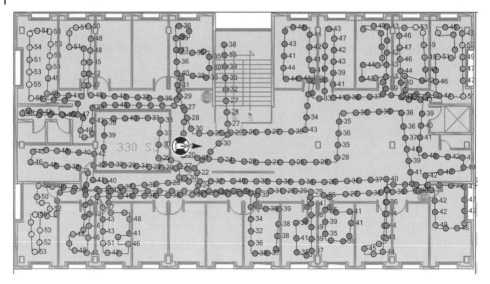

Figure 7.15 Design survey example.

in particular locations, which is difficult to do with complete confidence with any propagation model. While a good macrocell prediction model may achieve an 8 dB standard deviation of error, well-conducted and calibrated measurements should be repeatable with a standard deviation as low as around 3 dB.

Similarly, in an indoor environment, it is often important to conduct surveys from some or all of the proposed antenna locations in order to validate initial assumptions made concerning the properties of the building materials or the detailed building construction geometry. It is also important to check the coverage of the existing systems, particularly for licence-exempt bands such as those used by Wi-Fi systems, for which no central record of potential interferers exists.

An example of a design survey is shown in Figure 7.15. Each point represents a signal strength sample and the colour bar indicates the signal strength level. The tested antenna is shown with an arrow indicating the direction of maximum radiation (boresight).

7.5.2 Existing Coverage

Ongoing measurements are conducted by network operators to determine the performance of their network. These may be compared to previous measurements, revealing the impact of network changes intended to optimize performance or the impact of changing user traffic patterns. They may also be used to benchmark network performance against alternative operators or network technologies. Such measurements may simply be of signal strengths, but more usually they involve measuring a large number of parameters simultaneously in order to determine the experience of a real customer and to relate this to the network parameters, both RF and logical.

For example, it is common to place repetitive voice calls using mobile phones in an engineering mode and to determine the proportions of such calls that are dropped or subject to blocking. It is increasingly common to also measure the data performance that may be experienced by users in the form of measurements of throughput, latency and other quality-of-service indicators. Simultaneously, the measurement system must log the network state and signalling traffic, to determine the network messages such as handover commands and mobile measurement reports, which will help to diagnose the causes of any problems and suggest actions to be taken to optimize performance in the future.

7.6 Guidelines for Effective Radio Measurements

We have appreciated the importance of measurements for in-building design work, including many aspects within the process: for tuning and validating propagation models, for determining existing coverage in a new site, to identify coverage holes, to decode channels to find potential pilot polluters in 3G, to assess network performance of recently deployed in-building cellular systems or simply to optimize parameters. The issue now has to do with how to conduct such measurement campaigns in order to get the most benefit from them, having as a result accurate measurements that could be used. This involves planning the measurement campaign and, once a plan is in place, conducting such measurements efficiently. This topic aims at reviewing these two aspects in radio measurements: planning and execution (conducting).

Many of the issues related to outdoor measurements also apply indoors, but there are further considerations relating specifically to the methods of navigation and choice of measurement locations. Since this book is totally dedicated to indoor wireless systems, we will focus our attention on guidelines for effective indoor radio measurements. For a more detailed discussion on the issues and guidelines related to outdoor measurements as well, refer to Saunders and Aragón-Zavala (2007).

7.6.1 Planning Your Measurements: The MCP

In Section 7.2.1 and as part of the indoor radio design methodology, a measurement campaign plan (MCP) was mentioned. A measurement campaign plan is a document that establishes the requirements and planning needed for the propagation testing to be performed. This plan encompasses essential issues such as motivation, general measurement requirements, base station locations, testing routes, testing guidelines, workplan and required equipment checklist. Some of these issues will be discussed in more detail in this section.

The importance of having a measurement campaign planned in advance relies on more than being a simple checklist document. The aims and goals of what is to be achieved after the testing are often overlooked and should not be minimized. Time is sometimes critical, especially when the measurements are performed in public areas where permission is given only for a limited timeframe. Therefore, careful planning is essential if successful results are to be obtained.

The MCP aims to give a brief introduction to the measurement campaign, including some background and scope, which is required here to stress the importance of conducting measurements as part of a project. To support this, a standard documents background can be included for reference. This assists the building owner, for example, who grants permission to access the building site for the measurements, a general overview (without too much technical information) of the measurement campaign and why it is important.

7.6.1.1 Introduction

The *introductory* part normally establishes a baseline for the measurements, but is often not enough. An explanation of what drives engineers, planners, scientists or whoever is to conduct the measurements is also required, to justify the main 'motivation' for the measurements. This is usually a statement that describes and explains the purpose of the measurements. It should be precise, clear and catches the attention of the reader, as it often justifies the need for the measurement campaign.

7.6.1.2 Objectives

As part of the MCP, there is also a need to define clear, achievable and measurable *objectives*. These objectives will be used as a basis for future evaluation of the measurement campaign. Often building owners want to make sure that whoever is conducting the measurements has clear objectives in mind – in a way, they want to understand that the measurement campaign is 'worthwhile' and that it must be conducted to achieved specific targets.

As an example of the definition of motivation and objectives in an MCP, let us suppose that the project consists of determining the existing coverage of four GSM operators in an airport, for benchmarking purposes, and that the survey is conducted on behalf of operator A (B,C and D are the remaining three GSM operators). A way to write the main motivations for this work could read like this: 'Establish a coverage comparison amongst the four GSM operators in the airport, to assess market penetration of operator A, in order to take further actions to improve their coverage at certain areas.' This brings the attention of the airport authorities as they may see the benefit for airport users in achieving better signal quality.

In terms of objectives, they could be as follows:

- Determine existing coverage levels for operator A in critical areas within the airport, for GSM, to identify coverage holes.
- Establish a comparison of the extent of coverage with the other operators (B,C and D) to identify the market penetration for operator A in the airport.

7.6.1.3 Requirements

The *requirements* section in an MCP establishes testing requirements depending on many factors, such as technology, equipment availability, design requirements, operator's requirements, and what is to be done with the measurements. When specifying requirements in the MCP, for example, a specific test equipment to use is specified, how repeatable should the measurements be, sampling and averaging requirements (especially if fast fading is to be gathered) as well as the type of sampling that is needed (distance or time).

To understand these requirements for the measurement campaign better, let us take our benchmarking existing coverage exercise for operator A in an airport example. Technical requirements for these measurements could be:

- Accurate measurements are needed for propagation modelling, so CW measurements are to be used.

- GSM and UMTS sites are to be deployed, thus transmitters and receivers at these frequency bands are required.
- The operator has specified to take samples in 1 m bins and to average using linear power; thus receiver equipment must be selected to guarantee that this requirement is fulfilled so that sufficient samples are taken, and averaging is performed in postprocessing.
- Repeatability is needed for the measurement campaign, to assess cross-correlation factors. Therefore, a line track on the floor is created and followed for each measurement.

7.6.1.4 Antenna Locations

Prior to the measurements, *test antenna locations* must be selected according to the type of campaign and the scope that is desired for this. These antenna locations are chosen according to knowledge of the building under consideration – a visit to the building is strongly recommended to identify potential test antenna locations.

On the day of the measurements, these antenna locations should be maintained within a minimum error margin, which is established according to campaign accuracy requirements. Also, test antenna heights and types of antennas to be used must also be selected here.

For the example shown here, note that five antennas have been selected to be used in this building, as depicted in Figure 7.16. The first antenna, a directional one, has been located there to illuminate the corridor and to gather information about wall losses from different materials. Antenna 2 has been placed at the entrance of the building to illuminate the lobby area. It will be used to determine leakage levels outside the building and assess the backlobe performance of this directional antenna. Omni antennas 3 and 4

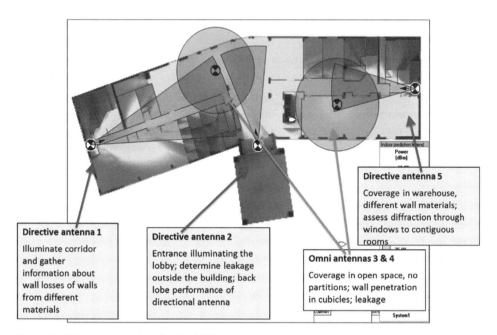

Figure 7.16 Antenna locations for the MCP example.

have been placed at the areas shown in the figure to provide coverage in open spaces, with no partitions, as well as to gather wall penetration in cubicles and to determine leakage levels. Finally, antenna 5, also directional, provides coverage in the warehouse, assesses diffraction through windows to contiguous rooms and also gathers information about other wall materials.

7.6.1.5 Walk Test Routes

For *walk test routes*, to have a plan is essential, especially for saving valuable time when conducting the measurements. The areas to measure must be clearly identified, even if the very specific paths are not exactly marked – it is often sufficient to indicate the rooms and types of walk routes to be followed (straight lines, zigzag, circles, etc.).

It is important at this stage to emphasize that it is not critical to walk in all the building. The most representative rooms and spaces, such as corridors and atriums, have to be selected. Also, bear in mind that, whenever possible, to obtain floor penetration loss information, walk at least a floor above and/or below the tested antenna.

Finally, the extent of the walk should include all areas until the received signal strength has reached the noise floor and should not be too close to the tested antenna so as to saturate the receiver.

7.6.1.6 Workplan

The *workplan* deals with the timing of the MCP: what needs to be done and when. A workplan is very useful in planning activities, resources and times, and is usually performed using a Gantt chart. Here, task dependencies can easily be identified as well as milestones, and adjustments can be made if a task is at risk of delaying the whole measurement campaign. A priority here is to finish the measurement campaign on time and within budget!

7.6.1.7 Implications of Not Having an MCP

The implications of *not* having a measurement campaign plan or having an *incomplete* version include:

- Delays in conducting the measurements. Sometimes this can make the difference between finishing on time or leaving some areas untested, which may result in additional expenses as a second visit may be required.
- Impossibility to conduct *some* or *all* measurements. If permission was not granted, or not properly negotiated, or a piece of testing equipment was not planned and included in the checklist, it may be the case that the measurement campaign would need to be cancelled.
- Erroneous measured data. Failure to carefully design the type of data that needs to be gathered, or the channels to be tested, or the walk routes that need to be performed, or the receiving equipment setup may result in collecting data that is wrong or contains errors that cannot be fixed during postprocessing.

7.6.2 Choose a Suitable Navigation System

As signal strength samples are recorded, they need to be linked to spatial coordinates and time stamps in order to establish a spatiotemporal frame for measurements. Propagation

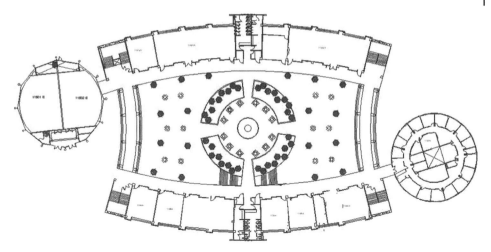

Figure 7.17 Example of an available floorplan for in-building data collection.

models are distance-dependent, as path loss varies with distance, and temporal variations are also to be recorded. Spatial variations are extremely important since they determine to a great extent the cell dimensioning as well since three dimensions are required for in-building work – unlike for most outdoor projects, where horizontal distances are much greater than heights.

There is a major limitation in navigation and positioning for indoor scenarios as it is a fact that the signal from the GPS (global positioning system), which is greatly used for drive tests and outdoor measurements, cannot penetrate buildings. Thus, alternative methods for in-building navigation and positioning should be examined, one of which is described next.

The most common approach to navigation indoors is called *way-point navigation*. This method has been widely used in many commercial in-building collection tools developed by several radio data collection software and hardware manufacturers.

The approach consists of having a digital representation of parts of the building in consideration, which may be split by floors or regions. This representation is typically a bitmap image showing the floor layout, as shown in Figure 7.17, but some collection tools are capable of navigating using CAD formats, which provide vector representations of walls and other building features. The user interacts with the data collection software using a 'pen' input device on the touch-screen of the collection computer. This input selects their position at a given point in time on the floor layout displayed on the collection system's graphical user interface. The user's entry of their start position begins recording regular samples from the RF receiver hardware and data collection continues as the user walks in a straight line to a new position within the building and then stops. The user then indicates this end-point to the collection software on the floor layout using the touch-screen.

The collection software then uniformly assigns a position to each discrete sample collected over the period between the start time and end time of the walk segment the user has just completed, assuming the user has walked at a constant speed. This process is repeated segment by segment, as the user moves around the building, until all required

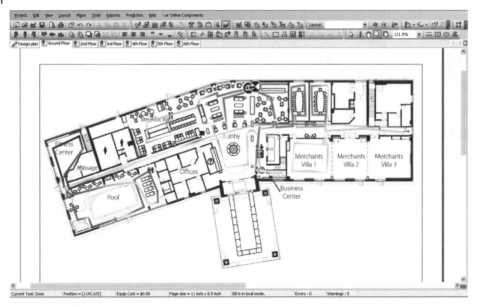

Figure 7.18 Typical in-building measurement route and signal strength.

areas of the building have been surveyed. This is the approach used in the example shown in Figure 7.18; note that the route is composed of straight-line segments due to the way-point approach. A skilled user can achieve accuracies of the order of 1 m from way-point navigation.

This way-point approach has various disadvantages. The use of high-resolution maps is normally required, since the user must determine his position with reasonable accuracy. This may not often be available, especially for large facilities with a complicated building layout. Also, in cases where a measurement route should be repeated, way-point navigation can provide neither the repeatability nor the accuracy required. Nevertheless, way-point navigation is very simple to implement and is sufficiently accurate for most common applications (Aragón-Zavala, 2003).

Both antenna locations and walk routes need to be sufficiently accurate. Note that repeating the same walk route will give different results, even when conducted carefully, as depicted in Figure 7.19, and this contributes to the 'in-built' uncertainty of the whole measurement/prediction/verification process.

When this positioning accuracy is required, other indoor positioning and navigation systems have been employed and suggested using mobile robots with a 'fifth wheel' mechanism, which samples at equally spaced intervals (Aragón-Zavala and Saunders, 1999; Radi *et al.*, 1998). This configuration is often desired:

- When measurements are to be isolated from the human body.
- If repeatability is required; for example when time variations in the channel are analysed and different runs over exactly the same route are to be performed.
- In hazardous environments for the field engineer; for example exposure to high levels of radiation.

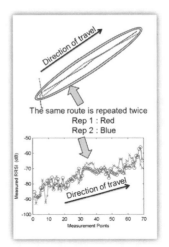

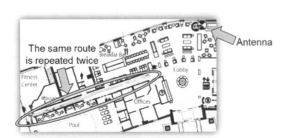

Figure 7.19 Impact of positioning.

- If high accuracy in position is necessary at the expense of a longer time to conduct the measurements, with requirements for absolute position updates and limitations on the surface on which the robot can navigate.
- Whenever autonomous navigation is needed, providing an accurate and inexpensive method of navigation for indoors.

Figure 7.20 shows two of the mobile robot platforms employed for these special measurement campaigns. One follows a route marked out on the floor with a high degree of accuracy, while the other is pushed by the user and uses odometry and gyroscopes to determine its position. It becomes a very interesting project for students to build a specialized robot responsible for the navigation and positioning of the radio transmitters or receivers within a measurement campaign – this certainly provides students with a broader picture of practical applications of robotics in the wireless industry!

7.6.3 Signal Sampling and Averaging Considerations

While the requirements of the measurement system will depend in detail on the specific application for the measurements, there are some generic issues related to the way in which mobile signals must be sampled, which are common to all applications.

In modern receiver systems, signals are recorded by digital *sampling*, producing a series of discrete samples rather than a continuous signal record. The available sample rate is typically limited by the speed of the associated analogue-to-digital converters, the available storage space and, in the case of scanning multiple channels, the retuning rate of the receiver. It is then important to determine a sample rate that represents the signal sufficiently accurately for the application in hand.

For path loss modelling, the interest is mainly in determining the local mean of the signal, removing the fast fading component while providing a high-confidence estimate of the underlying power associated with the overall path loss and shadowing processes. This implies that all of the samples gathered must be taken within a time period over

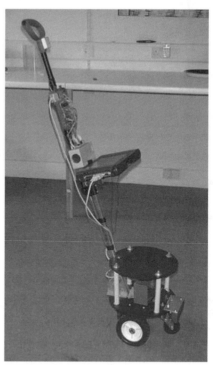

Figure 7.20 Mobile robots employed for indoor channel measurements.

which the mobile receiver is well within the shadowing correlation distance (the distance taken for the normalized autocorrelation to fall to 0.37); otherwise the local mean will not represent the shadowing variations adequately, which loses the detail of the system coverage that the test is aimed at revealing. There must also be enough samples so that the receiver noise floor does not excessively affect the estimate. On the other hand, it is important that the samples taken are not so closely gathered that they have a high probability of being in a fading null, or peak, which will produce a significant over- or underestimation of the local mean. In this application only the signal amplitude is of direct relevance, although sampling of both in-phase and quadrature components simultaneously will effectively produce twice as many independent samples of noise and of fast fading within the same distance.

The local mean is an *average* of the received signal strength, often used for path loss modelling. For this, we are not interested in the exact rapid variations of the mobile signal – this is handled statistically, as the nature of mobile radio signals is random. The description of these fading characteristics is beyond the scope of this book. For further details, refer to Saunders and Aragón-Zavala (2007).

The example plots shown in Figure 7.21 illustrate the effects of undersampling and oversampling a signal when the local mean is to be obtained. The message here is that the signal must be sampled at the right sampling rate – much quicker sampling rates will overload the receiver's memory and fast fading (very rapid signal variations) will also be captured and must be removed. If not enough samples are taken, shadowing (slow

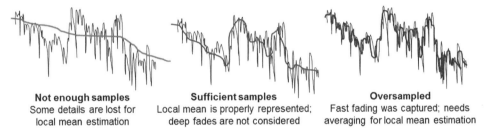

Not enough samples	Sufficient samples	Oversampled
Some details are lost for local mean estimation	Local mean is properly represented; deep fades are not considered	Fast fading was captured; needs averaging for local mean estimation

Figure 7.21 Signal sampling cases for in-building mobile radio signals.

fading) characteristics will be lost and therefore there is a high risk of not producing an accurate enough model.

While oversampling is normally not considered to be a serious issue affecting the measurements, since the excess of samples and consequently the fast fading can be removed by applying suitable averaging methods, undersampling can be a problem, as there is no way to get shadowing effects if enough samples were not taken. For the latter, a revisit to the site needs to be done and more measurements at a higher sampling rate would be required.

For estimating the local mean signal power for path loss modelling, there are various estimators that can be used, as follows:

- Median, which consists of taking the 50% value of all the data set within the interval of interest.
- Decibel mean, where averaging is performed in decibels directly. For example, two samples at −85 dBm and −93 dBm would be averaged in dB as simply: $-(85 + 93)/2 = -89$ dBm.
- Linear voltage mean, which involves performing the average in volts.
- Linear power mean, where the averaging is performed in watts. For example, for the previous samples, to perform the average, both values must first be converted to watts: −85 dBm is 3.162 pW whereas −93 dBm is 0.501 pW. Thus, the average in watts is: $(3.162 + 0.501)/2 = 1.8315$ pW. If we convert this figure back to dBm we have −87.4 dBm.
- Optimal estimator, which uses a weighted average of samples. This is slightly more complex than the previous estimators but seems to yield the highest accuracy.

Figure 7.22 shows a comparison of these methods to estimate the local mean, which shows how the error on either side of the mean for each estimator varies with the number of samples used for 90% confidence. It should be recalled that all of these values assume Rayleigh fading statistics and it is assumed that all samples experience statistically independent fading. Further details of the above-mentioned averaging methods can be found in Saunders and Aragón-Zavala (2007).

Table 7.2 shows the minimum electrical distance for each of the estimator types. In some cases this electrical distance may be typically some tens of metres for outdoor environments. In such cases, which are most likely to occur at lower frequencies, selection of a good estimator is important to avoid averaging the detail of the shadowing information. For in-building environments, where the shadowing correlation distance is of the order of a few metres, it may be impossible to meet this criterion and a

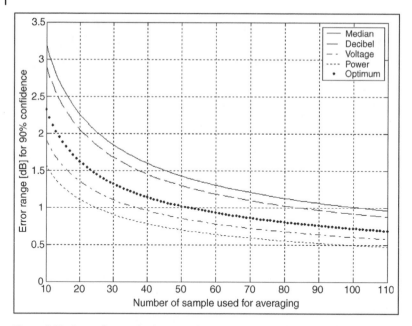

Figure 7.22 Averaging methods comparison.

Table 7.2 Independent sampling limits.

Estimator	Average length for independent sampling	
	90% confidence	**95% confidence**
Mean	39λ	55λ
Decibel mean	32λ	46λ
Linear voltage mean	14λ	20λ
Linear power mean	10λ	11λ
Optimal	20λ	28λ

compromise between averaging fast fading and not disturbing shadowing must be reached. One approach is to average all the samples in a small area rather than along a line, thus defining what is known as a *bin*.

Another approach to determining appropriate sample rates for channel sampling is to consider the first *Nyquist criterion*. It can simply be stated as follows: for a band-limited signal waveform to be reproduced accurately, each cycle of the input signal must be sampled at least twice. In the context of a narrowband fading signal, this implies that at least two samples per wavelength are required. Under such conditions, an interpolation procedure can be applied to calculate the values of the original, continuous signal with any degree of accuracy.

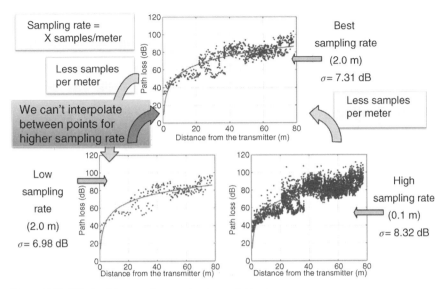

Figure 7.23 Effects of incorrect averaging in model tuning.

There is a relationship between the maximum speed v_{max} at which data are collected and the sampling frequency f_s:

$$f_s = \frac{v_{max}}{0.38\lambda} \qquad (7.5)$$

For accurate results, this maximum speed should not be exceeded. For in-building work, this is rarely an issue; however, for outdoor data collection this has to be taken into account by limiting the driving speed.

To assess how important it is to consider the appropriate data averaging window on model tuning accuracy, Figure 7.23 shows three examples of the same collected data but averaged in a different way. Notice how if proper sampling has been done, fast fading is removed without eliminating shadowing variability (given by σ). The power law model, shown as a red line, is the same for all figures as the best fit is based on the average path loss.

7.6.4 Documentation

It is essential to document the survey in a form that will allow the measurements to be used at a much later date, perhaps years later, with all important parameters and conditions recorded. A survey completion form is typically designed for the measurement campaign to facilitate this. It should typically include:

- Site details, contact names and access arrangements. This is usually handy for future surveys, which may need to be conducted at the same site.
- The type of survey being conducted.
- Details of which floors are to be surveyed and the corresponding data filenames.
- The purpose of the survey; for example propagation model calibration, existing coverage, design survey, etc.

- The building description; for example corporate office building, airport, railway station, etc.
- The on-site dates and time of surveys along with which floor they were on and the names of the associated data files and photographs should be completed.
- A complete list of the serial numbers of the measurement equipment should be completed.
- The form should be signed and dated by the radio field engineer.
- The survey completion form should be scanned and stored with the project data.

Bear in mind that, as for any other project in engineering, the habit of documentation is essential; especially for radio measurements that you conduct on a site that you are just visiting for a few hours, this becomes even more important. Suppose you are doing measurements in an airport hundreds of kilometres away from your company premises; if you did not document all in detail, how do you expect to obtain specific details of antenna positions, heights, building characteristics, etc.? You will need to go back! This is not too bad if it is just a matter of travelling back to the site, but what if you were only granted access for that day? You might need to wait weeks and in some cases months to go back!

7.6.5 Walk Test Best Practice

In order to properly characterize a building, the walk routes should be carefully selected. The following guidelines may help:

- The walk route should include both line-of-sight and non-line-of-sight situations.
- Data should be collected over paths that include penetration through multiple walls of various construction types and also through floors above and below the test antenna location.
- The walk should cover the area of interest fairly evenly, walking all the way across rooms where practical, rather than only close to the edges.
- The walk should be in straight lines given the way-point navigation and multiple routes should be conducted in large open areas, typically with around 10–20 m between routes in a grid or zigzag pattern.
- Walk through as many doorways as possible; that is if a room has two different entrances, walk through each door at least once
- Walk smoothly at a constant pace to minimize the navigation error.
- If it is desired to directly compare multiple instances of the same route, then consider placing markers on the floor to act as an accurate reminder of the way-points used.
- Survey antennas should be low-gain omnidirectional antennas to illuminate the whole environment, unless a specific directional antenna characteristic is of interest.
- An outdoor measurement route should usually be conducted to determine signal leakage from the building.
- Survey for different types of environment (e.g. cluttered, open, densely populated, etc.).

There are occasions in which, although a predefined route has been chosen, during the day of the measurements it is not possible to have access to a specific area in the building and therefore the route plan needs to be adjusted accordingly. For example, suppose you are doing radio testing in an international airport and it turns out that during the day you

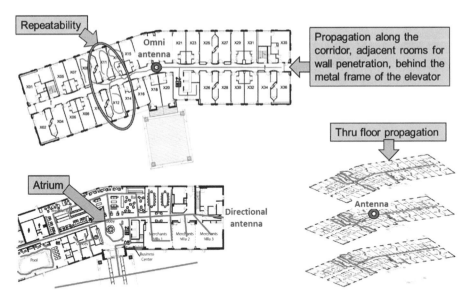

Figure 7.24 Walk route selection example.

had been granted access, an airline VIP lounge is closed and there is no way to measure there. The radio engineer should be capable of making decisions quickly and on-the-move should any circumstance similar to this occur. A viable alternative would be to choose a contiguous room or space where a similar environment exists and therefore measurements in this area could be of great value.

See in the example of Figure 7.24 how walk test routes have been selected for the same floor and for through-floor propagation. Walk routes have been included in many contiguous rooms to indicate repeatability and to have more samples when characterizing wall losses. An atrium in the ground floor (indicated below) shows how the walk route included an atrium. A walk along a long corridor has also been included. Finally, for through-floor propagation, by having an omnidirectional antenna in the middle floor two walk routes have been included in the floors above and below.

The use of insufficient or inappropriate walk test routes may bring as a consequence errors in the model tuning, leading to signal prediction errors. This is shown in Figure 7.25 where an 'only far' route and an 'only close' route have been selected. Note how signal prediction can significantly change depending on which of the routes is chosen, if the model is to be tuned using these measurements. The best approach here is, if possible, to select walk routes that are representative for the different environments in the building in order to gather valuable propagation information. This is sometimes not possible as access may not be granted to all desired test locations, but at least awareness of the consequences of not testing there needs to be made.

7.6.6 Equipment Calibration and Validation

Equipment calibration and system validation are essential ingredients for any in-building measurement campaign, which should not be overlooked, and a fair amount of time and resources must be dedicated to both if accurate measurements are to be produced.

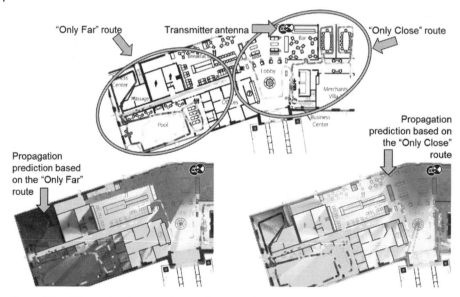

Figure 7.25 Effects of an inappropriate walk route selection.

Equipment *calibration* is a procedure that aims to suppress the impact of measurement inaccuracies due to several factors that may affect operating specifications of equipment. Calibration is very important for any accurate propagation work, and if not performed periodically and prior to measurements, errors cannot be measured or accounted for, leading to totally incorrect results. Calibration is a periodic procedure that is typically done *at least* every 12 months, and could be more frequent depending on the type of equipment that is to be calibrated.

On the other hand, equipment is *validated* by an accuracy assessment of the system integration operation; this means that all the equipment is validated when interconnected to other elements within the measurement system. *Validation* is the process of ensuring that the complete measurement system, from transmitter via cables and antennas to the receiver, and the data logging system deliver reliable and accurate results. It is typically done at least every 3–6 months depending on how often the equipment is used, or before very important measurement campaigns.

Amongst the most important issues related to calibration/validation of radio equipment, we can mention:

- Transmitters (output power, frequency, power supply, operation in long periods of time)
- Receivers (signal injection, dynamic range, frequency, fading)
- Passive elements (cable loss, antenna pattern, attenuators, connectors, splitters, etc.).

This means that special care must be taken with these issues and calibration and validation routines must include these elements. This applies to both outdoor and indoor measurement campaigns.

Calibration is an essential procedure to be performed for every measuring equipment since all of the propagation model accuracy evaluation and tuning processes depend on a good set of measurements, which otherwise would not be possible to get without the use of carefully calibrated equipment. There is no point in having a good model if the measurements used to evaluate or calibrate it are flawed – its parameters will be wrong anyway!

One of the risks in using a logarithmic scale is that belief of 'only a few decibels' will not have any impact on the accuracy of the work. Just to give an example, 1 dB offset in measurement equipment can mean 12% more or less antennas in a design – and you will never know which it is, more or less!

Calibration aims at ensuring that individual equipment is working within specifications given by the manufacturer. For example, if an 11 dBm output transmitter is used for CW testing, it is expected that the transmitter outputs 11 dBm within a tolerance of 0.1% (assuming this is the specification given by the manufacturer). If the equipment has not been calibrated, the transmitter may transmit 10 dBm (even if the display in the transmitter indicates 11 dBm). This 1 dB of error will be carried to the actual measurements and affect the results greatly, and the worst effect is that engineers will not notice it! Likewise, if the transmitter indicates a carrier frequency of 913.8 MHz and it has not been calibrated, it might be the case that it actually transmits a carrier at 913.45 MHz, well out of the expected frequency and of the tolerance limits given by the manufacturer.

For an in-building system integrator, these specifications will usually exceed the manufacturer's ones, making sure the equipment is not 'pushed to operate in the limits'.

The main issue with validation is to ensure that signal strength is accurately reported, both given a transmit power setting and a receiver reading. To confirm this, very well calibrated power meters are needed for the transmitter, and a calibrated signal source (typically a signal generator) for the receiver is also employed. The sources need to be calibrated to a higher accuracy than is required from your measurement equipment.

On the other hand, frequency accuracy is also validated, within a predefined range according to specifications, since frequency drift may be present under certain atmospheric conditions (an increase in the temperature, for example).

Since measurement campaigns tend to last from a few minutes to several hours, stability over time is also of interest to be validated. Typically, transmitters change their power significantly in the first 10–30 minutes after switch-on. This needs to be characterized and taken into consideration. A common practice is to switch the transmitter on and leave it for at least an hour for its output power to stabilize.

Finally, reflections from passive components characterized by the VSWR need to be validated, to estimate the amount of power that is being reflected. This applies for antennas, feeders, connectors, etc.

There are some components that cannot be validated. For example, antennas and moving parts of the system are not included in the validation exercise. Dynamic issues like sampling rates and averaging are also not possible to be validated, and even sometimes the entire system chain (transmitter and receiver) cannot be validated in one go. For these cases, and to overcome the limitations of not validating the system while ensuring we minimize measurement errors and uncertainties, set the equipment up exactly as used in practice, including all software options and equipment setting, in a clear open environment; for example big car parks with no cars can work well. Next, repeatedly walk a simple well-characterized route and derive the appropriate offsets and adjustments required.

7.7 Model Tuning and Validation

Indoor propagation *model tuning* consists of adjusting model parameters that depend on specific characteristics (materials, partitions, clutter, etc.) to account for building or site details that otherwise could not be modelled. This process is performed using radio measurements on the areas and places of interest, obtaining through the use of these measurements valuable information related to the site. Bear in mind that although this is necessary in the process, measurements are not always available or it can be impractical to perform an excessive amount of them, for which radio designers should have a very clear picture of how many measurements are required.

Tuning empirical path loss models is usually the starting point for the rollout of any network. For example, to find the parameters n and K in the power law models described in Chapter 5, measurements can be conducted and, after applying linear regression to these, the values for the path loss intercept and path loss exponent can be found for a specific building or site.

By carefully using radio measurements to tune propagation models, we can reduce the number of measurements needed to design a site and hence produce a higher precision result than 'default' modelling alone – it is difficult to generalize unique parameters that could apply to a model without tuning the model with measurements. Thus, the result is an optimum trade-off between the benefits and costs of both approaches, saving time and money in the design and implementation.

On the other hand, *model validation* is the process to validate signal strength predictions made using propagation models against measurements. This is often performed after the system has been designed and base stations have been deployed around the site, to verify that signal levels and coverage targets are met. It is important to emphasize that different sets of measurements should be used for tuning and validation.

In summary, the model and validation process consists on the following aspects:

- Import measurements to be used for tuning and validation.
- Trace adjustments.
- Select appropriate measurements for tuning, since erroneous data can yield incorrect predictions.
- Some measurements are reserved for testing the accuracy of the tuned model, since it is not correct to use the same measurements for tuning and validation.
- Manual tuning of the wall materials, as they strongly impact the propagation mechanisms involved.
- Manual tuning of the floor materials, for through-floor predictions.
- Perform tuning of model parameters.
- Examine the tuned coefficients.
- Validate model accuracy with the test measurements; that is the data set reserved for validation only.

So far, the pros and cons of modelling versus measurements in the design process have been examined and all the benefits that propagation models bring to an in-building design have been highlighted. It is clear that although propagation models can yield more accurate predictions, if erroneous data are used to 'tune' the model parameters, significant errors can also be produced in the predictions.

By carefully using radio measurements to tune propagation models we can reduce the number of measurements needed to design a building, and hence produce a higher precision result than 'default' modelling alone – it is difficult to generalize unique parameters that could apply to a model without tuning the model with measurements. Thus, the result is an optimum trade-off between the benefits and costs of both approaches, saving time and money in the design and implementation.

7.7.1   Measurements for Model Tuning

There is no need to walk absolutely everywhere in the building; this would take an enormous amount of time and resources and would not necessarily bring more benefits to the predictions. However, sufficient measurements need to be collected in order to tune the model accurately. Note that a separate set of measurements is required for model tuning and for model validation (Aragón-Zavala *et al.*, 2006). The same data set cannot be used for both, as biased results would then be obtained, leading to an 'almost perfect' propagation model tested against the areas for which the same measurements were used for tuning.

The question here is: how much is enough? Some guidelines would be:

- Represent as many environments (materials, partitions, etc.) as possible within the building.
- Extract valuable information from challenging environments such as atriums and corridors – model tuning will need these data to represent these effects more accurately.
- Gather data to model wall and floor loss factors.
- Sampling rate should be adequate to represent local mean statistics (if only this is required).

Once the measurements have been imported from the collection tool, it is very important to perform a measurement data check to make sure the levels that are visualized are congruent and seem correct. Depending on the postprocessing and design software, different approaches can be followed. If using the iBwave design (iBwave Solutions Inc., 2016), this can be normally performed visually using a colour palette indicating signal strength values and the walk route, as depicted in Figure 7.26. If erroneous samples are detected in a large area, measurements should be repeated there.

For *empirical model tuning*, detailed measurements are made of the path loss for sites in a range of environments. This is particularly important for in-building design due to the large variability of building environments and materials. Then models of the propagation path loss are fitted to these measurements for later applications within radio planning tools.

Figure 7.26 shows a scatter plot of measured data with best-fit approximations for a power law model and for a free-space loss model. The spread about the mean value in the plot, also known as shadowing, as seen in Chapter 4, is due to the non-distant-dependent losses, such as:

- Transmission loss from walls
- Reflections
- Diffractions from corners
- Variabilities in the measurement process.

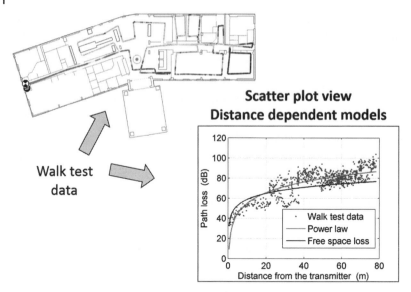

Figure 7.26 Visualization of the walk test data.

7.7.2 Factors Affecting Model Tuning

There are many factors that can affect model tuning performance, which are described as follows:

- Propagation mechanisms, which may affect especially the areas that could not be surveyed and therefore make the model underestimate or overestimate path loss in these zones.
- Antenna locations, as their choice has a deep impact on the amount of useful information that can be extracted to calibrate the model (wall/floor factors, building materials, etc.).
- Extent of surveyed data, as when insufficient locations were collected, there is not sufficient information to gather all the relevant parameters of the model.
- Averaging, for oversampled data, as when not done correctly, can 'smooth' the data in excess or leave fast fading still present.
- System calibration, as when uncalibrated equipment was used to collect the measurements, it is impossible to determine measurement errors and inaccuracies and to quantify them.

7.7.3 Impact of Having Insufficient Measurements for Tuning

To assess the impact of having an insufficient number of measurement samples in model tuning, Figure 7.27 shows various examples of outcomes having a different number of measurements. The figures correspond to:

- All survey data used for tuning are included.
- Only one measurement route is included (21% of the total survey data).

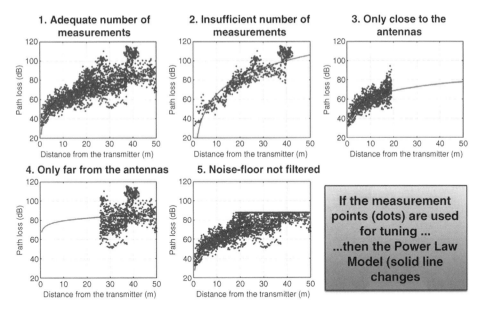

Figure 7.27 Examples of measurements used for model tuning.

- All survey data used for tuning close to the transmitter are included.
- All survey data used for tuning far from the transmitter are included.

All survey data used for tuning are included. Noise-floor (minimum signal level that can be received) was assumed at 87.5 dB. Noise-floor was not reached in any of the survey data. However, this figure shows how the scatter plot would look if the noise-floor were reached at 87.5 dB. To produce this figure all points with a greater path loss were manually moved to the noise-floor path loss, at 87.5 dB.

The effects of insufficient measurements for model tuning also can be seen by a visual inspection of the predicted signal strength plots. The upper plot in Figure 7.28 shows a prediction with a sufficient number of measurements; we see all the areas with reasonable values of signal strength. Now let us examine the plot below with an insufficient number of measurements. Refer back to the scatter plots in Figure 7.27. Clearly the NLOS exponent in this case is much higher and therefore a very low signal strength is measured on these locations – as expected since there was no way to determine a more approximate value for such an exponent.

If measurements are taken *only close to the tested antenna* and not far away, the path loss exponent obtained is rather optimistic and does not account for additional wall losses for far away locations; therefore, the predictions obtained overestimate the signal strength values that would have been obtained having sufficient samples also far away from the antenna. Refer to the scatter plots in Figure 7.27.

On the other hand, if measurements are taken *only far from the tested antenna* and not close to it, the propagation prediction far from the tested antenna in some areas is similar to the prediction if survey data were available close and far from the antenna. However, note that the prediction differs in other areas far from the tested antenna. The propagation far from the tested antenna is dependent on the signal strength in areas

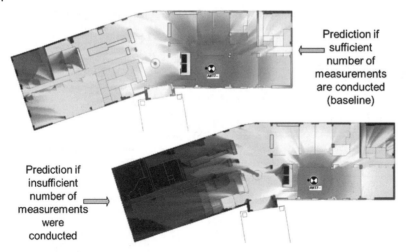

Prediction if sufficient number of measurements are conducted (baseline)

Prediction if insufficient number of measurements were conducted

Figure 7.28 Visual impact of insufficient measurements for model tuning.

close to the transmitter antenna. Therefore, for an accurate prediction we need to tune the propagation model with survey data from both close and far from the tested antenna.

The noise-floor is another important aspect to take into account when performing model tuning. Sometimes, measurements are collected at far distances from the transmit antenna that have reached the receiver noise-floor. These samples therefore are of no use as they do not represent real values, as seen in the last plot shown in Figure 7.27.

7.8 Conclusion

Radio measurements are an essential ingredient of the overall in-building design process. It is rather difficult to visualize their importance if the benefits and need for measurements have not been analysed in depth. A clear understanding of the value of measurements for in-building design projects has been developed through the review of the different types of in-building measurement systems and their characteristics, appreciation of the impact of inaccuracies while performing measurements and the various guidelines and recommendations for in-building testing, which will assist the in-building radio engineer in his or her design.

During this chapter, the basics of propagation measurements and their classification, according to their characteristics and the type of equipment utilized for these, have been presented. There are important issues to take into consideration, such as measurement inaccuracies, sampling and navigation, that were reviewed to make sure they are accounted for when planning and conducting a measurement campaign. A brief overview of equipment for in-building measurements was also given, discussing some technical details of relevant components and equipment, such as transmitters, receivers, feeders and antennas. The role of having a well-documented measurement campaign was also discussed, especially when postanalysis of the collected data is to be performed: there is a need to have a good reference. Finally, tips and tricks for effective measurements were presented, along with some walk route guidelines and recommendations.

There is a very relevant issue not to be overlooked here, and it corresponds to calibration. For most cases, the measurements reported by the equipment out of the box will not be correct and some offsets will need to be applied, no matter what specifications are required from manufacturers. This happens not because the equipment is faulty but due to some of the many variables affecting the measurements, which lead to uncertainties difficult to predict and measure. Thus, simple 'correction factors' can apply to any given measurement system at a given frequency and configuration. There is then some important organizational work to ensure that these values are carefully recorded and applied, and that values for all given equipment can be tracked over time.

Prior to any propagation measurement campaign, make sure all transmit and receive equipment chains have been validated and that individual key devices have been calibrated. Although calibration and validation certainly take time, they can save time afterwards and definitely will guarantee that the measured data are as accurate as possible.

References

Aragón-Zavala, A. (2003) In-building cellular radio system design using measurements, PhD dissertation, University of Surrey, UK.

Aragón-Zavala, A., Brown, T.W.C. and Castañón, G. (2016) Polarization and effects on hidden node/shadowing margin for TVWS, *IEEE Transactions on Broadcasting*, **62** (1).

Aragón-Zavala, A., Jevremovic, V. and Jemmali, A. (2014) Auto-correlation and cross-correlation analysis for sport arenas at 850 MHz and 2.1 GHz, in 8th European Conference on Antennas and Propagation (EuCAP), Amsterdam, Netherlands, pp. 428–432, April 2014.

Aragón-Zavala, A. and Saunders, S.R. (1999) Autonomous positioning system for indoor propagation measurements, in Proceedings of the 50th Vehicular Technology Conference, VTC Fall 99, Amsterdam, Netherlands, pp. 785–789, September 1999.

Aragón-Zavala, A., Belloul, B., Nikolopoulos, V. and Saunders, S.R. (2006) Accuracy evaluation analysis for indoor measurement-based radio-wave-propagation predictions, *IEE Proceedings on Microwaves, Antennas and Propagation*, **153** (1), 67–74.

iBwave Solutions Inc. (2016) URL: http://www.ibwave.com.

Monk, N. and Wingbier, H.S. (1956) Communications with moving trains in tunnels, *IRE Transactions on Vehicular Communications*, **7**, 21–28.

Radi, H., Fiacco, M., Parks, M. and Saunders, S.R. (1998) Simultaneous indoor propagation measurements at 17 and 60 GHz for wireless local area networks, in Proceedings of the 48th Vehicular Technology Conference, Ottawa, Canada, pp. 510–514.

Saunders, S. and Aragón-Zavala, A. (2007) *Antennas and Propagation for Wireless Communication Systems*, 2nd edition, John Wiley & Sons, Ltd, Chichester. ISBN 0-470-84879-1.

8

Capacity Planning and Dimensioning

Capacity determines to a great extent the service availability and hence impacts on the quality-of-service of a radio network. For example, to be able to establish a call or a successful data connection anytime, anywhere in a building, from a user's point of view, is a subjective measure of network quality. Nowadays with the increasing growth of data traffic, this necessity can be aggravated if users are not able to establish reliable connections at the speed they require, thus affecting a user's experience. Both situations are definitely reflected in an operator's revenue, as more satisfied customers will attract more users and, likewise, poor quality-of-service, measured as blocked calls, does not encourage customers to continue utilizing the network.

   The calculation of the number of channels needed to cater for the capacity requirements of the users in a specific building is to be reviewed here, once the number of users and type of users is determined, as well as the traffic profile that is expected. This approach has also to account for possible system expansions and expected growth in the number of users or subscribers over time. There are special considerations that need to be made for indoor systems and that differ from those used in outdoor networks, in particular for those challenging venues having a large number of users for peak time periods (stadiums, airports, etc.). These will be emphasized in this chapter.

   In general, every indoor wireless technology has its own characteristics and approaches to plan for capacity. This chapter aims at giving an overview for capacity dimensioning for voice and data traffic and planning for any indoor wireless technology.

8.1   Introduction

For indoor scenarios where hotspots of population are expected, capacity dimensioning plays an essential role in network performance. For voice, as a limited number of channels are available for use within the cell, if the system is not properly dimensioned, additional users to those that can be served will not be allowed to connect to the network or will be blocked. Moreover, for UMTS networks, if capacity is incorrectly estimated, more "unexpected" users will generate more interference to that accounted in the link budget and hence reduce the cell radius significantly, as the noise rise will increase in the uplink. On the other hand, for the downlink, if more users are served than those dimensioned for, the cell radius will also be reduced as more power is needed to be distributed amongst a higher population. For LTE networks, which have much more intensive use of data traffic, capacity requires a careful dimensioning and proper data

Indoor Wireless Communications: From Theory to Implementation, First Edition. Alejandro Aragón-Zavala.
© 2017 John Wiley & Sons Ltd. Published 2017 by John Wiley & Sons Ltd.

offloading to alleviate excessive peaks that, for example, a cellular network may not be able to cope with.

Capacity is also a very important performance parameter for Wi-Fi networks. Here, the number of users has a great impact on network throughput, due to the protocol that Wi-Fi devices use to send packets over the air interface: sense the channel for activity, then transmit if the channel is free, wait for acknowledgement and retransmit after a certain time if no acknowledgement is received. Obviously, the more users are in the network, the more likely that collisions may occur and hence even though the data rate can be apparently fast, the actual throughput is greatly reduced.

The layout of the building has also a significant effect on the way capacity is planned. Walls and partitions may help indoor *zones* to be created for which a proportional number of resources can be allocated according to traffic demands. A different approach clearly needs to be taken when planning capacity for a corporate office building, an airport or a stadium. However, one common consideration that is taken is where the *hotspots* of traffic are located. Although this is a term often employed in Wi-Fi networks, the same principle holds for other wireless systems.

There is a relevant consideration to be made depending on the type of services offered to the customer when designing for capacity. Voice traffic is considered a real-time service, that is to be delivered with the minimum possible delay. The quality of service is greatly affected not only if the voice call is rejected to enter the network but also if the signal is greatly delayed – little tolerance is allowed. On the other hand, data traffic normally is not affected if data are delayed; for example you do not really care too much if an email arrives a few seconds or even minutes late. Thus, different methods need to be employed depending on the type of traffic considered – traffic categories and some characteristics are further explained in Section 8.3.5.

Therefore, when designing an indoor network, an accurate knowledge of the required number of channels can prevent the system from having high blocking and additional unexpected noise rise, as well as the propagation characteristics of the building, all as an integrated approach. The methods involved in determining the necessary number of channels in accordance with the building population and the distribution of these users within the building are defined as *capacity planning*.

Bear in mind that, in the past, estimation of spectrum requirements were considered as a framework focusing on a single system and market scenario, something that nowadays is no longer applicable (ITU, 2013). New models have to be used that allow for consideration of spatial and temporal correlations among telecommunication services, taking into consideration the market requirements and network deployment scenarios. This has been mainly motivated by the convergence of mobile and fixed telecommunication and multinetwork environments as well as the support of attributed systems such as seamless interworking between different complementary access systems.

8.2 An Overview On Teletraffic

Teletraffic is the application of traffic engineering theory to telecommunications, using the basic knowledge of probability and statistics including queuing theory, the nature of traffic, practical models, measurements and simulations to make predictions and to

Figure 8.1 A.K. Erlang, Danish mathematician who developed teletraffic theory.

assist in the planning of telecommunication networks. This applies to both wired and wireless networks.

The field of teletraffic was created after the work of a Danish statistician, Agner Krarup Erlang, shown in Figure 8.1, in the beginning of the twentiethcentury. Erlang published his work "The Theory of Probabilities and Telephone Conversations" in 1909, where he was able to calculate how many lines were required to service a specific number of users, given a certain availability of the lines for the users.

Erlang's work and teletraffic was originally developed to be used in circuit-switched networks, but in recent times it has also been extended to packet-switched networks. Although teletraffic is applied for both, the theory has evolved to account for the particularities of broadband networks, where voice is not the only service to be offered, but many other services and connections need to be available. Alternative traffic models for broadband in addition to those proposed by Erlang had therefore to be proposed, which include parameters such as *mean demand* and *traffic variability*. These models are out of the scope of this book. For a more detailed study on teletraffic, refer to Schwartz (1987).

8.2.1 Trunking

In modern communications, and specifically in telephony, *trunking* is a concept by which a communications system can provide network access to many users by sharing a set of channels (or lines) instead of providing them individually, on a dedicated basis. In addition to cellular, this concept also applies to private mobile radio systems (PMRs), in which many users have access to a limited number of channel resources. This is the key of the success of many telecommunication systems, such as cellular and PMR, since only a limited number of channels are required to service an area (e.g. a cell) on the assumption that the probability of all users utilizing the network at the same time is low. In fact, networks are dimensioned to account for some of those users to be rejected or blocked, as will be explained later.

Those wireless communication systems that make use of trunking are also called *trunked radio systems* (TRSs). These systems provide greater efficiency than conventional radio systems, at the expense of greater management overhead. TRSs pool all the channels into one group and use a site controller that assigns incoming users to free channels as determined by the TRS protocols. Depending on the specific wireless technology, these entities (site controller, channels, etc.) may have different names and functions.

8.2.2 Loss and Queue Networks

There are two ways in which traffic can be handled in a network, depending on how incoming users are handled when all channels are busy:

Loss systems, where calls that cannot be handled are given an equipment 'busy tone' and are rejected. For example, in cellular systems, there are occasions where the mobile phone indicates a full coverage level and there is sufficient signal strength but when a user attempts to make a phone call, a 'busy tone' is heard. Loss systems are normally those used for voice traffic.

Queue systems, where calls that cannot be handled immediately are queued and delays are specified, after which the call is rejected if no available channels are in place. Data traffic can be handled in queue systems.

8.2.3 Busy-Hour

Capacity dimensioning for wireless networks, and in particular for cellular systems, is performed during the times of the day where most traffic is likely to occur. In a communications system, the *busy-hour* is then referred to as the sliding sixty-minute period during which the maximum total traffic load occurs in a given 24-hour period.

It is expected that the 'peak' in traffic occurs during the busy-hour and therefore sufficient channels should be allocated for this purpose to guarantee a given QoS.

For highly-populated environments such as airports, to account for capacity dimensioning, the busy-hour is chosen according to airport daily activity as well as season of the year. Summer is normally a very busy season and thus planning should be done for busy-hour traffic during this period. There are other environments that exhibit a more uniform behaviour, such as corporate buildings, for which traffic is expected to be fairly regular throughout the year. Shopping centres, on the other hand, have Christmas and special seasonal dates as the busiest times of the year.

There are other wireless systems that exhibit a similar 'busy-hour' kind of behaviour and for which special capacity considerations should be made. For example, Wi-Fi systems in a university campus are busier during term times and become more congested in specific periods of time, such as exams. Regardless of the system in consideration, the same design principles apply, and should be employed to account for sufficient radio resources.

8.3 Capacity Parameters – Circuit-Switched

In the design of a circuit-switched network for capacity there are various parameters that should be used, which are described in this section. Note that these parameters apply for voice traffic only and cannot be used for data traffic.

8.3.1 Blocking

As we have discussed earlier, for trunked systems capacity is determined by a finite number of channels available for all users within a building on the basis that not all users make a phone call at the same time and thus utilize all channels. In the event that this happens, the cell is said to be *blocked* and a degradation in the quality-of-service to users is experienced. In most modern cellular systems, when a call is rejected to access the network, a busy tone is sent to the user and the call is immediately terminated.

Blocking represents a major problem, especially for indoor systems. As an example, suppose an international airport has various macrocell sites around and relies on macrocell penetration to provide service to customers. Even if the received signal strength is sufficient in most of the areas of the airport according to operator require-ments, capacity may not be enough to accommodate all users, especially during busy hour times or seasons. Not only is the airport left with blocked calls but users outside and in the surroundings of the airport will also suffer from blocking as well. This is one of the reasons why dedicated indoor cells need to be deployed in such busy venues, even though coverage may be considered as sufficient from outdoor cells.

8.3.2   Grade of Service

Also denoted as GoS, grade-of-service is the probability of a call in a circuit-switched network being blocked or delayed for more than a specific interval. The GoS is normally expressed as a percentile or a decimal fraction; for example 2% or 0.02. GoS is always referenced to the busy-hour or when traffic is more intense.

To determine the GoS of a network when the traffic load and the number of channels are known, the *Erlang-B* formula is used:

$$P_r = \frac{\dfrac{A^C}{C!}}{\sum_0^K \dfrac{A^k}{k!}} \tag{8.1}$$

where P_r is the GoS, A is the total offered traffic in erlangs and C is the total number of available channels. This formula is idealized as it does not attempt to make assumptions about the call request rate and duration, but it is a useful start point for estimating the number of channels required. Erlang-B assumes Poisson arrivals, call durations are either at a fixed length or exponentially distributed and blocked calls are not retried immediately after receiving a busy signal.

It is important to recall that Erlang-B should be used when estimating the number of channels in a cell. The decision of the cell borders should come from radio channel modelling considerations, which account for propagation characteristics.

An Erlang-B graph (Figure 8.2) is often provided to quickly estimate the required number of channels for a given GoS with an offered traffic figure. Note that this graph is only given as an approximation and exact values should be obtained using Equation (8.1).

8.3.3   Traffic per User

Traffic is measured in erlangs: one erlang (E) is equivalent to one user making a call for 100% of the time. A typical cellular voice user generates around 2–30 mE of traffic during the busiest hour of the system: that is, a typical user is active for around 0.2–3.0% of the time during the busy hour. These figures tend to increase for indoor environments, fluctuating around 50–60 mE.

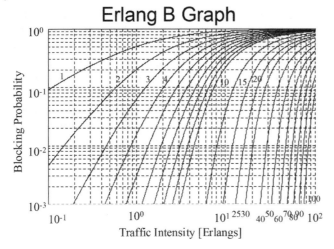

Erlang B Graph

Figure 8.2 Erlang-B graph.

Traffic per user A_u is required if the traffic per cell is to be computed. Therefore, a user traffic profile is often described determined by different traffic categories, as specified in Section 8.3.4, in which an average mobile phone user makes λ calls of duration H during the busy hour, where λ is known as the call request rate and H is the holding time. Hence, the average traffic per user is

$$A_u = \lambda H \tag{8.2}$$

8.3.4 Offered and Carried Traffic

A distinction should be made here between *offered* and *carried* traffic, for which a different use of the Erlang unit is made.

When used to describe *offered traffic*, a value followed by 'erlangs' represents the average number of concurrent calls that would have been carried if there were an unlimited number of circuits; that is if the call attempts that were made when all circuits were in use had not been rejected or blocked.

When used to represent *carried traffic*, a value followed by 'erlangs' represents the average number of concurrent calls carried by the circuits or other service-providing elements, where that average is calculated over some reasonable period of time. The period over which the average is calculated is often one hour (the so-called busy hour for indoor capacity calculations), but shorter periods (e.g. 15 minutes) may be used where it is known that there are short spurts of demand and a traffic measurement is desired that does not mask these spurts. For example, if a radio channel is occupied for one hour continuously it is said to have a load of 1 erlang. Thus, for U users in the cell, the total carried traffic A is given by

$$A = UA_u \tag{8.3}$$

The relationship between offered traffic and carried traffic depends on the design of the system and user behaviour. Three common models are (a) callers whose call-attempts are rejected go away and never come back, (b) callers whose call-attempts are rejected try

Table 8.1 Estimated user loads for voice traffic.

User type	Estimated traffic per user [mE]
Extreme	200
Heavy	100
Normal office	50
Private	20

again within a fairly short space of time and (c) the system allows users to wait in queue until a circuit becomes available.

8.3.5 Traffic Categories

Indoor radio users exhibit distinct behaviour depending on a given *traffic profile*. This traffic profile specifies the way each user will load the wireless system and since these profiles normally change on a daily basis, average values should be taken. Depending on the type of user, some user loads for voice traffic can be considered as indicated in Table 8.1 (Tolstrup, 2011).

This is one useful way to categorize the various types of traffic that can be utilizing radio resources in a building, but it is worth pointing out that wireless service providers and cellular operators may have their own traffic profiles to dimension the network for capacity.

In general, users in a wireless network are distinguished by several factors:

- Service levels (voice, email, web browsing, video, streaming, . . .)
- Technologies (UMTS, HSPA, LTE, Wi-Fi, . . .)
- Spectrum bands (cellular, PCS, AWS, ISM, . . .)
- Traffic (expected usage of network resources).

When defining service types, the characteristics of each technology should be taken into account, in order to establish reasonable traffic profiles. For example, for cellular systems, the radio designer should recall that:

- GSM does not support any data service – perhaps only very constrained and limited using GPRS.
- HSPA does not support voice, only data.
- LTE does not support circuit-switched voice, only VoIP.

Having this in mind, it is possible to create a *subscriber mix*, which is defined as the average percentage of users that use a particular service type. Therefore, for example, for an LTE subscriber mix, it can be assumed that 25% of the traffic is voice, 10% is email, 25% is web browsing, 8% is video conferencing, 12% is video streaming and 20% is data download.

A more detailed classification and explanation will be given in the next section, related to data capacity, which inherently considers, for example, very low data rate services in which voice may be included.

8.4 Data Transmission Parameters

Much like GoS is used for circuit voice user capacity, as described in Section 8.3, there are other capacity parameters used for data user capacity, which are described here. Capacity planning for data traffic needs to be contemplated, especially for UMTS and LTE cellular systems, where a significant amount of the total network traffic is data. For this, Erlang-B calculations cannot be employed, but we must be aware that data traffic will take up capacity that could have been used for voice. If dimensioning for data, data rates for uplink (UL) and downlink (DL) are normally specified, and this depends on the way, for example, operators define typical user profiles for data services; for example 384 kbps DL, 128 kbps UL for a user who normally downloads more data than uploads. Since each operator or wireless provider defines specific traffic profiles for data according to marketing studies and other factors, typical user rates are hard to exemplify and capacity dimensioning needs to be done for very specific cases and scenarios.

8.4.1 Delay

Delay covers the time required to establish the service when a request is made and the time required when information is received. Delay could have a big impact on the user satisfaction depending on the types of service and applications.

Delay or latency contributors are the user equipment, network and the server. Digital mobile signals require a significant amount of signal processing in the terminal and the network before a service is delivered. There are exchanges of signalling between the devices and the network in order to establish a secure and reliable connection. The radio access network (RAN) and the core network (CN) can be optimized and configured for minimum latency. The MNOs are constantly carrying out surveys to reduce the round trip time (RTT) where possible. These are comparatively static latency contributors, which can be optimized by reconfiguration of the network when a new service is introduced.

A significant contributor to latency is the call setup. Depending on the state of the mobile, the call setup can take from 200 ms to 2000 ms depending on the network configuration. Mobile applications development should take account of the way the mobile communicates with the network to minimize the amount of exchanges required before a request is completed.

8.4.2 Throughput

Another important dynamic contributor is the network throughput, that is the capability to handle large amounts of traffic. Throughput depends on the network bandwidth, which is affected by the hardware capacity (the number of channel cards), the backhaul transmission capacity (leased line or Ethernet) and the network noise rise due to traffic load. These have to be properly planned and catered for during the network planning process.

8.4.3 Latency

There are some services that require the delivery of packets with a minimum delay in order not to degrade the quality of service. One of the measures or key performance indicators (KPIs) that is frequently used for this is *latency.* Network latency is a measure of how much time it takes for a packet of data to get from one designated point to another. For example, for service classes with critical timing such as voice, to keep latency to a minimum is essential. In fact, with the imminent deployment of 5G networks in the near future, latency is one of the key parameters that guarantees the expected quality of service for many types of service to be offered, since this parameter is expected to be lower than previous standards. Real-time services are therefore expected, for example, for connected factories (or factories of the future) and thus special attention should be paid to this parameter.

8.5 Capacity Limits

There are two types of capacity limits that a wireless technology may have:

- Hard limit
- Soft limit.

Hard limit is technology limit that constrains capacity and is determined by the number of:

- Time slots per carrier (GSM)
- Walsh codes per carrier (UMTS)
- HSPA codes per carrier (HSPA)
- Resource blocks per carrier (LTE).

When the hard limit in a network is reached, the system will refuse connection until resources become available. These resources are normally categorized as time slots, codes or resource blocks depending on the specific technology.

On the other hand, *soft capacity limit* is influenced by factors related to power: transmit power limitation and interference. Thus, to achieve a certain service level, the user equipment must maintain a certain $SINR$, E_b/N_0 or C/I for that service level. Therefore, for a given $SINR$ at the user equipment, the base station should allocate adequate power to that user, whereby two cases may occur:

- Let us assume there is strong interference close to the UE, so to achieve the target $SINR$ there may not be enough power.
- Even if interference is low, the UE may be quite far away from the base station (Node B), in which case the transmit power may also not be available.

One of the consequences of having a soft capacity limit is that when the QoS target is not guaranteed, the UE is assigned a lower data throughput rate.

8.6 Radio Resource Management

Radio resource management (RRM) is a set of algorithms used for optimal utilization of various parameters such as transmit power, channel allocation, handover criteria, modulation scheme, error coding scheme, etc., to control co-channel interference and other radio transmission characteristics in wireless communication systems. The ultimate objective of RRM is the use of limited radio spectrum resources and radio network infrastructure as efficiently as possible.

RRM deals with multiuser and multicell network capacity issues, to achieve maximum capacity when several users and adjacent base stations share the same frequency channel. These capacity enhancements can supersede those provided by advanced source and channel coding schemes, in particular using dynamic RRM strategies.

For wireless systems such as cellular, WLAN and broadcasting networks, where the limiting factor is often co-channel interference rather than noise, RRM plays an essential role in optimizing the use of spectrum.

RRM is a wide topic that certainly is out of the scope of this book. In the context of various wireless technologies and those related to network planning and dimensioning, an excellent reference is Stasiak *et al.* (2011).

8.7 Load Sharing: Base Station Hotels

One strategy that has been employed in recent years to optimize capacity resources for buildings is *load sharing*, where various network resources can be shared by a collection of neighbouring buildings and hence maximize resourcing. The principle behind this sharing is that different venues and buildings will have distinct traffic profiles that depend on the time of the day, day of the week and the season, and thus load profiles can be combined to increase trunking gain. For example, a shopping centre will have low traffic until opening hours and no traffic at night, having more peaks during weekends than in weekdays. On the other hand, an office building exhibits a similar behaviour to that of a shopping centre on weekdays, but for weekends no traffic is experienced. If radio resources are shared amongst the two buildings, load profiles can be combined and thus the resources of the cell can be much better used.

A *base station hotel* is the technique where coverage and capacity is provided by a centralized base station to buildings located in the same area. This solution provides a low-cost alternative for areas where high capacity demands exist and for which the number of cells need to be increased. The connection between the base station equipment room and each remote site is connected normally using a single optical fibre.

In the centralized base station hotel multiple services can be fed to a central main DAS unit, from which optical fibre is fed to the individual antenna locations connecting remote units to the DAS. Note that both the DAS master and remote units have modular capabilities for new cards or amplifiers to be added if required should network expansion be neededTolstrup (2011).

One of the greatest advantages of this concept is that the central base station hotel could be located in an existing equipment room, thus resolving potential lack of space for remote locations that support multioperator and/or multitechnology solutions. The use of optical fibre allows distances between the central base station hotel and the remote units of more than 15 km to be reached.

In summary, with the use of a base station hotel, capacity can be extended on demand without the need to invest in additional infrastructure at the remote locations but instead add functionality to the centralized base station and easily deploy remote units at the required locations.

8.8   Traffic Mapping

It is common practice for operators to assume even user density across a building, which is sufficiently valid if the building is reasonably small; however, for denselypopulated buildings such as airports and shopping centres, another method to distribute this user density is required. For example, an airport has many areas where user loads can be clearly observed to change as a plane lands and large bursts of passengers walk through the piers towards immigration control, in many cases making an extensive use of their mobile phones. Clearly, this traffic profile should be different to that in an arrivals lounge where people are waiting for passengers to arrive or to baggage reclaim zones. For Wi-Fi networks, throughput can be substantially reduced if high bursts of traffic in wireless hotspots are not accounted for, especially in those environments having uneven user distribution, making use of network resources extensively, such as airports, train stations, shopping centres, etc.

A technique to distribute the building population and traffic was derived from mobility models, producing traffic maps according to user density (Aragón-Zavala *et al.*, 2009). This suggests a much better use of network resources and a better estimation of the required number of channels to avoid blocking, which by no means can be derived simply by examining network statistics and previous traffic per existing cell. The difficulty for allocating enough channels per cell and moreover to determine the number and location of such cells is essential and, along with coverage predictions from the building in consideration, this can be performed.

Traffic mapping was conceived as a mapping exercise that was developed to allocate radio resources according to traffic demands, and how channels are allocated in accordance with such uneven user distributions across a building, mainly for voice in GSM systems. It was successfully used at the beginning of the millennium (2000–2003), especially to dimension traffic resources for airports. Nowadays with the increase of data traffic, these methods are insufficient and other methodologies need to be used, which will be described in Section 8.9.

8.9   Capacity Calculations

Different approaches should be followed to dimension a network for capacity, considering many aspects such as:

- Circuit-switched (voice) or packet-switched (data) services
- Wireless technology (cellular, Wi-Fi, etc.)
- Standards (GSM, UMTS, LTE, HSPA, etc.)
- Traffic forecasts to account for future growth
- Deployment scenario.

The specific method may slightly vary from one technology to another, but in general techniques for voice and techniques for data have similarities regardless of other technical details.

Recently the ITU-R has published recommendations and a methodology to calculate spectrum requirements for IMT based on the new cellular standards, which carry a significant amount of data traffic (ITU, 2013). The methodology is applicable to packet

Table 8.2 Service categories. Taken from ITU (2013). Reproduced with permission of International Telecommunication Union.

Service type	Conversational	Streaming	Interactive	Background
Super-high multimedia	SC1	SC6	SC11	SC16
High multimedia	SC2	SC7	SC12	SC17
Medium multimedia	SC3	SC8	SC13	SC18
Low data rate and low multimedia	SC4	SC9	SC14	SC19
Very low data rate	SC5	SC10	SC15	SC10

switch-based traffic and can accommodate multiple services. Based on this methodology, some guidelines are given in this section.

8.9.1 Service Categories

According to ITU (2013), a service category (SC) is defined as a combination of service type and traffic class as shown in Table 8.2. The ITU lists these service categories using specific numbers (e.g. SC4), and for each service type there are four *traffic classes*: conversational, streaming, interactive and background. These traffic classes are explained in the following section.

8.9.1.1 Service Types
Peak bit rates are used to categorize the service types, for which service demanding similar peak rates can be grouped together. The following service types can be identified:

- Very low data rate, requiring peak bit rates of up to 16 kbps. Speech and simple message services fall within this service type. Applications related to sensor networks or low data rate telemetry are examples of this service type.
- Low data rate and low multimedia, supporting data rates up to 144 kbps. This service type accounts for all pre-IMT-2000 data communication applications.
- Medium multimedia, with a peak data rate of 2 Mbps. Recall that IMT-2000 data communications such as those defined in 3G support this data rate and therefore this service type aims at sustaining compatibility with these applications.
- High multimedia, accommodating high data rate applications up to 30 Mbps including video streaming services, those that are provided with xDSL services for fixed wired communications.
- Super-high multimedia, supporting super-high data rates multimedia applications, currently provided using fibre-to-the-home (FTTH) services in fixed wired networks. Data rates vary from 30 Mbps to 100 Mbps/1 Gbps.

8.9.1.2 Traffic Classes
This methodology applies the traffic classes described in ITU (2003), for which four quality-of-service classes are defined from the user perspective:

- *Conversational*, such as telephony speech. Other applications may require this scheme, such as voice-over-IP (VoIP) and videoconferencing tools. Real-time conversation is characterized by a low transfer time, hence restricting the limit for acceptable transfer delay. If this maximum delay is exceeded, the result may be an unacceptable lack of quality in the conversation or videoconference.

- *Streaming*, which occurs when the user is looking at real-time video or listening to audio. This real-time data flow always aims at a live (human) destination, not a machine. Data flow must be preserved, although there are no requirements on low transfer delay. The end-to-end delay variation must be limited, to preserve the time relation between information entities of the stream. Normally, the highest acceptable delay variation over the transmission media is related to the capability of the time alignment function of the application; thus delay variation is much greater than the one given by human perception.
- *Interactive*, when the end user (machine or human) is online requesting data from remote equipment, such as in web browsing, database retrieval, server access, etc., for human interaction with remote equipment, or polling measurement records and automatic database enquires for machine interaction with remote equipment. For interactive traffic, at the message destination, there is an entity expecting a message (response) within a certain time and therefore round-trip delay is of importance here. The content of the packets must be transparently transferred, having low BER.
- *Background*, when the end user (typically a computer) sends and receives data in the background, as is the case for e-mail, SMS, download of databases and the reception of measurement records. The absence of a parameter at the destination expecting to receive the data within a certain time limit characterizes this scheme, although there is still a delay constraint since data are effectively useless if received too late for any practical purpose. Thus, this scheme is more or less delivery time insensitive.

As can be seen above, the key difference amongst them is how delay-sensitive the application is. For these four traffic classes, conversational and streaming are served with circuit switching whereas interactive and background are served with packet switching.

8.9.1.3 Service Category Parameters

SCs are characterized by parameters that are obtained either from market studies or from other sources. Relevant parameters are (ITU, 2006b):

- User density, in users per km^2
- Session arrival rate per user, in sessions/s per user
- Mean service bit rate, in bps
- Mean session duration, in s/session
- Mobility ratio.

The mobility parameter is used in traffic distribution whereas the other parameters characterize the demand of different service categories. Terminal mobility is related to application usage parameters and is defined in ITU (1999) as:

- In-building
- Pedestrian
- Vehicular.

Mobility requirements depend upon the speed of the mobile stations, for which mobility classes are categorized as follows (ITU, 2006a):

- Stationary, 0 km/h
- Low mobility, speed between 0 km/h and 4 km/h
- High mobility, speed between 4 km/h and 100 km/h
- Super-high mobility, speed between 100 km/h and 250 km/h.

These range limits should be related to typical characteristics of cellular radio networks. Therefore, for application of the mobility classes in the methodology, the mobility classes from market studies are redefined as follows:

- Stationary/pedestrian, speed between 0 km/h and 4 km/h
- Low mobility, speed between 4 km/h and 50 km/h
- High mobility, speed greater than 50 km/h

8.9.2 Service Environment

A service environment is defined in ITU (2013) as a combination of a service usage pattern and teledensity, and represents common service usage and volume conditions.

A *service usage pattern* is a common user(s) behaviour in a given service area and is categorized according to an area where users exploit similar services and expect similar QoS. The following service usage patterns are used:

- Home
- Office
- Public area.

On the other hand, population density and the number of devices per person are relevant factors when considering service environments. Thus, geographical area is divided according to these factors into *teledensity* categories. Each teledensity parameter is characterized by population density and communication device density. Teledensity is categorized into the following:

- Dense urban
- Suburban
- Rural.

Service environments are defined for the combinations of teledensity and service usage patterns as shown in Table 8.3. Examples of user groups and applications of service environments are as follows (ITU, 2013):

SE1. *User groups*: private user, business user. *Applications*: voice, Internet access, games, e-commerce, remote education, multimedia applications.

SE2. *User groups*: business user, small and medium enterprise. *Applications*: voice, Internet access, video conferencing, e-commerce, mobile business applications.

SE3. *User groups*: private user, business user, public service user, tourist, sales people. *Applications*: voice, Internet access, video conferencing, mobile business applications, tourist information, e-commerce.

SE4. *User groups*: private user, business user. *Applications*: voice, Internet access, games, e-commerce, multimedia applications, remote education.

Table 8.3 Identification of service environments. Taken from ITU (2013). Reproduced with permission of International Telecommunication Union.

Service usage pattern	Dense urban	Teledensity Suburban	Rural
Home	SE1	SE4	SE6
Office	SE2	SE5	
Public area	SE3		

SE5. *User groups*: business user, enterprise. *Applications*: voice, Internet access, e-commerce, video conferencing, mobile business applications.

SE6. *User groups*: private user, farm, public service user. *Applications*: voice, information application.

When designing for capacity, spectrum requirements should first be calculated separately for each teledensity. The final spectrum requirements are calculated by taking the maximum value amongst spectrum requirements for the three teledensity areas defined above.

8.9.3 Radio Environment

Radio environments (RE) are defined by the cell layers in a network consisting of hierarchical cell layers, that is macro, micro, pico and hot-spot cells. The cell areas for the different radio environments are used in capacity calculations by the method specified in ITU (2013).

In addition to the limits on cell sizes related to network deployment costs and the spectrum requirement, technical limits exist. The upper technical limit is determined by the propagation conditions, terminal transmit power limitations and delay spread. The lower technical limits for the cell sizes are determined by an increase in unfavourable interference conditions; however, this limit is negligible compared to the limit imposed by deployment costs.

As the deployment of micro, pico and hot-spot cells do not greatly vary between teledensity areas, the assumption of taking the same 'maximum' cell area is made for the spectrum calculation method. Macro cells are different, since the teledensity has an impact on the targeted cell area as well as on the deployment of base stations. Therefore, the cell area of a macro cell is made teledensity-dependent in spectrum requirement calculations. Examples of maximum cell areas per RE are included in ITU (2013) for various teledensity areas, as denoted in Table 8.4.

Since the availability of REs depends on the service environment, the total area of a particular service environment is only covered to a certain percentage X by each radio environment. Therefore, example coverage percentages that are used in distributing the traffic amongst REs are given in ITU (2013) as follows:

- SE1: macrocell, 100%; microcell, 0%; picocell, 0% and hotspot, 80%
- SE2: macrocell, 100%; microcell, 0%; picocell, 20% and hotspot, 80%
- SE3: macrocell, 100%; microcell, 80%; picocell, 20% and hotspot, 10%
- SE4: macrocell, 100%; microcell, 0%; picocell, 0% and hotspot, 80%

Table 8.4 Example maximum cell area per RE (km^2). Taken from ITU (2013). Reproduced with permission of International Telecommunication Union.

Radio environment	Dense urban	Teledensity Suburban	Rural
Macro cell	0.65	1.5	8
Micro cell	0.1	0.1	0.1
Pco cell	0.0016	0.0016	0.0016
Hot spot	0.000065	0.000065	0.000065

- SE5: macrocell, 100%; microcell, 20%; picocell, 20% and hotspot, 20%
- SE6: macrocell, 100%; microcell, 0%; picocell, 10% and hotspot, 50%.

8.9.4 Radio Access Technology Groups (RATGs)

The RATGs that have been considered in the methodology contemplate four groups as follows:

Group 1. Pre-IMT systems, IMT-2000 and its enhancements
Group 2. IMT-Advanced systems
Group 3. Existing radio local area networks (LANs) and their enhancements
Group 4. Digital mobile broadcasting systems and their enhancements.

The first two RATGs are used in all the methodology, whereas the last two only are considered for the first steps.

Required radio parameters for each group are presented in ITU (2013) and are used as an example for the subsequent calculations. The use of more realistic figures for various indoor environments is presented in Chapter 11.

8.9.5 Methodology Flowchart

The generic flowchart for the spectrum requirement calculation methodology is shown in Figure 8.3. In summary, the methodology is as follows:

Step 1. Present the different definitions used in the methodology, as given in Sections 8.9.1, 8.9.2, 8.9.3 and 8.9.4.
Step 2. Analyse the market data, obtained from ITU (2006a).
Step 3. Values for the methodology are computed.
Step 4. Distribute traffic to different RATGs and radio environments inside the RATGs.
Step 5. Determine the required system capacity to carry the offered traffic. Capacity calculation algorithms are given separately for circuit-switched and packet-switched service categories.
Step 6. Calculate the spectrum requirements of RATG1 and RATG2.
Step 7. Apply necessary adjustments to take into account practical network deployments.
Step 8. Calculate aggregated spectrum requirements.
Step 9. Provide the spectrum requirements for RATG1 and RATG2 as outputs.

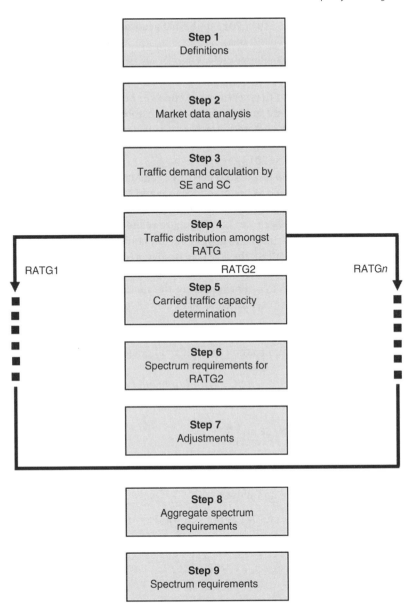

Figure 8.3 Flow chart for spectrum requirement calculations. Taken from ITU (2013). Reproduced with permission of International Telecommunication Union.

8.9.6 Market Data Analysis

From the flowchart in Figure 8.3 and once all variables and terms have been properly defined, obtain market data from surveys and questionnaires, which included items such as services and market survey for existing mobile services, service and market forecast, driving forces for future markets, other views of future services, key market parameters, etc. Data are provided for the years 2010, 2015 and 2020 according to the responses in such questionnaires included in ITU (2006a).

In order to perform an exhaustive market data analysis, the following steps should be followed:

List up applications and services, so that services are chosen not to be overlapped from an application that can be general and essential enough to categorize all collected services properly. As an example, if Internet access is selected as one service, traffic attributes such as the mean service bit rate should be specified, for example 10 Mbps. More examples are listed in ITU (2013).

Specify traffic attribute values for each service. These are related to traffic characteristics of the service. Attribute values such as mean service rate and average session duration need to be specified for each service defined earlier.

Specify market attribute values for each service. These are related tousers' perspectives. The idea here is no longer to have fixed and geographically equal amounts of spectrum allocated to each RATG but to allow the RATGs to give a spectrum to each other during times when it is unused. This provides a time-varying and regionally-varying nature of traffic for different RATGs. For analysing the market data, values of user density and session arrival rate per user for each service on each environment and time interval need to be specified.

Map services into a service category table for each SE. Each service can be mapped into a table composed of service type and traffic class. This table needs to be developed for each service environment.

Compute market attribute values for each SC, SE and time interval. This is fully defined in the next section.

8.9.7 Traffic Demand Calculation by SE and SC

From the previous section, the last step in market data analysis consists of computing market attribute values for each SC, SE and time interval. The relevant parameters to be considered in the methodology are as follows:

- *User density* (users/km^2) of a certain SC is the sum of user densities of each service mapped into the SC:

$$U_{m,t,n} = \sum_{s \in n} U_{m,t,s} \tag{8.4}$$

where $U_{m,t,n}$ and $U_{m,t,s}$ represent the user density of SC n and the user density of service s inside SC n.

- *Session arrival rate per user* (sessions/suser) of a certain service category is the weighted average of session arrival rate per user of each service mapped to this service category. The weight of each service is the user density. Thus:

$$Q_{m,t,n} = \frac{\sum_{s \in n} U_{m,t,s} Q_{m,t,s}}{U_{m,t,n}} \tag{8.5}$$

where $Q_{m,t,n}$ and $Q_{m,t,s}$ represent the session arrival rate per user of service category n and the session arrival rate per user of service s inside SC n.

- *Average session duration* (s/session) of a certain SC is the weighted average of an average session duration of each service mapped to this SC. The weight is the session arrival rate per area. Thus:

$$\mu_{m,t,n} = \sum_{s \in n} w_{m,t,s}\mu_{m,t,s} \tag{8.6}$$

and

$$w_{m,t,s} = \frac{U_{m,t,s}Q_{m,t,s}}{U_{m,t,n}Q_{m,t,n}} \tag{8.7}$$

where $\mu_{m,t,n}$ and $\mu_{m,t,s}$ represent the average session duration of service category n and the average session duration of service s inside SC n.

- *Mean service bit rate* (bps) of a certain service category is the weighted average of the mean service rates of each service mapped to this SC. The weight is the sum of the average durations of all sessions that arrive during a unit time per area. Thus:

$$r_{m,t,n} = \sum_{s \in n} \overline{w}_{m,t,s}r_{m,t,s} \tag{8.8}$$

and

$$\overline{w}_{m,t,s} = \frac{U_{m,t,s}Q_{m,t,s}\mu_{m,t,s}}{U_{m,t,n}Q_{m,t,n}\mu_{m,t,n}} \tag{8.9}$$

where $r_{m,t,n}$ and $r_{m,t,s}$ represent the service data rate of service category n and the service data rate of service s inside SC n.

- *Mobilty ratio* of a certain service category is the weighted average of each mobility ratio for a user of SC of each service mapped to the SC, and this mobility ratio is assumed to be not time-dependent. Thus:

$$MR\_market_{m,t,n} = \sum_{s \in n} \overline{w}_{m,t,s}MR\_market_{m,s} \tag{8.10}$$

where $MR\_market_{m,t,n}$ and $MR\_market_{m,t,s}$ denote the mobility ratio of service category n and the mobility ratio of service s inside SC n, and is applied to all mobility cases. The mobility ratios MR_market for stationary (sm), low (lm), high (hm) and super-high mobility (shm) need to be mapped into the methodology mobility ratios MR for stationary/pedestrian (sm), low (lm) and high mobility (hm) as follows:

$$MR\_sm_{m,t,n} = MR\_market\_sm_{m,t,n} + MR\_market\_lm_{m,t,n} \tag{8.11}$$

$$MR\_lm_{m,t,n} = J_mMR\_market\_hm_{m,t,n} \tag{8.12}$$

$$MR\_hm_{m,t,n} = (1 - J_m)MR\_market\_hm_{m,t,n} + MR\_market\_shm_{m,t,n} \tag{8.13}$$

Note that J_m are factors that depend on the service environment, as given in Table 8.5.

8.9.8 Traffic Distribution Amongst RATGs

Each service environment is supported by one or more RATGs, so traffic per service environment can be distributed to traffic per RATGs.

Table 8.5 Example *J*-values for mapping mobility classes in SE. Taken from ITU (2013). Reproduced with permission of International Telecommunication Union.

Service environment	J_m-value
1	1
2	1
3	1
4	1
5	0.5
6	0

8.9.8.1 Distribution Ratios

The distribution ratios $\xi_{m,t,n,rat,p}$ are used to distribute session arrival rates into RATGs and REs. These ratios need to be derived separately for different SCs in different SEs and time intervals for uplink and downlink traffic. The distribution ratios are defined in three phases:

- *Phase 1*, which determines which combination of RATG and RE cannot support a given service category in a given SE. Therefore, zero distribution ratios apply for:
  - RATG4 for unicast SC
  - REs that do not exist in the considered SE
  - REs that are not supported by a given RATG
  - Combination of RATG and RE for which the application data rate from RATG definitions is smaller than the required data rate of a particular SC
  - Macrocell RE for those RATG that do not support the entire range of velocities associated with the high mobility classes defined before.
- *Phase 2* distributes traffic between RATGs. The distribution ratio depends on the available RATG in each RE and SE, as specified in Phase 1. Table 8.6 shows an example of distribution ratios amongst available RATGs as suggested in TU (2013).
- *Phase 3* distributes the traffic amongst the radio environments based on mobility ratios and coverage percentages. Since the methodology defines the mobility classes stationary/pedestrian, low and high, the mapping of mobility classes to radio environments should be done, as indicated in Table 8.7. This mapping is assumed to be the same for all RATGs, and the velocity ranges and parameter maximum supported velocity of each RE are chosen accordingly.

Traffic distribution follows the principle to use the RE with the lowest mobility support that just satisfies the requirements. According to this, all stationary/pedestrian traffic would go to hotspots and picocells, all low mobility to microcells and all high mobility to macrocells. In practice, the total area of a particular SE is only covered a certain percentage X by each RE.

Table 8.8 defines the population coverage percentage of each radio environment as an example given in ITU (2013). This population coverage is independent of the RATG, but if a particular RATG does not support a certain RE at all then the corresponding cell edge

Table 8.6 Example of distribution ratios amongst available RATGs. Taken from ITU (2013). Reproduced with permission of International Telecommunication Union.

Available RATGs	% RATG1	% RATG2	% RATG3	% RATG4
1	100	0	0	0
2	0	100	0	0
3	0	0	100	0
4	0	0	0	100
1,2	20	80	0	0
1,3	20	0	80	0
1,4	10	0	0	90
2,3	0	20	80	0
2,4	0	10	0	90
3,4	0	0	10	90
1,2,3	20	20	60	0
1,2,4	10	10	0	80
1,3,4	10	0	10	80
2,3,4	0	10	10	80
1,2,3,4	10	10	10	70

Table 8.7 Mapping of mobility classes to RE.

Mobility class	Radio environment
High mobility	Macrocell only
Low mobility	Micro- and macrocell
Stationary/pedestrian	All RE

Table 8.8 Example population coverage percentage of the RE in each SE. Taken from ITU (2013). Reproduced with permission of International Telecommunication Union.

Service environment	Radio environment			
	Macrocell	Microcell	Picocell	Hotspot
SE1	100	0	0	80
SE2	100	0	20	80
SE3	100	80	20	10
SE4	100	0	0	80
SE5	100	20	20	20
SE6	100	0	10	50

data rate of this RATG/RE combination shall be set to zero, so that the Phase 1 of the traffic distribution will force the corresponding distribution ratio to zero.

Using the population coverage percentages X_{hs}, X_{pico}, X_{micro} and X_{macro}, the algorithm distributes the following traffic proportions to the hotspot, pico, micro and macro radio environments:

$$\xi_{pico\&hs} = \min\left(X_{pico} + X_{hs}, MR\_sm\right) \tag{8.14}$$

$$\xi_{micro} = \min\left(X_{pico}, (MR_{sm} + MR\_lm) - \xi_{pico\&hs}\right) \tag{8.15}$$

$$\xi_{pico\&hs} = \min\left(X_{pico} + X_{hs}, MR\_sm\right) \tag{8.16}$$

Recall that

$$MR_{sm} + MR_{lm} + MR\_hm = 1 \tag{8.17}$$

Finally, between hotspots and picocells the traffic is distributed according to the relation of the population coverage ratios of hotspots and picocells as follows:

$$\xi_{hs} = \xi_{pico\&hs} \cdot \frac{X_{hs}}{X_{pico} + X_{hs}} \tag{8.18}$$

$$\xi_{pico} = \xi_{pico\&hs} \cdot \frac{X_{pico}}{X_{pico} + X_{hs}} \tag{8.19}$$

8.9.8.2 Distribution of Session Arrival Rates

The session arrival rate per area of SC n and SE m distributed to RATG rat and RE p in time interval t, denoted as $P_{m,t,n,rat,p}$, is computed from the distribution ratio $\xi_{m,t,n,rat,p}$, the user density $U_{m,t,n}$ and the session arrival rate per user $Q_{m,t,n}$ calculated in Equation (8.5) as follows:

$$P_{m,t,n,rat,p} = \xi_{m,t,n,rat,p} \cdot U_{m,t,n} \cdot Q_{m,t,n} \tag{8.20}$$

Recall that

$$\sum_{rat} \sum_{p} \xi_{m,t,n,rat,p} = 1 \tag{8.21}$$

Thus

$$\sum_{rat} \sum_{p} P_{m,t,n,rat,p} = U_{m,t,n} \cdot Q_{m,t,n} \tag{8.22}$$

In order to get the traffic accumulated from all users in a cell, the session arrival rate/cell is computed as

$$P'_{m,t,n,rat,p} = P_{m,t,n,rat,p} \cdot A_{d,p} \tag{8.23}$$

where $A_{d,p}$ represents the cell area in km^2 of RATG rat in teledensity d and RE p. Note that d is determined by m, as seen in Table 8.3.

8.9.8.3 Offered Traffic

The offered traffic needs to be computed for each service category. The conversational and streaming classes are served with circuit switching, whereas the background and interactive classes are served with packet switching.

For *circuit switching*, the aggregate values of the product of session arrival rate per cell and average session duration for different teledensities d are collected to the offered traffic $\rho_{d,t,n,rat,p}$. This represents the sum of average durations of all sessions of SC n that arrive per unit time in a cell with teledensity d, RATG rat and radio environment p in time interval t:

$$\rho_{d,t,n,rat,p} = \sum_{m \in d} P'_{m,t,n,rat,p} \cdot \mu_{m,t,n} \tag{8.24}$$

The aggregate values of the mean service bit rate $r_{d,t,n,rat,p}$ (bps) for teledensity d are as follows:

$$r_{d,t,n,rat,p} = \frac{\sum_{m \in d} P'_{m,t,n,rat,p} \cdot \mu_{m,t,n} \cdot r_{m,t,n}}{\rho_{d,t,n,rat,p}} \tag{8.25}$$

For *packet switching*, the offered traffic $T_{d,t,n,rat,p}$ is given as the aggregate offered traffic over the SE, which belong to the same teledensity. Thus:

$$T_{d,t,n,rat,p} = \sum_{m \in d} P'_{m,t,n,rat,p} \cdot \mu_{m,t,n} \cdot r_{m,t,n} \tag{8.26}$$

8.9.9 Carried Traffic Capacity Determination

The required system capacity (bps) is determined separately for circuit-switched (denoted as $C_{d,t,rat,p,cs}$) and packet-switched ($C_{d,t,rat,p,ps}$) traffic. The number of circuit-switched service categories is denoted as N_{cs}, while the number of packet-switched service categories is N_{ps}. Note that the total number of service categories N is given by $N = N_{cs} + N_{ps}$.

8.9.9.1 Circuit-Switched Traffic

For *circuit-switched traffic*, the Erlang-B formula can be extended to the multi-dimensional case, which also allows simultaneous occupation of several channels by each call. Let the system state be $i = (i_1, i_2, \ldots, i_{Ncs})$ where i_m is the number of calls of class m currently using channels v. The steady-state probability mass function is then

$$P(i) = G(v)^{-1} \prod_{m=1}^{Ncs} \frac{(\rho_m)^{i_m}}{i_m!} \tag{8.27}$$

and

$$G(k) = \frac{1}{k} \left[\sum_{j=0}^{k-1} G(j) + \sum_{m=1}^{Ncs} v_m \rho_m G(k - v_m) \right] \tag{8.28}$$

where $G(k) = 0$ for $k < 0$. This algorithm yields the blocking probabilities for systems with up to v channels all at once and is applicable when the total number of channels v is given. If the total number of channels needs to be calculated, the system capacity is obtained by multiplying the required total number of channels by the bit rate of the channel:

$$C_{d,t,rat,p,cs} = \kappa \cdot r \tag{8.29}$$

where κ is the required number of channels per cell and r is the bit rate per channel, in bps. Therefore, κ is obtained as the smallest v that satisfies the following conditions (for a maximum allowable blocking probability π_n):

$$B_n(v) < \pi_n, \quad 1 \le n \le N_{cs} \tag{8.30}$$

$B_n(v)$ is the blocking probability for calls of class n and is given by

$$B_n(v) = \sum P(i) = 1 - \frac{G(v - v_n)}{G(v)} \tag{8.31}$$

8.9.9.2 Packet-Switched Traffic

For *packet-switched traffic*, the system capacity needs to fulfil each SC's mean delay requirement, which can be done using a queuing model applicable for independent arrival times of packets and an arbitrary distribution of packet size. Since the complexity of this algorithm is out of the scope of this book, we leave for the reader to investigate further the detailed calculations for the packet-switched capacity in ITU (2013). Case study examples have been included in Chapter 11 for completeness.

8.10 Wi-Fi Capacity

The best way to illustrate Wi-Fi capacity considerations is through an example. Since Chapter 11 case studies focus mainly on cellular network deployments, a Wi-Fi example is presented here for reference.

8.10.1 The Challenge

Let us assume a large shopping mall has a floor area of $140000 \, \text{m}^2$ and 45 million visitors annually, which makes it a very attractive venue for potentially mobile Internet and other high-speed, high-volume applications. It has a high percentage of customers using smartphones due to a large number of technology stores and youth attractions. A distributed antenna system (DAS) was built to provide cellular coverage, but is running out of capacity, and there is no good business case for upgrading it. We wish to use Wi-Fi to offload the excess capacity – we need to design the Wi-Fi network to deliver this without excessive cost. Figure 8.4 shows the layout of the shopping centre.

8.10.2 Facts and Figures

To start with, let us have a look at some relevant figures that were given to us at the beginning of the project:

- There is a market share of 30%.
- The penetration of data-capable mobile devices for the demographic in the shopping mall is expected to reach and stabilize at 80% in the next two years.

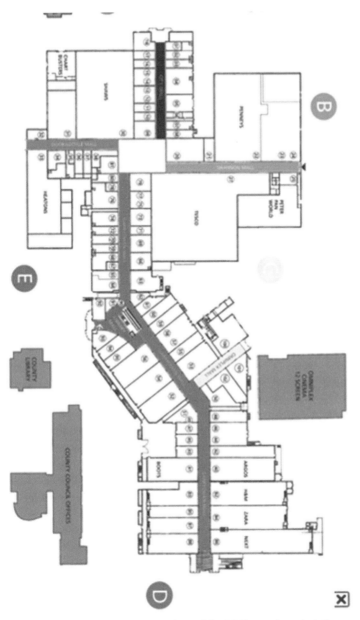

Figure 8.4 Shopping centre example used for Wi-Fi capacity calculations.

- We are experiencing a doubling of data volumes annually.
- We want to design our Wi-Fi system to offload excess demand for the next 3 years.
- Currently our DAS network, which has four HSPA sectors, is running 10% over capacity.
- There are lots of other access points in the mall, causing an effective noise rise of 3 dB throughout the 2.4 GHz band.
- We are aiming for a cell edge throughput of 10 Mbps.

8.10.3 Coverage Design

Coverage and capacity in a Wi-Fi network are closely related. Thus, designing for coverage needs to be accomplished first. For coverage purposes, our aim is to have 10 Mbps at the cell edge over a floor area of 140000 m^2, according to the given figures related to the mall. Therefore, we need to decide which Wi-Fi technology to use, which as yet has not been specified. Although 802.11 n delivers higher throughputs than 11 b/g, it requires wider MIMO antennas and support in the mobile device, which is currently unusual; therefore, we need to select 802.11 b/g for this purpose, and by looking at some graphs for 10 Mbps, a cell-edge SINR of 6 dB is required. If interference causes a noise rise of about 3 dB, we require a signal level of:

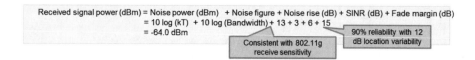

Received signal power (dBm) = Noise power (dBm) + Noise figure + Noise rise (dB) + SINR (dB) + Fade margin (dB)
= 10 log (kT) + 10 log (Bandwidth) + 13 + 3 + 6 + 15
= -64.0 dBm

Consistent with 802.11g receive sensitivity

90% reliability with 12 dB location variability

Given our calculated target received signal, we find the maximum path loss from our link budget. The following assumptions are made:

- 20 dBm EIRP
- Client devices have antenna gain of −5 dBi (including some body loss)

Hence:

Max path loss (dB) = Transmit power EIRP (dBm) − Required receive power (dBm) + Receive antenna gain (dBi)
= 20 − (-64) + (-5)
= 79 dB

Using the ITU-R P.1238 propagation model described in Chapter 5, the range is computed as:

$$L = -28 + 10\,n\,\log(r) + 20\log(f)$$

$$\text{Hence } r = 10^{\frac{L+28-20\,\log f}{10n}} = 10^{\frac{79+28-20\,\log(2400)}{2.8}} = 26\text{m}$$

Path loss exponent 2.8 in ITU-R model

Having a coverage area of 140000 m^2, this is equivalent to a square area of 374 × 374 m, approximately. Therefore, if we assume an overlap of 30% of the cell radius is enough for handovers (Figure 8.5), then

$$o = 0.3d$$

$$d = 2 \times 26 = 52 \text{ m}$$

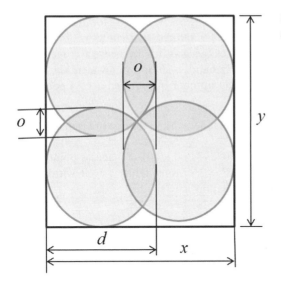

Figure 8.5 Parameters used for a number of AP calculations.

$$N \approx \left(\frac{x-d}{d-o}+1\right) \cdot \left(\frac{y-d}{d-o}+1\right)$$

$$= \left(\frac{374-52}{52-0.3\times52}+1\right) \cdot \left(\frac{374-52}{52-0.3\times52}+1\right)$$

$$= 101 \text{ access points}$$

8.10.4 Capacity Design

Next, let us determine our capacity demands for the system. In general, from the facts and figures, although we have information related to market share and the penetration of smartphones used in the mall, what we really need to estimate our current capacity is the fact that the cellular DAS is running 10% over capacity with four HSPA carriers. To project for three years time, the doubling of data volumes annually is taken and the fact is that we will be using Wi-Fi to offload this extra capacity.

Therefore, assuming one HSPA sector has a realistic capacity of 1.5 Mbps, then

$$\text{Current demand} = (3 \times \text{ sector capacity}) + 10\%$$
$$= 3 \times 1.5 \times 1.1$$
$$= 4.95 \text{ Mbps}$$
$$\text{Demand in three years} = 4.95 \times 2 \times 2 \times 2 = 39.6 \text{ Mbps}$$

If this demand is spread evenly over the area, the demand per access point is 39.6 Mbps/ 101 = 392 kbps.

8.10.5 Additional Challenges

Some assumptions were made during our design process, which may need to be revised at this stage. For example, we have dimensioned our system assuming users are uniformly distributed over the mall floor area. This may not be the case for many areas in the mall, and in fact there is a food court area of $10000\,\text{m}^2$ that contains 20% of the shoppers at lunchtime, who they use their devices twice as heavily as the average. Also, the food court has more devices in the 2.4 GHz band, resulting in an increase in the noise-floor. We therefore need to determine the number of required access points to provide capacity for this area as follows:

- This implies $2 \times 20\% = 40\%$ of the nominal total demand in three years is in the food court; that is $0.4 \times 39.6\,\text{Mbps} = 15.84\,\text{Mbps}$.
- This area is covered by approximately $10\,000/140\,000 \times 101 = 7$ access points.
- Therefore, each access point serves $15.84/7 = 2.3\,\text{Mbps}$. This is well within the capacity of an access point so no extra installation is needed.
- However, a 5 dB increase in the noise-floor requires more APs.

As you can see in the calculations above, these new requirements imply that the nominal total demand is now increased to 15.84 Mbps. If we estimate the number of APs that are considered for this area with the current estimations, 7 APs are to be used here, and each AP serves about 2.3 Mbps, which is well within the capacity of an AP and therefore no need for additional installation is envisaged. However, the noise rise increase makes this increase in APs inevitable.

The reason why more APs are needed if the noise rise is increased is due to the cell-edge requirements, to guarantee the minimum SINR specified – if the noise is increased, the signal levels have also to be increased to maintain the SINR.

Using our propagation model, we can determine that 5 dB in path loss corresponds to a ratio of previous and new radius of

$$\frac{r_1}{r_2} = 10^{5/10n}$$

If we say the ratio of the number of APs is inversely proportional to the square of the distances, then

$$\frac{A_2}{A_1} = \left(10^{5/10n}\right)^2 = 2.3$$

Therefore, the food court requires $2.3 \times 7 = 16$ access points.

8.11 Data Offloading Considerations

An extensive growth in mobile data traffic has been seen in the last few years, especially with the launch of more sophisticated devices such as smartphones, capable of handling much more data volumes, and it is expected that for the next five years there will be an exponential growth of mobile data traffic. Interestingly enough, the number of subscribers will not follow the same trend – it is expected to grow at a much slower pace. Nowadays, more devices are getting connected to the Internet, and a very large increase in traffic indoors, of about 80%, has been reported. This brings new challenges to operators and service providers, in order to have enough network resources to satisfy such increasing demands.

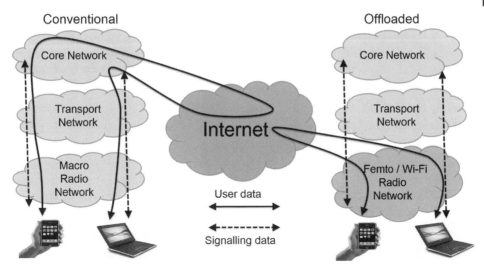

Figure 8.6 The data offloading scenario for capacity.

Since one of the elements that has influenced most of the growth in mobile data traffic has been mobile device evolution, mobile terminals and handsets have evolved in the last years, commencing with dongles in 2007, towards the advent of smartphones in 2009 and tablets in 2010. Therefore, mobile broadband demand has already been through at least three major evolutions arising from devices in the last few years. Interestingly enough, parallel growth in data revenue and data volume is not observed; whereas historically in the UK revenues were growing more or less aligned with volume, the forecast shows that available revenue per bit is declining steeply, even as user expectations of connectivity and speed are rising. On the other hand, traffic growth is expected to growth exponentially, which brings major concerns as to finding ways to rapidly reduce the cost per bit.

Conventionally, mobile data traffic has been handled towards the macro radio network, which through the transport and core networks connects to the Internet and brings this traffic back to the mobile portable devices, thus imposing a very large traffic demand in the macro network. On the other hand, the concept of an *offloaded network* uses a femto/Wi-Fi radio network to carry the traffic generated by portable devices and in this way offloading transport and core networks since access to the Internet is now possible directly, unlike in conventional networks. This allows the transport and core networks to only deal with signalling data and all user data are conveyed towards the Femto/Wi-Fi radio network, hence maximizing network resources, as depicted in Figure 8.6.

In Figure 8.7, offload of traffic from smartphones and tablets is shown for the next five years, where it was expected to gain an increasing share and volume of mobile device traffic to be offloaded to fixed networks via Wi-Fi or femtocells. In August 2011, 37% of mobile device access in the US was over Wi-Fi, up three percentage points in only three months, something that suggests that the use of WLAN for handling mobile traffic is increasing. On the other hand, many phones are now equipped with Wi-Fi connectivity, of which 74% of people having this feature in their phones use it. Moreover, Wi-Fi connectivity in mobile phones and other portable devices is becoming increasingly

Petabytes per Month

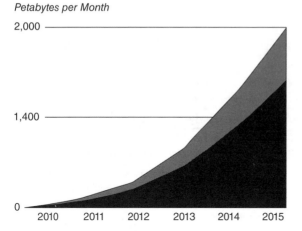

Figure 8.7 Offload forecasts.

popular: 44% of smartphones had Wi-Fi in 2011 and by 2014 it was expected that 90% would be equipped with Wi-Fi. Finally, and to fully have the picture of the increasing use of Wi-Fi in portable devices, Wi-Fi enabled mobile phone shipments per year reached 141 million in 2009 and grew up to 520 million by 2014!

Figures 8.6 and 8.7 suggested a definite use of Wi-Fi for mobile applications, but recall from the figures therewere still a small percentage of all phones in the market and Wi-Fi were often not switched on or not always accessible.

According to the Wireless Broadband Alliance, 47% of mobile operator believe Wi-Fi hotspots are very important or crucial to enhance customers' experience, offload busy mobile broadband and provide a value-added services platform. This is a very interesting perspective that has radically changed, as in the early days of Wi-Fi and 3G, this was seen as a competitor for cellular!

Major operators around the world are now adopting Wi-Fi offload, supported by industry initiatives to ease the process. In some cases they are building new custom Wi-Fi networks, in others they are acquiring existing Wi-Fi providers while others are simply opportunistically using existing third-party networks.

The following benefits are identified:

- Improves user experience where it really matters: typically indoors.
- Reduces the cost per bit, by making use of available backhaul and unlicensed spectrum, and third party (including customer) infrastructure.
- Offloads the traffic where little operator value is added, while retaining control of added value services and customer information; for example voice, SMS, MMS, location-based services not offloaded, while video streaming is offloaded.

On the other hand, some risks are involved:

- Offloading data could lead to offloading of customer revenue and reduced loyalty.
- Use of third party infrastructure makes it hard to manage end-to-end customer experience: who does the customer call when something goes wrong?
- User experience of Wi-Fi is very variable – great when it works, but seamless access, authentication and speed are patchy.

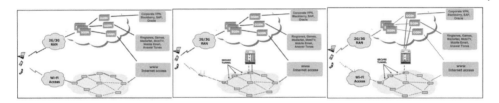

Network Bypass	Managed Network Bypass	Integrated Network Offload
Straightforward but loses control	*Allows metering and monitoring, but not carrier-managed content over Wi-Fi*	*More seamless experience and carrier-managed content*

Figure 8.8 Data offload via Wi-Fi.

8.11.1 Data Offload Using Femtocells

One of the possibilities that was introduced to offload traffic from the macro network is the use of femtocells. Femtocells offload radio access network and backhaul – these constitute the vast majority of cost in mobile networks, since access to the Internet can be achieved without going through the RNC of the macro network. In fact, with Selected IP Traffic Offload (SIPTO), femtocells can also offload the core network, further reducing cost and increasing scalability.

8.11.2 Data Offload Using Wi-Fi

The second alternative is the offload of traffic via Wi-Fi. Three options are presented in Figure 8.8, as follows:

- **Network bypass.**This method can be done straightforwardly, but a loss in control can be experienced.
- **Managed network bypass.**This is similar to the previous one, but now metering and monitoring are allowed. As a drawback, this approach does not support carrier-management content over Wi-Fi.
- **Integrated network offload.**This could be seen somehow as a combination of the previous two, thus showing more seamless experience and carrier-managed content.

8.11.3 Femtocell versus Wi-Fi

Having provided an overview of the two possible alternatives for macro network traffic offload, a comparison follows.

8.11.3.1 Wi-Fi

- Unlicensed spectrum: cheap but variable and prone to interference from other in-band technologies
- Device support improving, but variable and user may disable
- Large installed base of access points, but not all suitable
- Risks reducing customer loyalty
- Seamless authentication, access and mobility are very variable.

8.11.3.2 Femtocells

- Licensed spectrum: good quality but need to manage
- Radio always on: addresses all devices
- Access point deployment growing fast but still in early stages
- Keeps customer firmly associated with operator, building loyalty
- Uses existing authentication, mobility and billing mechanisms for seamless extension of existing network.

8.11.4 Carrier Wi-Fi

In order to achieve efficient data offload by using Wi-Fi, it has to be considered as part of the mobile operator network. This ensures the delivery of acceptable and consistent service, which otherwise could not be guaranteed. For example, suppose the Wi-Fi service and the operator are not connected or linked whatsoever and the user is entering a Wi-Fi zone for which the WISP does not have an agreement with the mobile operator. How could the user be connected to such a network as part of an integrated, seamless service, without the hassle of having to authenticate every time he joins in the Wi-Fi zone?

Ways for achieving this integration have been proposed by many vendors (Ruckus Wireless, Cisco, etc.) using reference architectures with advanced radio technology and interference rejection techniques, comprehensive end-to-end management and fast inexpensive backhaul links. Bear in mind that all of these require specific support in the handset and the network, with some of them requiring support in the AP as well – this should be considered when this integration is being planned.

8.11.5 UMA/GAN

There are two terms often used in Wi-Fi, which are used indistinctively and representing almost the same technology. UMA, or Unlicensed Mobile Access, is a technology that provides roaming between cellular networks and Wi-Fi. Under this context, the 3GPP implementation of UMA is known as GAN, or Generic Access Network, a telecommunication system that extends mobile voice, data and IP Multimedia Subsystem/Session Initiation Protocol (IMS/SIP) applications over IP networks. Unlicensed Mobile Access, or UMA, is the commercial name used by mobile carriers for external IP access into their core networks.

The UMA/GAN technology works as follows:

- UMA-enabled handset comes in range of UMA-enabled Wi-Fi.
- Handset connects to Wi-Fi and contacts UMA network controller (UNC).
- User gets authenticated and approved. Core network updated and voice and data traffic is routed via Wi-Fi instead of cellular.
- Roaming and handover between 2G/3G and GAN can in principle be transparent and seamless, but there are risks of interruption and call quality degradation

The major challenge that UMA/GAN faces is the fact that relatively few phones (e.g. not iPhones) support UMA and can involve deep changes. However, Blackberry devices do support it and some Android devices can add UMA support via a user-installed

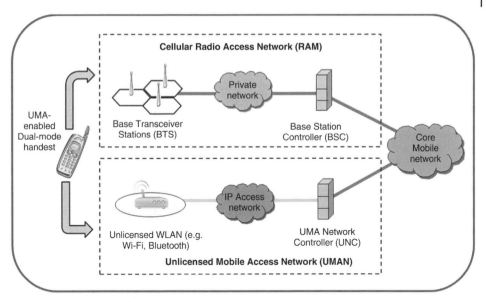

Figure 8.9 UMA/GAN architecture.

application – in summary, there are ways to go around this problem but still need further development to allow devices to support this feature without too much complexity, especially from a user's perspective.

Figure 8.9 depicts a simple architecture in which the core mobile network integrates the cellular radio area network (RAN) with an unlicensed mobile access network (UMAN) and how a UMA-enabled dual-mode handset can access both networks and roam from one to another. This is essential for the integration of Wi-Fi technology in mobile operator networks.

8.11.6 Seamless Authentication

Seamless authentication is a standard feature of 2G/3G/4G cellular, taken for granted by users and operators alike. In Wi-Fi, however, this is not so straightforward and is seen by operators as the biggest challenge for adoption. Authentication is the top barrier to wider adoption and usage of Wi-Fi amongst end users.

The most favoured approach to this is now the 802.11u standard, which includes support for SIM-based authentication via the IETF's Extensible Authorization Protocol (i.e. EAP-SIM/EAP-AKA for GSM/UMTS, respectively). This allows the subscriber's profile to be carried over from cellular into Wi-Fi and deliver similar security to the cellular network.

8.11.7 Turning Wi-Fi into an Operator Network

Several additional initiatives are working to ensure Wi-Fi services can be managed by operators as part of their network, including: WBA Next Generation Hotspot, WBA Roaming and WFA Hotspot.

8.11.7.1 WBA Next Generation Hotspot

This initiative is based on the use of roaming via 802.11 u, 802.1× and EAP-SIM standards. It promises seamless interoperability via WFA certified hotspots (http://www.wi-fi.org/for_operators.php). Wi-Fi certified handsets mean service providers can be confident that the devices on their network are going to deliver the very best experience to their subscribers. The hundreds of phones that have been certified work interoperably with thousands of other Wi-Fi certified networking devices and consumer electronics. This certification is done by the Wi-Fi Alliance.

8.11.7.2 WBA Roaming

This initiative promotes the interconnection between international operators similar to cellular roaming arrangements via various specifications, such as WRIX and WISPr 2.0.

8.11.7.3 WFA Hotspot

This initiative somehow mirrors the Next Generation Hotspot (NGH) suggested by the Wireless Broadband Alliance (WBA) and covers the equipment aspects of 802.11 u, using WPA2-Enterprise and EAP-based authentication.

8.11.7.4 ANDSF

Access Network Discovery and Selection Function (ANDSF) is an entity within an evolved packet core (EPC) of the system architecture evolution (SAE) for 3GPP compliant mobile networks. The purpose of the ANDSF is to assist user equipment (UE) to discover non-3GPP access networks – such as Wi-Fi or WIMAX – that can be used for data communications in addition to 3GPP access networks (such as HSPA or LTE) and to provide the UE with rules policing the connection to these networks.

This is a 3GPP specification (TS 24.312 and some related information in TS23.402 and TS 24.302) for Access Network Discovery and Selection. It allows 3GPP equipment to discover non-3GPP equipment and to follow rules to police and prioritize the connection between these networks based on operator-set policies.

8.11.7.5 I-WLAN

This is a 3GPP spec allowing carrier managed services over Wi-Fi. It is specified in 3GPP TS 23.234. It operates by setting a secure (IPsec) tunnel. However, in a managed Wi-Fi network this leads to double encryption; that is IPsec and Wi-Fi link-layer security, thus decreasing battery life.

8.11.8 Discussion

A very interesting perspective for many carriers and mobile operators has been to reconsider the role of Wi-Fi as a viable strategy to provide a high-speed alternative to users who may not require full mobility, that is 'nomadic' users, as they consume data on their mobile device. Wi-Fi offload is not only a technically viable strategy but also seems commercially attractive and sound. As an offload technology, Wi-Fi thus brings three main benefits to operators:

- It is cost effective.
- It is widely available.
- It is widely used.

Even though a large number of subscribers can access Wi-Fi for free, only a portion of these actually use Wi-Fi (slightly more than 50%, according to some studies). We must recall that the data itself is the one to be offloaded, not the customer! On the other hand, focus is placed only in data offload, but what about voice traffic? Some reports of frustrations and concrete blocking factors can be found in many Wi-Fi forums that have made the use of Wi-Fi sometimes more of a pain than a useful service – mainly through inconsistent user experiences. In summary:

- Client software, which needs to be downloaded to the mobile device. The way it is executed brings an enormous frustration to users, since it requires long and often hard configurations, as well as authenticating to many portals in order to get the service that is required.
- Discovery, referred to as a great problem for the moble consumer. Issues like whether the Wi-Fi network they can "see" is a partner of their mobile operator or not; hotspots whose SSID is reported as 'Public Wi-Fi' and prompt a screen to secure payment before the service can be used. There is a great deal of confusion on these aspects.
- Authentication is highly variable – different methods and options are available for each operator, some of them not working properly; forgotten passwords, etc.
- Service continuity, since some customers sometimes experience poor service, even though the mobile device is officially connected to the Wi-Fi network. Usernames/passwords, cookies and device IDs are not sufficiently secure for many carriers to open access to premium services such as music or mobile TV – disappointing for customers!
- Need to liaise with many other players if non-integrated – or build a whole network, reducing value of Wi-Fi ubiquity in the first place.

8.12 Conclusion

Capacity dimensioning and spectrum requirements calculation is an essential part of any in-building wireless network, especially to determine the necessary spectrum resources required to provide sufficient QoS for the distinct services and applications that nowadays wireless networks offer.

We have seen that in past times capacity planning was a very simple exercise of just allocating enough channels for a certain blocking probability that the system required, given that the main service was only voice, with very limited data usage. However, in recent years, this has been reversed and now more data communications demand a different design approach, in order to guarantee the QoS for each type of service and application. A careful classification of service environments, categories, etc., is mandatory for a proper design, and examples of this have been included in Chapter 11 for further reference.

Finally, the deployment of 5G networks in the near future will require a user to upgrade this methodology as many more devices (thousands and perhaps millions with the Internet of Things) will be deployed, with much higher data usage. This needs to be revised and formulated, but, for now, let us focus on current wireless technologies and let us have those thoughts for future editions of this book.

References

Aragón-Zavala, A., Cuevas-Ruiz, J.L., Castañón, G. and Saunders, S.R. (2009) Mobility model and traffic mapping for in-building radio design, in Proceedings of theIEEE International Conference on Electrical, Communications and Composition, CONIELECOMP 2009, Cholula, Mexico, pp. 46–51, February 2009

ITU, (1999) International Telecommunication Union, ITU-R Recommendation M.1390-0: Methodology for the calculation of IMT-2000 terrestrial spectrum requirements, Geneva.

ITU (2003) International Telecommunication Union, ITU-R Recommendation M.1079-2: Performance and quality-of-service requirements for International Mobile Telecommunications-2000 (IMT-2000) access networks, Geneva.

ITU (2006a) International Telecommunication Union, ITU-R Recommendation M.2072-0: World mobile telecommunication market forecast, Geneva.

ITU (2006b) International Telecommunication Union, ITU-R Recommendation M.2079-0: Technical and operational information for identifying spectrum for the terrestrial component of future development of IMT-2000 and IMT-Advanced, Geneva.

ITU (2013) International Telecommunication Union, ITU-R Recommendation M.1768-1: Methodology for calculation of spectrum requirements for the terrestrial component of International Mobile Telecommunications, Geneva.

Schwartz, M. (1987) *Telecommunication Networks: Protocols, Modelling and Analysis*, 1st edition, Adison Wesley, USA. ISBN 978–0201164237.

Stasiak, M., Glabowski, M., Wisniewski, A. and Zwierzykovski, P. (2011) *Modeling and Dimensioning of Mobile Networks: From GSM to LTE*, 1st edition, John Wiley& Sons, Ltd, Chichester. ISBN 978–0470665862.

Tolstrup, M. (2011) *Indoor Radio Planning: A Practical Guide for GSM, DCS, UMTS, HSPA and LTE*, 2nd edition, *John* Wiley & Sons, Ltd, Chichester. ISBN 978–0470710708.

9

RF Equipment and Distribution Systems

When designing an indoor radio system, one of the most important parameters to consider is to provide enough coverage in the building, as part of the requirements stated in Chapter 3. This is applicable to UMTS and LTE systems as well, although the design rules may change. This signal needs to be distributed across the building in the most suitable, efficient and economical way, ensuring the highest quality-of-service for all users.

In the early days of cellular radio, coverage inside a building used to be achieved by placing a macrocellular base station sufficiently close to it and downtilting the antennas to rely on outdoor-to-indoor penetration of the signal. This approach may still be applicable for small buildings in rural and suburban areas, and this is of course the easiest way to achieve in-building coverage. However, this method is limited only to buildings with sufficient LOS conditions with the serving cell, which do not attenuate the signal excessively. The RF signal is attenuated as it penetrates through the building and is affected by constructive and destructive rays adding together, or multipath propagation. Building penetration losses are often sufficiently large and thus produce unacceptable coverage in many areas. As well as providing poor coverage in many situations, this approach also imposes a further constraint on the amount of channels that can be allocated to the building – capacity from the external cell is taken to handle the in-building traffic at the expense of an increase in blocking.

The concept of the use of an internal cell or picocell to provide dedicated coverage and capacity inside a building is suggested for such demanding buildings, but the choice of the most suitable way of distributing this power across the building depends on system requirements and building structure. A comparison of the different types of distribution systems is made in this chapter and recommendations as to which to choose in different situations are also made. The costs and performance of the indoor system will depend critically on making the choice; therefore it is important to have a clear understanding of this before attempting to optimize any indoor system. Finally, the main characteristics of the RF equipment and components used in such distribution systems are also discussed.

9.1 Base Stations

Base stations are the transmitting element in many indoor wireless networks, such as cellular, Wi-Fi, DECT, etc. The use of different names vary according to the technology and some architecture issues; for example for Wi-Fi the name *access point* is employed,

Indoor Wireless Communications: From Theory to Implementation, First Edition. Alejandro Aragón-Zavala.
© 2017 John Wiley & Sons Ltd. Published 2017 by John Wiley & Sons Ltd.

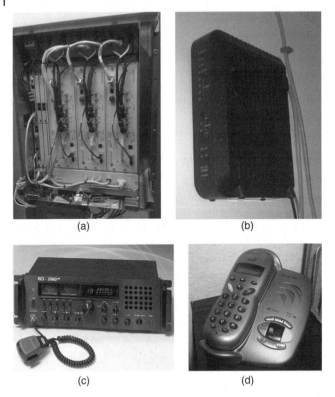

(a) (b)

(c) (d)

Figure 9.1 Examples of wireless base stations: (a) cellular eNodeB; (b) Wi-Fi access point; (c) PMR base station; (d) DECT cordless phone base station.

for DECT the term *base* is used and for PMR the name *infrastructure* is often encountered. Examples of base stations are depicted in Figure 9.1.

A base station can be located inside a building or outdoors, especially for cellular networks. Sometimes Wi-Fi systems also may have an access point that is mounted on an exterior wall of a building, especially to provide coverage to a corridor. In cellular, if the base station is employed outdoors, a suitable housing equipped with power supply and ventilation is often enough, as well as with the necessary security precautions to avoid any vandalism on the site. A backhaul network may become an issue in these cases, especially if the site is located in rural areas with little infrastructure. For indoor base stations, these requirements are similar but normally the backhaul connection is easier since the building has sufficient infrastructure to accommodate these needs.

While low levels of RF power are usually considered to have negligible effects on health, national and local regulations restrict the design of base stations to limit exposure to electromagnetic fields. Technical measures to limit exposure include restricting the radio frequency power emitted by the station, elevating the antenna above ground level, changes to the antenna pattern and barriers to foot or road traffic. For typical base stations, significant electromagnetic energy is only emitted at the antenna, not along the length of the antenna tower.

Site acquisition is the procedure that needs to be completed in order to deploy a new cellular base station at a certain location. For airports, for example, this procedure may involve the construction of new facilities to house all the base station equipment provided with enough ventilation, cooling and power. Also, the site would need to be properly secured and locked. For Wi-Fi, the base station or access point is much more reduced in size and therefore its installation does not require a special cabinet or room, but only an area where power is available and Ethernet backhaul connection can be easily obtained.

For cellular base stations, technical specifications may vary according to the standard (GSM, UMTS, LTE, PCS, etc.), but in general we may summarize the most important as follows:

- Operating frequencies
- Electromagnetic compatibility requirements
- Electrical safety requirements
- Cooling and ventilation requirements
- Size, weight and dimensions
- Power supply requirements.

9.2 Distributed Antenna Systems

A *distributed antenna system* (DAS) is a system of indoor antennas throughout the building connected together via coax or fibre cabling and routed back to a central base station in the building. This base station is in turn backhauled into the operator's core network. This base station and the DAS then appears as a distinct cell site on the operator's network. Depending on the capacity requirements of the network smaller DASs may be fed by a picocell or repeater rather than a full macrocellar base station.

A very simple way to understand the DAS concept is to see the DAS as a means for distributing radio signals over a large and complex coverage footprint, as shown in Figure 9.2. The number of antennas varies from several to hundreds depending on the size of the coverage area. In essence, it is a system of indoor antennas throughout the building connected together using coaxial or fibre cabling and routed back to a central base station in the building, which is in turn backhauled into the operator's core network. This appears as an extra separate cell site on the operator's network.

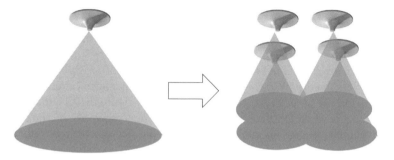

A single antenna radiating at high power ...   ... is replaced by a group of *N* antennas to cover the same area

Figure 9.2 The distributed antenna system concept.

DAS can offer the advantage of a centralized base station installation that segregates the maintenance of the operator's equipment from that of the DAS. This may be important where security is a main concern, such as airports, for example.

Distributed antenna systems are normally used for medium to large enterprises for which the benefits of their use can be easily justified. If the end-user is *not* the building owner they will see the benefits of DAS as an improvement to their existing cellular service.

In the case of the business or building owner they may have to cover part or all of the cost of the DAS installation but for major corporate customers the operator is generally willing to pay all the costs incurred. However, there will inevitably be a lot of disruption with a DAS installation if appropriate cabling and antennas were not built into the building from when it was first built, and this is rarely the case.

Technically DASs can support multiple operators but this is cheaper to achieve in passive than active systems. In public venues a multioperator DAS tends to be required to minimize the requirement to deploy multiple systems. In large corporate DASs, particularly when installed by the operator, this will be limited to a single operator for commercial reasons. In fact, there have been cases of DASs being taken out and completely new systems being installed when corporations have changed operator – this large investment by operators/corporations can sometimes make changing operators less attractive on commercial grounds. In Australia, building owners have gone as far as approaching the government and getting legislation passed to require operators to keep each other informed of DAS in-building deployments to encourage collaborative DAS deployments in an effort to try to avoid these situations.

Once installed the indoor service improvements seen by consumers from a DAS will include:

- Better availability and quality of cellular voice, SMS and data services in the building
- Better cellular data rates
- Improved mobile phone battery life
- Depending on the deployment the DAS may support all operators.

In terms of types of DAS, there are passive, active or hybrid DAS. These are explained in more detail below.

9.2.1 Passive DAS

Passive DAS is made up of coaxial cables, splitters and couplers. It is normally used for small to large buildings with a relatively simple layout. It is the preferred DAS option because of its wideband nature and also because power handling is not normally an issue as long as the antenna radiated power complies with ICNIRP guidelines on exposure limits (ICNIRP, 1998). On the other hand, due to the large noise figure that is added by passive elements, its use is often restricted to smaller buildings, especially for noise-limited systems in the uplink, such as in cellular 3G. An example of a passive deployment is shown in Figure 9.3.

Passive DAS consists of the head-end, an intermediate passive network and remote antennas. At the head-end, signals from multiple base stations are received and combined into a single signal. This composite signal is then split via power dividers into several parallel composite signals that are sent to remote antennas via a passive

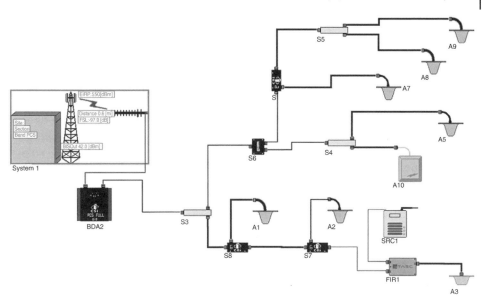

Figure 9.3 Overview of a typical passive DAS installation.

network. Passive networks consist of coaxial cables, splitters and power dividers. A block diagram of a passive DAS is shown in Figure 9.4.

Base stations used in this configuration are usually macrocell base stations with a transmit power of around 20 W (43 dBm). A high transmit power is needed to overcome passive losses between BTS and remote antennas, as there are no intermediate RF amplifiers in the network. Coaxial cables used in passive networks are usually ½ or ¾ inch in diameter; the latter have a lower attenuation per metre, but are more expensive and more difficult to install as they are bigger and less flexible.

While passive DASs are less expensive to deploy and maintain than active or hybrid DASs, their major drawback is their uplink signal-to-noise ratio (SNR) degradation. Poor SNR causes a poor uplink data rate, which is a major problem for sporting events, where uplink traffic often exceeds downlink traffic.

9.2.2 Active DAS

In cases where a building has a complex layout and the space available for the installation of coaxial cables is limited or difficult to install, an active DAS is used. An active DAS is

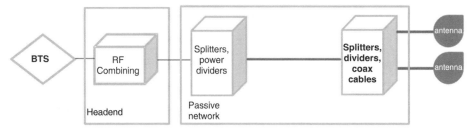

Figure 9.4 Passive DAS block diagram.

normally made of a master unit that interfaces with the base station and distributes the radio signal to remote amplifiers installed near the target coverage areas. The connection between the master unit and the remote units can be in the form of CAT5/6 cables, multimode or singlemode optical fibres. Thus, the master unit has the function of converting the radio signal from the base station into a lower frequency signal or optical signal, depending on the medium used. The remote amplifier converts the signal back to the radio signal for radiation. Therefore an active DAS requires careful system dimensioning to ensure that the power handling is within the capabilities of the active elements. The coverage capability of each antenna can be engineered for the intended service requirements.

One of the characteristics of an active DAS is that the amplifiers balance out any transmission losses that may occur in the link and also have a much better noise performance than a passive DAS. They are also very flexible, which means that zoning and different antenna configurations are possible with the simple reinstallation of the remote antenna units. On the other hand, they are much more expensive and complex to deploy and install than passive components.

In terms of architecture, the active DAS consists of a headend, an intermediate active network and remote antennas. As is the case with the passive DAS, the headend receives signals from several BTS and combines them into a single RF signal. On the other hand, the active network receives the RF signal, converts it to optical, splits the optical signal into several parallel optical signals and sends them via fibre cables further down the network. When the optical signal gets in the vicinity of the target coverage area, it is converted to a digital signal and sent via cable TV (CATV) to digital/RF power amplifiers, which are also known as remote units (RU). At the RU the digital signal is converted to RF and amplified. Remote antennas are located next to RUs and connected to them via short RF jumper cables (0.5–1 m). A block diagram of a typical active DAS is shown in Figure 9.5.

The maximum distance between RF/optical and optical/digital converters is determined by an *optical link budget*. This link budget depends on the type of fibre cable (multimode versus single mode) and the type of optical connectors used, but its maximum range is in general of the order of a few kilometres. The maximum distance between an optical/digital converter and an RU depends on the type of CATV cable used, as some CATV cable types have less attenuation per metre than others. Typical CATV

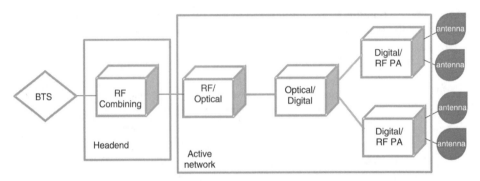

Figure 9.5 Active DAS block diagram.

choices for active DAS are RG-6, RG-11 and RG-59. The system gain between the optical/digital converter and an RAU can, to a great extent, cancel out the CATV loss between the two.

Since remote antennas are very close to power amplifiers, the EIRP from the antennas is fairly uniform, which makes active DAS design and planning easier. On the other hand, as a power amplifier reamplifies the RF signal, the BTS signal need not be as high as it needs to be for a passive DAS. This is an important aspect, especially for LTE indoor radio networks.

There are two major disadvantages for an active DAS. First, the cost to deploy an active DAS and maintain the network is high, especially compared to passive DAS solutions. The cost to deploy the first RU in a DAS is high, while adding subsequent RUs makes the DAS progressively less expensive. The consequence is that active DAS are rarely cost effective to deploy in small venues. The second drawback is the power requirement for RUs, as provisioning AC power at remote locations adds to the total cost of buildout. In some cases, instead of AC power, RU can get DC power over copper, if composite fibre/copper cable is deployed. The maximum distance over which DC power can be effectively sent to an RU depends on the copper wire gauge and RU DC power requirements. Typical copper wire gauge values are 12, 14, 16 and 18 AWG, while typical power requirements range from 12 VDC to 75 VDC.

9.2.3 Hybrid DAS

A hybrid DAS is a mixture of passive and active DAS making the best use of the base station output power. This configuration is normally used for buildings with either existing passive installations and for which coverage needs to be easily extended or for areas where it is difficult to have an active DAS, for example due to power requirements, and therefore passive elements are easier to add to existing active DAS deployments. These systems, like all DAS installations, require expert detailed design, planning and installation.

A hybrid DAS consists of the headend, intermediate hybrid network and remote antennas. As is the case with the previous DAS networks, the headend receives signals from several BTS and combines them into a single RF signal. The active network receives the signal and converts it to optical, then splits the optical signal into several parallel optical signals and sends them via fibre cables further down the network to a remote unit. At the RU a digital signal is converted to RF, amplified and split into a few parallel RF signals using power dividers. These RF signals are sent via coaxial cables to remote antennas. In some venues, such as tunnels, radiating cables may be used instead of coaxial cables and remote antennas. An example of a hybrid DAS is shown Figure 9.6.

The presence of passive elements (coaxial cables, radiating cables) between antennas and RUs adds attenuation to the output signal from the RU, making EIRPs at the antennas unequal. However, the elimination of optical/digital conversion makes the hybrid DAS simpler, with fewer active network elements. It also separates the RU from the antenna, which allows RUs to be mounted in telecommunication cabinets, further away from antennas. A typical RU–antenna distance ranges from 10 to 50 metres, depending on the RF propagation environment.

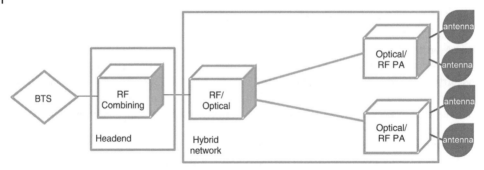

Figure 9.6 Hybrid DAS block diagram.

9.2.4 Installation

The installation of DASs is generally challenging and requires a skilled team to be designed and installed successfully. Design and installation constraints include:

- Cable routing to stay within maximum bend radius
- Management of intermodulation between carriers in multioperator systems
- Management of handover and interference with neighbouring macrocells
- Restrictions on antenna location due to aesthetic or physical access reasons
- Restrictions on cable routes due to access restrictions.

Installation is extremely costly and therefore this type of solution tends to be used in high-capacity large buildings only; for example airports, shopping centres, large enterprises, or stadiums, whether as active, passive or hybrid deployments. Figure 9.7 shows an example of a typical DAS installation.

9.3 RF Miscellaneous – Passive

We have spent some time in Chapter 7 discussing relevant characteristics related to measurement equipment, such as transmitters, receivers, cables, connectors, etc. Most

Figure 9.7 Behind the scenes of a typical DAS installation.

of the principles presented there are also applicable to installed in-building radio networks, since some of this equipment is utilized here as well, with minor changes. For example, transmitters in radio measurements are employed to characterize propagation behaviour and estimate coverage in the building, whereas for network installations they are employed to provide service and coverage; for example in base stations. The particular characteristics of a transmitter employed for measurements and one used for base stations certainly differ, but similarities exist in the way both types should be looked after and maintained. We will focus here more on RF equipment that is often used for distribution systems, presenting a brief overview of each component. To begin with, we will focus only on passive networks.

9.3.1 Cables

Coaxial cables are amongst the most popular transmission lines used in distribution systems, although not the only ones – optical fibre in conjunction with CAT5 cables are often employed in an active DAS and certainly in Wi-Fi deployments. For passive installation coaxial is the most popular choice. The parameters that should be carefully observed when selecting a cable are:

- *Cable type and diameter.* Bulky cables often have less losses but may represent difficulties when installing them in certain venues. Cable diameters vary, but the most popular ones are $\frac{1}{2}''$, $\frac{1}{4}''$, $\frac{7}{8}''$, $1\frac{1}{4}''$ and $1\frac{5}{8}''$.
- *Losses.* Any cable has losses that vary with distance, and these are often quoted for every 100 m in dB and for various frequency bands. At higher frequencies the lumped model for a cable (containing resistance, inductance, capacitance and reactance) changes its characteristics and acts as a low-pass filter, for which its cutoff frequency varies with the cable design. Therefore cable losses normally increase with frequency.
- *Bandwidth.* Cables operate over a certain frequency range within which their performance is as expected. Beyond this maximum frequency they act as filters and should not be used.
- *Impedance.* As with many other passive elements in an RF network, impedance should match the rest of the network elements (transmitter, load, etc.) for minimum reflections to occur. For radio systems, the most common impedance is 50 Ω.
- *Minimum bending radius.* RF cables should not be bent beyond a specified bending radius, otherwise they can be damaged and thus produce undesirable reflections. Their radius is specified in mm and varies depending on the diameter and construction of the cable.
- *Connectors.* For radio frequencies, suitable connectors should always be used to avoid mismatches and reflections. This also applies to Wi-Fi cables.

Twisted-pair cables are also used in indoor networks, especially as part of an active DAS to interconnect remote units with expansion units. One of the most popular ones used is CAT-5, used also for Ethernet installations.

Finally, a radiating cable is also used as part of a hybrid DAS or a purely passive DAS, since coverage can be extended in certain areas where it may be difficult to install antennas. These are the most relevant parameters:

- *Coupling loss*, in dB, also specified as a percentile and for which a coupling loss distance is given.
- *Longitudinal loss*, in dB/100 m, which is very similar to that of coaxial cables. This loss also varies with frequency.
- *Impedance*, in ohms, which should be matched with the termination load connected at the end of the radiating cable as well as with the transmitter.
- *Maximum frequency*, in MHz, beyond which the radiating cable should not be used.

Recall that radiating cables should be installed not in close proximity to metallic structures since their radiation characteristics may be greatly affected and undesired reflections may be produced, thus reducing the cable performance.

9.3.2 Splitters/Combiners

Splitters and combiners are often associated together since they perform the inverse operation: *splitters* divide the signal coming from a cable in two, three or more derivations; on the other hand, *combiners* add many signals coming from many sources into a single cable. In distributed antenna systems, the use of splitters is very common, as from the main base station power needs to be distributed accordingly inside the building. In summary, a splitter that is connected with the outputs as inputs and vice versa is a combiner. Figure 9.8 shows a commercial two-way splitter and its symbol.

Figure 9.3 shows an example of a building for which the signal is being divided to distribute it to many areas, as part of a passive DAS installation. An approximate value of 3 dB loss is associated with every power split. For example, from Ericsson, the KRF 102 246/2 has a 3 dB coupled loss (two-way splitter), whereas the KRF 102 246/3 has a 4.8 dB coupled loss (three-way) and the KRF 102 246/4 has a 6 dB coupled loss (four-way). Note that in addition to this coupled loss, there is an insertion loss of around 0.1 dB that needs to be considered for every port. If a port is not used, it must be terminated using a dummy load to avoid reflections.

A very useful way to calculate the total loss L_{splitter} from a splitter (dB) having n_p ports and $L_{\text{insertion}}$ insertion loss (dB) is as follows (Tolstrup, 2011):

$$L_{\text{splitter}} = 10 \log(n_p) + L_{\text{insertion}} \tag{9.1}$$

The first term $10 \log(n_p)$ is the so-called *coupled loss* and is also given in dB.

9.3.3 Antennas

As discussed earlier in Chapter 6, antennas constitute one of the basic and essential elements for any indoor network, thus requiring special attention. We will avoid giving a

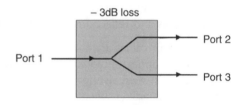

Figure 9.8 Splitter/combiner and its symbol.

thorough description of the characteristics of these elements – this was done in Chapter 6. Instead, we will focus on special considerations for in-building installations, as follows:

- *Installation constraints.* Antennas need to be installed only at places where their radiation patterns can be used efficiently in order to maximize coverage at designated spaces. For example, although it may be the best option to choose an omnidirectional antenna to provide coverage in the middle of a fall, if it is not possible to install ceiling-mounted antennas then directional panels mounted on walls or pillars need to be chosen. Antenna size can also be an issue if space is very constrained.
- *Aesthetics.* Cosmetic requirements restrain the use of indoor antennas that may visually 'pollute' a space inside a building. On the other hand, these antennas should be as discrete as possible in appearance so that users do not notice their location.
- *Frequency band requirements.* Some deployments require multiband antennas, which should be capable of operating in the designated bands. In general, multiband antennas are nowadays readily available and allow cost and space savings.
- *SISO or MIMO.* For modern cellular and other wireless standards, the use of MIMO is gaining popularity and therefore consideration of this needs to be taken into account.
- *Accessibility.* For areas where antennas are difficult to install, such as underground mines or tunnels, radiating cables are used. Special consideration should be taken to provide adequate maintenance and replacement (if this is the case) to these elements.

9.3.4 Directional Couplers

Their function is similar to that of splitters, but directional couplers only couple power flowing in one direction. Power entering the output port is coupled to the isolated port but not to the coupled port. Figure 9.9 shows an example of a directional coupler and its symbol. They can be used for separating transmitted and received signals in a system, for example.

Important parameters for directional couplers are:

- *Coupling factor.* It takes normally negative values and in practice should not exceed −3 dB since more than this would result in more power output from the coupled port than that from the transmitted port. This coupling varies with frequency, and the quality of a directional coupler is often specified in terms of variations of this loss

Figure 9.9 Directional couplers.

around its center frequency. The coupling factor $L_{3,1}$ in dB for a directional coupler having input port power p_1 and coupled port power p_3, both in W, is given by

$$L_{3,1} = 10 \log\left(\frac{p_3}{p_1}\right) \quad \text{dB} \tag{9.2}$$

- *Insertion loss.* The main line insertion loss from the input port to the transmitted port is

$$L_{i2,1} = -10 \log\left(\frac{p_2}{p_1}\right) \quad \text{dB} \tag{9.3}$$

Part of this loss is due to some power going to the coupled port, which is known as the *coupling loss.* This is given by

$$L_{c2,1} = -10 \log\left(1 - \frac{p_3}{p_1}\right) = L_{3,1} \quad \text{dB} \tag{9.4}$$

Note that the coupling loss and coupling factor represent the same thing – Equations (9.2) and (9.4) yield an identical answer.

For practical directional couplers, the insertion loss consists of a combination of other losses, named coupling loss, dielectric loss, conductor loss and VSWR loss.

- *Isolation.* For a directional coupler, isolation is defined as the difference in signal power levels in decibels between the input and the isolated port when the other ports are terminated in matched loads. Therefore, for example, for a four-port coupler, isolation between input ports 1 and port 4 is

$$I_{4,1} = -10 \log\left(\frac{p_4}{p_1}\right) \quad \text{dB} \tag{9.5}$$

Isolation is normally specified by some manufacturers as the *coupler loss* (be careful not to confuse it with the coupling loss). Typical values can vary from a few dB to many tens, as for the directional coupler from AlanDick, the COU-F0825-50A, which has an isolation of 50 dB and a maximum insertion loss of only 0.15 dB.

9.3.5 Tappers

A tapper or uneven splitter are used to divide the power from one into two lines, but the power, unlike the splitter, is not equally divided among the ports. For indoor deployments, this can be used to tap small portions of the signal from a passive DAS installation of a very long cable and use this to feed antennas along the cable. Thus there is no need to install many parallel heavy cables whilst maintaining the losses reduced.

A tapper would be in RF equivalent to a variable resistor in electronics design, since voltage levels can be adjusted and balanced in a circuit by adjusting the resistance of the device, while power levels can also be adjusted and balanced in a building. Figure 9.10 shows an example of a tapper.

Tappers normally have a port with high losses and one with low losses. The high-loss port is often specified in the tapper by its 'coupler loss', whereas for the low-loss port a 'coupled loss' is reported. For example, for CommScope tappers, such as the 255580, the coupled loss is around 0.3 dB whereas the coupler loss is 10 dB.

Figure 9.10 Tapper

9.3.6 Attenuators

For some applications, especially in passive DAS, signal levels need to be reduced and adjusted to avoid signal spillage in some areas, for which a signal attenuator is often used. Also, attenuators can be used in an active DAS to reduce high power levels and protect amplifiers from not being overdriven. Figure 9.11 shows an attenuator usually employed for in-building installations.

As done for resistance in electronics, when a specific value can be obtained by connecting resistors in series and/or in parallel, attenuators can be cascaded to obtain the desired attenuation value. For example, three 10 dB attenuators connected in series will produce a 30 dB total attenuation.

Figure 9.11 Attenuator.

9.3.7 Circulators

A circulator is a transmission line or waveguide device that allows energy to pass easily (i.e. with low attenuation, < 0.5 dB) in the forward direction but offers great opposition (losses > 20 dB) to signal flow in the reverse direction.

Circulators have been placed in a circular format, as depicted in Figure 9.12. The ports are arranged so that a signal entering at one port exits at the next. For example, a signal entering port A is absorbed in matched port B. Similarly, energy entering port B is absorbed in matched port C. Signal energy from port B cannot go to port A because of isolator action between these ports. By the same principle, signal entering port C travels to port D but not port B.

Circulators are used to protect a source or device where the mismatch or VSWR in a system is very high or where the load may be inadvertently disconnected, as is the case when an antenna may be accidentally disconnected from a radio transmitter. Figure 9.13 shows the use of a circulator to protect the output of a 900 MHz radio transmitter.

Circulators have an advantage over matching pads or attenuators because they offer little attenuation to the signal in the conducting direction. However, they suffer from the disadvantage of being relatively bulky and heavy in coaxial and waveguide form. Newer technologies have allowed the reduction in size and weight for circulators, which is the case for the one shown in Figure 9.13.

A common application for circulators is to use it to separate transmit and receive directions from combined TX/RX ports, but it needs to be used mainly for low power applications to reduce PIM issues in the circulator.

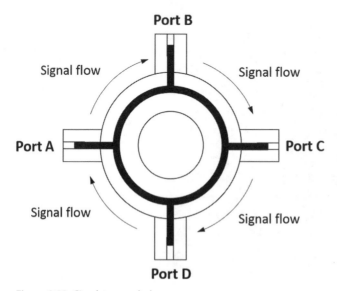

Figure 9.12 Circulator symbol.

Figure 9.13 Circulator used to protect a 900 MHz amplifier output.

9.3.8 Terminations/Dummy Loads

Terminators, such as that shown in Figure 9.14, are used as matching loads on a transmission line or in an open port on any other component (circulators, splitters, etc.) to avoid reflections. They are also known as dummy loads, and radiating cables should also be terminated using these devices or antennas.

Figure 9.14 Terminator.

Figure 9.15 Triplexer.

9.3.9 Duplexers

These devices are used to separate a combined transmit/receive signal into separate lines, for example when the same antenna is being used in the receiver and the two separate lines are required. The two most relevant parameters to check with duplexers are insertion loss and maximum power rating.

9.3.10 Diplexers/Triplexers

Duplexers are commonly confused with diplexers, but indeed we are talking about two different devices. A *diplexer* combines signals from different bands; for example 900 MHz and 1800 MHz could be combined in a single 900/1800 MHz output coming out from a diplexer. If three bands are involved, then a *triplexer* is used.

Note that diplexers/triplexers, as was the case for splitters, can be used in either direction to combine or separate signals from different frequency bands.

Figure 9.15 shows a triplexer used to combine various signals in TV and satellite applications.

9.4 RF Miscellaneous – Active

Active devices are widely used nowadays for many indoor wireless deployments. Two of the most relevant ones are cellular and Wi-Fi, and a brief description of key active components follows.

9.4.1 Amplifiers

Signal amplification is often required, especially for large facilities, and therefore special care should be taken when selecting amplifiers for this purpose. In general terms, amplifiers are employed to extend the coverage in a specific area without increasing the capacity of the system.

Bidirectional amplifiers are used when a two-way amplification is required. A very common device used to extend the coverage in cellular systems from an outdoor cell inside a building is known as a *repeater*. Due to its relevance and importance, Section 9.5 is dedicated to repeaters.

Relevant aspects to check when selecting an amplifier include: gain, noise figure, automatic gain control (AGC) availability, operating frequency, gain step size, impedance, maximum output power, compression and intercept points (related to linearity), size and weight. Remember that amplifiers add noise to a system and potentially could create IMD (intermodulation distortion), which affects the quality of the signal being amplified. Also make sure that for the selected frequency band the amplifier is linear – otherwise distortion will occur at out-of-band frequencies.

9.4.2 Active DAS Components

Active DAS architectures include various elements that convert the signal to/from RF from/to optical, which will be explored in the following sections. For a deeper analysis of active DAS, refer to Tolstrup (2011).

9.4.2.1 Main Unit

The main unit is the first element in the architecture of an active DAS system and connects the base station equipment with expansion units, which are the next elements in the network. These connections are done using coaxial cable, and main units have RF-to-optical converters as well as A/D converters to transport signals using optical fibre over long distances (a few km depending on the type of fibre; i.e. multimode or monomode) and thus minimize losses. This unit generates and controls internal calibration signals in the system together with internal amplifiers, thus adjusting gains and levels to the different ports to compensate for the variance of internal cable losses between all the units.

9.4.2.2 Expansion Unit

The next element is the network is the expansion unit. These devices are distributed throughout the building and are placed in central cable risers. The expansion unit is connected to the main unit using optical fibre, with separate lines for uplink and downlink. The expansion unit then converts the optical signal coming from the main unit into an electrical signal for the remote unit.

9.4.2.3 Remote Unit

The last element in an active DAS is the remote unit, which is installed close to the antenna in order to keep losses due to cables to a minimum. The remote unit converts the signal from the expansion unit from an electrical signal to an RF signal in the downlink and the RF signal from the mobile units into an electrical signal in the uplink. The remote unit is normally fed with power from the expansion unit in order to avoid an expensive (or unavailable) local power supply at each antenna point.

9.5 Repeaters

A *repeater* is a mobile signal amplification device or 'booster', which requires an adequate 'donor' signal from the macrocell network to be present outside the building to be received, amplified and retransmitted into the building. Unlike many other dedicated in-building solutions, repeaters do not require a broadband connection. As repeaters simply boost the 'donor' mobile signal level they do not add capacity to the network and as such tend to be used in areas of poor coverage rather than in high traffic areas.

By using a radio repeater, as shown in Figure 9.16, basically a bidirectional amplifier with a high-gain and narrow-beam BTS-facing antenna and one or several antennas serving the mobiles, the geographical coverage can easily be extended into in-building areas, with the repeater gain overcoming the building penetration loss. The RF signal received by the repeater can be distributed across the building.

9.5.1 Repeater Deployments

Repeater solutions tend to be deployed in the following three ways: operator deployed repeaters, traditional consumer repeaters and intelligent repeaters. These are explained in this section.

9.5.1.1 Operator-Deployed Repeaters

These typically consist of a rooftop or wall-mounted antenna on the outside of the building to pick up the 'donor' mobile signal from a macrocell. This is then amplified and retransmitted into the building, which can either be done wirelessly or over cables to antennas in the building (which may be a DAS). These systems need skilled installation and planning to minimize interference to the donor macrocell and with deployments carefully monitored by operators for interference. Typically an operator will deploy one of these solutions on request in large buildings for large corporate customers with service issues as part of their service agreement. Operators also deploy repeaters outside this in

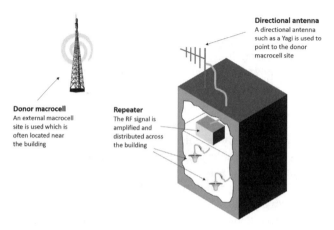

Directional antenna
A directional antenna such as a Yagi is used to point to the donor macrocell site

Donor macrocell
An external macrocell site is used which is often located near the building

Repeater
The RF signal is amplified and distributed across the building

Figure 9.16 RF repeater.

low traffic areas where coverage is problematic over a small area, such as tunnels and roads in rural areas. The general view of operators in stakeholder discussions was that repeaters are only deployed as a last resort and are not a preferred solution.

9.5.1.2   Traditional Consumer Repeaters

These tend to be small, low cost, bidirectional amplifier units that the user can place on their window ledge to pick up a 'donor' mobile signal, which is then 'boosted' and retransmitted into the building. In-car repeater units are also available. In the UK these units can be purchased but are illegal to deploy without permission from the operator, which is difficult to obtain. This is due to interference concerns to the macrocellular network, which are much greater in these devices than in operator deployed repeaters as:

- These units tend to be broadband units working across a number of operators.
- The operator has no direct control over where the unit is used and its settings.
- The low cost and small form factor of the unit means that the same level of skilled installation and setup associated with operator deployed repeaters to minimize interference cannot be achieved.

Generally the low cost and small form factor of these units means that there is not good isolation between the two opposite-facing antennas in the unit, which in turn limits the amplification that can be achieved without having feedback issues. This combined with the placement of the unit on a window ledge rather than having an antenna external to the building will limit the levels of service improvement achieved.

In the case of consumer deployed repeaters, in the UK at least, while there have been no direct barriers to vendors producing consumer repeater products, and indeed a range are available to purchase online, the deployment of these has been limited by the regulatory rules put in place on the use of such products to protect operators from interference in their licenced spectrum. These currently require consumers to gain approval for their repeater product from their operator prior to deployment. This is generally difficult to obtain as UK operators remain to be convinced that traditional consumer repeater products will not cause interference to their networks.

In the US the opposite has been true and repeaters have been permitted to be deployed without specific permission from operators, leading to their widescale use there. However, recently the FCC position on repeaters has been revised notably to:

- More tightly specify the functionality and performance of commercial repeaters
- Require all users of repeaters, both legacy users and new users, to register their device with their operator and gain their approval for its operation.

This move towards tighter specification of the technical functionality and performance of consumer repeaters means that this allows operators to confidentially approve deployment of particular product ranges. Indeed, in the US all four nationwide carriers are reported to have consented to the use of consumer repeaters on their networks for products within the new regulations. However, it is likely that operators will always have a preference for knowing and controlling where repeaters are deployed.

This tighter regulation may in the short term limit availability of consumer repeaters to perhaps even just the intelligent repeater solution currently available from Nextivity but there are no barriers seen on why other vendors could not upgrade existing product ranges to offer similar features to meet these tighter technical requirements.

9.5.1.3 Intelligent Repeaters

These are a recent update to traditional consumer repeater products and consist of two boxes (i.e. separate donor and mobile units) with integral patchy antennas that are linked by a 5 GHz radio link (i.e. the same licence exempt band as used by the most recent Wi-Fi devices). The outdoor 'donor' signal is picked up by the donor box, digitized and transmitted to the mobile unit via the 5 GHz link and then retransmitted by the mobile unit. This allows the donor unit to be located near a window and the mobile unit to be positioned where the coverage improvement is required. This evolution of traditional consumer repeaters has notably been approved by some network operators and aims to minimize interference to the macrocellular network by:

- Using intelligent power control
- Having good isolation between the donor and indoor antenna as these are in separate boxes
- Repeating the signal from within the building rather than on the window, which isolates the indoor signal from the outdoor macrocellular network better
- Allowing the operator to remain in control of the unit by providing a simple management feature that allows operators to remotely signal to the intelligent repeater unit to switch off should the operator observe that the unit is causing interference.

Recently available LTE gateway products can be seen as extensions to intelligent repeaters. These are fixed consumer premise equipment units (CPEs) that provide a fixed broadband connection to a home from an outdoor LTE network. However, instead of providing an LTE signal indoors this broadband connection is distributed via Wi-Fi or a fixed Ethernet cable, but this limits services available to data services due to the limitations of Wi-Fi for providing cellular-like services. LTE relays represent the next evolution of these products and will include the ability to extend the coverage area of existing LTE cells. However, no LTE relay products are as yet available.

9.5.2 Disadvantages

With the use of repeaters to provide coverage inside a building, various disadvantages are envisaged. There is a need to ensure sufficient isolation to avoid feedback, as well as for cases where high gain is required to penetrate a building, as a rise in uplink noise becomes a problem. Although repeaters extend coverage, they do not increase network capacity, as they 'steal' the resources from the donor cell to be used indoors. This is a major limitation, as for dense-populated areas, a sacrifice in the capacity of the donor cell may lead to cell congestion if the resources are not properly allocated. Nevertheless, this option is still in use for low-cost easy-deployment situations with low-capacity demand and small coverage areas.

9.5.3 Installation Issues

For operator deployed repeaters, the end-user sees the benefits of improved mobile service without any action on their part. The building owner is likely to be involved in the physical installation of the unit and needs to tolerate some disruption while the system is installed. Disruption will be particularly high if the repeater is feeding into an indoor distributed antenna system, which requires a series of cables and indoor antennas to be installed throughout the building. The erection of the repeater antenna on the rooftop of the building may also require planning permission and cause additional inconvenience to the building owner in this way.

In the case of consumer repeaters these are straightforward plug-and-play units that in theory the user just needs to place on a window ledge and provide power to. However, installation can be complicated by the fact that the user needs to have a particular signal level outdoors before the repeater unit will work and, as there is no standard definition of a 'bar' of signal on a mobile handset, it can be challenging to work out if the outdoor signal is good enough and the unit should be working. Additionally, the move to better thermal insulation in buildings, and in particular the use of windows with a metallized coating, will make it more difficult for indoor repeater units to pick up the outdoor 'donor' signal without having an external antenna added or even the window left open.

For intelligent repeaters the user needs to place the donor unit on a window ledge and the mobile unit in the area where the service improvement is needed and provide power to both. Although the 5 GHz link between the two units does not require a clear line of sight, it cannot tolerate too much obstruction. The patch antenna in the donor unit also has some degree of directivity, which may require orientation to obtain the highest donor signal. This optimization process is supported by an indicator on the unit but may mean that the user needs to try out a few locations before finding the best setup. The issue of having a good outdoor signal applies, as discussed for the traditional consumer repeater case. The intelligent repeater product is also only available for UMTS currently so the consumer will need a UMTS handset for it to work. Additionally, heavy usage of the 5 GHz link by the repeater units may mean that Wi-Fi access points nearby may no longer achieve as high data rates as previously.

Distinct from the majority of other in-building solutions, repeater solutions do not require the user to have a broadband connection. This makes them the only solution for some niche areas where there is no broadband connection but a reasonably good outdoor cellular signal that is prevented from entering the building due to the building construction material or geometry (although an external antenna might be needed if thermal shielding in the windows is the main problem). As well as rural customers this may include those who wish to completely substitute their fixed line service with a mobile wireless service in their home or who may not be able to afford both a mobile phone and a fixed line service. As mobile service levels improve this is a category of users that could potentially increase with more consumers viewing mobile services as a complete substitute option for fixed line services.

Note that the fact that repeaters tend to be deployed in buildings with construction materials that prevent an outdoor signal reaching all parts of the building does not mean that the building will necessarily shield the outdoor macrocell network from interference from the repeater itself. In traditional consumer repeaters the repeater unit tends to be located on a window ledge and boosts the received signal in all directions, not just into the building, and so will leak the signal back towards the outdoor cellular network. Intelligent repeater solutions address this to a certain extent by locating the repeater unit inside the building rather than close to the outside of the building such as on a window ledge. However, in both cases the 'donor' outdoor signal is boosted in the building and so is likely to generate a signal outside the building at a higher level than the existing 'donor' signal from the outdoor network and hence cause interference issues.

Crucially all repeater products must be installed with the approval of operators due to UK policy on these devices. In the US the deployment of repeaters is much more open, which leads to them being more widely used.

9.5.4 Benefits

Once installed the indoor service improvements seen by consumers from repeater solutions will include:

- Better availability and quality of cellular voice, SMS and data services in the building (although this will be better through an operator deployed solution than the consumer installed options)
- Better data rates (although nearby Wi-Fi access points may see a reduction in their maximum data rates when the intelligent repeater is installed)
- Improved mobile phone battery life.

9.6 Conclusion

Network performance can be severely affected by the choice of RF components, thus a very careful selection is needed. For example, noise rise can degrade the performance of a cellular 3G system and there are potentially various sources for which noise can be increased, such as use of large lossy cables, using attenuators in the front end of the UL and thus contributing more to the overall noise figure and selecting amplifiers with high noise figures and relatively low gains. No matter how good we have done the propagation prediction work or how accurate our traffic demands are to account for capacity, if we have the wrong elements, performance will be greatly decreased! Remember, it is like a nice cooking recipe: the instructions and the execution to prepare the dish can be perfect, but if the ingredients are not the correct ones, we will not get the best taste!

References

ICNIRP 1998) International Commission on Non-Ionizing Radiation Protection, Guidelines for limiting exposure to time-varying electric, magnetic, and electromagnetic fields (up to 300 GHz), *Health Physics*, **75** (4), 494–522.

Tolstrup, M. (2011) *Indoor Radio Planning: A Practical Guide for GSM, DCS, UMTS, HSPA and LTE*, 2nd edition, John Wiley & Sons, Ltd, Chichester. ISBN 978–0470710708.

10

Small Cells

Prof. Simon R. Saunders

Adjunct Professor, Trinity College Dublin

Small cells are not a new concept, but over the last few years they have entered a new phase of growth and development, with a standards-based ecosystem, greater acceptance by operators and tens of millions deployed. As a foundational technology for network densification, they are a key element in advanced 4G and eventually 5G mobile technology.

In this chapter we chart the development of small cells, explaining some history and the technology principles and give an overview of market trends and the business case for small cells. We explain where small cells fit in the in-building wireless toolkit and explain how small cells are expected to contribute to the future evolution of the mobile network topology and to 5G.

This chapter draws on our previous book on femtocells (Saunders *et al.*, 2009) and our work with Small Cell Forum[1] and Real Wireless[2] and the interested reader is referred to those sources for further details.

10.1 What is a Small Cell?

There is no formal definition of small cells, but they can most easily be explained as any cell serving mobile traffic which is not a conventional macrocell. The Small Cell Forum provides the following definition:

> 'Small cells' is an umbrella term for operator-controlled, low-powered radio access nodes, including those that operate in licensed spectrum and unlicensed carrier-grade Wi-Fi. Small cells typically have a range from 10 meters to several hundred meters.

Small cells are usually fully fledged sources of both coverage and capacity – essentially mini base stations. Some people include the whole range of indoor coverage systems

---

1 www.smallcellforum.org.
2 www.realwireless.biz.

Indoor Wireless Communications: From Theory to Implementation, First Edition. Alejandro Aragón-Zavala.
© 2017 John Wiley & Sons Ltd. Published 2017 by John Wiley & Sons Ltd.

under the small cells banner, such as DAS, repeaters and Wi-Fi, but in this chapter we will focus solely on devices that autonomously deliver their own capacity and coverage using mobile standards.

Small cells incorporate significant intelligence to support simple incorporation into the operator network. They support self-optimization with little or no attention by the operator and with 'zero touch' installation on the part of the end user. Specific features vary by manufacturer, but they typically incorporate an ability to monitor the surrounding network and adjust themselves to provide good coverage over the desired area while avoiding interference beyond.

While these processes are conducted autonomously, the small cell remains always under the control of the network operator via a centralized management system.

Although small cells may be backhauled over the public Internet, they are secured via mutual authentication between the operator network and the cell, with all communication taking place over a secure, encrypted IPsec tunnel. The cell reports its location to the network and is only granted permission to transmit once the network checks it is authorized to do so, thereby meeting regulatory requirements on spectrum, emergency calls, etc.

10.2 Small Cell Species

Small cells have evolved to include several 'species'. These are briefly explained in this section.

10.2.1 Femtocells for Residential Environments

A femtocell will provide good coverage throughout a typical dwelling, with sufficient data and voice capacity for the needs of the occupants. They typically serve 4–8 simultaneous users and a coverage radius of tens of metres via transmit powers from a few milliwatts to around 100 millwatts. They are installed by the end user and backhauled via standard domestic fixed broadband and the Internet, as illustrated in Figure 10.1. Most femtocells are standalone units to be added to existing broadband routers, but increasingly operators are integrating femtocells within home routers, simplifying installation and providing opportunities for QoS management.

10.2.2 Picocells

Picocells or *enterprise small cells* provide service to office buildings and other work places, or to smaller public buildings. They may serve several offices or an office floor via transmit powers up to a few hundred milliwatts. They may be planned and installed by operator personnel via processes similar to those described in other chapters. However, their self-managing capabilities open the potential of much simple planning rules similar to those used for enterprise Wi-Fi systems, allowing IT personnel to deploy them as part of the enterprise infrastructure and to achieve far greater scale than has been possible with DAS. Enterprise small cells are backhauled via enterprise-grade broadband connections, although often these are dedicated rather than shared with existing enterprise connectivity to assure QoS, especially for voice. Within the building small

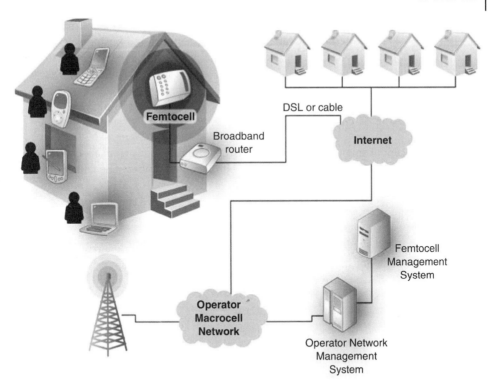

Figure 10.1 Residential femtocell concept.

cells will usually share existing LAN wiring, although usually segmented on a separate VLAN. Often the small cells in a building will be controlled and managed via a local controller, although the details of the segmentation of functions between the access points and controller, and the interfaces between them, are vendor-specific.

10.2.3 Metrocells

Metrocells or *microcells* are outdoor small cells, typically mounted on street furniture, such as lampposts or CCTV poles or else on the sides of buildings at typically 6–12 m above ground. They provide coverage of a focused area of a few hundred metres in diameter. They would usually be deployed in an area already covered by macrocells, with the intention of increasing the local system capacity, improving the user experience of data speeds and providing coverage deeper into buildings. Metrocells are backhauled either directly over fibre networks or else via intermediate wireless point-to-point or point-to-multipoint links.

10.2.4 Rural and Remote Small Cells

Rural and remote small cells (sometimes informally called 'meadow cells') are intended to provide coverage in areas where user density is too low to be economically served via macrocells or in special areas such as remote industrial sites or for special events. The technology used is essentially identical to metrocells, although backhaul may be via satellite links. The distinction is over the deployment model and business case.

In this chapter we will concentrate on indoor small cells, that is picocells and femtocells.

10.3 The Case for Small Cells

The basic concept of small cells is to deliver coverage and capacity from a location closer to the user than a conventional macrocell. This produces a number of benefits, which are briefly described here.

10.3.1 Capacity

By shortening the radio path to the user, interference can be better controlled, allowing denser reuse and hence higher capacity. This allows operators to support the enormous growth of demand seen in recent years in a sustainable fashion, even within limited spectrum resources.

10.3.2 Coverage

By placing cells indoor and in other confined spaces, coverage can be improved in a focused manner and targeted precisely and hence cost-effectively. Small cells come in many shapes and sizes, allowing them to be tailored to the needs of a given building.

10.3.3 User Experience

Small cells operate at higher signal-to-noise-plus-interference levels than macrocells and serve fewer simultaneous customers. As a result, typical data rates can be significantly higher. Since small cells offload congested macrocells, these benefits are seen for both the users on the small cells and for those remaining on the macrocell network.

10.3.4 Cost Effectiveness

By focusing coverage and capacity on exactly the places and users in greatest need, network investment can be reduced for a given level of investment. Small cells can be remotely managed, have self-optimizing capabilities and can make use of consumer- and enterprise-(rather than carrier-)grade backhaul, reducing operational expenditures. In the case of residential and (to some extent) enterprise femtocells they can be installed by users and installers without any RF background, reducing the initial capital expenditure of planning and installation.

10.4 History and Standards

Interestingly enough, the concept of small cells dates back to the last century, where mobile telephony standards were evolving. GSM picocell products were introduced in the mid-1990s and were used to serve some large-scale enterprises and some public buildings. However, these were largely cut-down macrocell equipment, with little cost advantage and requiring detailed RF planning, so deployments were sparse and operators tended to adopt an 'outdoors in always wins' approach.

However, by around 2007 3G had become established and was starting to be used more extensively for data applications, placing more emphasis on its coverage. Data applications tend to be used predominantly outdoors and 3G was deployed mainly at higher frequencies such as the 2.1 GHz band in Europe, so indoor coverage limitations were more apparent than for GSM.

The concept of femtocells was then introduced, building on several factors:

- The availability of low cost high-speed signal processing chips
- The increasingly widespread availability of domestic fixed broadband at reasonable data rates
- The take-up of compelling Internet applications
- Increasing adoption of smartphones
- The growth of importance of mobile data revenues for operators.

This combination of factors made small cells in their femtocell guise more practical than ever before. Vendors collaborated together, under the auspices of Small Cell Forum (originally known as Femto Forum), to agree technical architectures, management protocols and deployment best practices, which were subsequently standardized in partnership with bodies such as 3GPP and Broadband Forum.

In order to drive the adoption of femtocells with a good business case, the price point of femtocells needed to be at least two orders of magnitude lower than their macrocell cousins – that is less than around $100 – while supporting similar signal processing functions and more advanced self-optimizing capabilities. Meeting this challenge depended on the existence of open, interoperable standards, allowing vendors to produce a single device type to serve a wide operator audience, and giving operators an open choice amongst vendors.

Femto Forum drove these standards by gathering operator requirements and encouraging consistent vendor approaches, especially in terms of network architectures and management protocols. Building on this effort, 3GPP's Release 8 included a standardized architecture for 3G femtocells, known as Home Node B in 3GPP terminology. It also laid the foundations for small cells in LTE, including special features to be supported in the user devices, such as support for closed subscriber groups, easing the mobility management between the macrocells and small cell layers.

Therefore, while 3G small cell standards were added some 10 years after the initial standards, small cell concepts were incorporated in the very first release for LTE, although the details for LTE came in later releases (9 and beyond). Looking forward to 5G, small cells and other network densification techniques are foundational and essential if the very wide bandwidths available at millimetre waves are to be exploited.

Small cells have thus become a standard and essential element of the mobile toolkit, creating a heterogeneous network that can be better matched to the diverse needs of mobile data users than the inflexible macrocell-only architectures of the past.

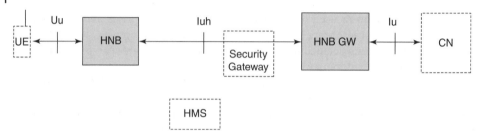

Figure 10.2 3G standardized small cell architecture.

10.5 Architecture and Management

The standardized architecture for 3G femtocells is illustrated in Figure 10.2. The femtocell is referred to in 3GPP specifications as a home NodeB or HNB. It communicates over the air interface (Uu) in a way that is entirely consistent with conventional 3G protocols, meaning that no changes to the user equipment (UE) are necessary. The HNB communicates with the operator core network (CN) over the Internet via the Iuh interface, where all the traffic is carried securely via an IPsec tunnel, following mutual authentication between the HNB and a security gateway. The HNB incorporates radio control functions similar to those undertaken by the radio network controller in conventional 3G networks, so Iuh is more like the Iu RNC-to-core interface than the conventional Iub base station-to-RNC interface. It is optimized to avoid unnecessary signalling traffic over the Internet when the HNB is idle. As well as being standardized, strenuous efforts in the industry have made this an open interoperable interface, permitting femtocells from multiple vendors to integrated into one network.[3] Iuh terminates at a specialized gateway (HNB GW) in the operator's core network, which includes the remaining functions that would be executed by a conventional RNC and then presents a conventional Iu interface to the operator's core. The femtocells are managed by a management system (HMS) that uses a standardized management protocol based on Broadband Forum's TR-069 management protocol, originally developed for managing home broadband modems and providing scalability to millions of devices.

Figure 10.3 shows the architecture for LTE small cells. While this is superficially similar to the 3G case, a significant difference is that the LTE small cell (HeNB) uses the same S1 interface as standard LTE base stations (eNB). The gateway (HNB GW) is simply a transparent concentrator that aggregates multiple S1 interfaces into a single S1. It may be omitted in the case of a relatively small deployment, giving the operator a more economical start-up route. See SCF025 for more details of the LTE architecture options.

These 3G and LTE architectures are applicable to individual femtocells deployed in the home or in small office environments where a single cell per premise is sufficient. In the case of larger environments, such as offices or public buildings, multiple cells are needed for coverage and capacity, and it becomes appropriate to handle interactions between these locally, including handover processes and local radio optimization. It may also be useful to integrate the small cells with local enterprise IT and telephony systems.

---

3 This was supported by "plugtests" conducted by Small Cell Forum and ETSI.

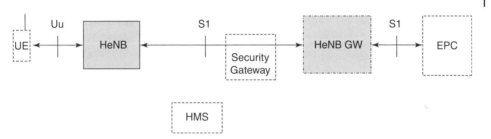

Figure 10.3 LTE standardized small cell/femtocell architecture.

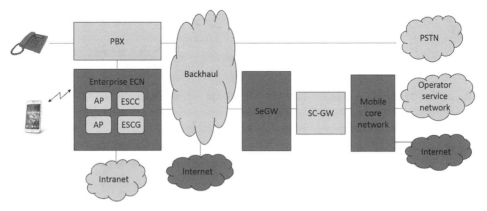

Figure 10.4 Enterprise small cells architectural framework. Taken from SCF067. Reproduced with permission of Small Cell Forum.

There is no standardised approach to this, as many different architectures may suit different buildings and different vendor concepts, but Figure 10.4 illustrates a general enterprise architecture framework. The local enterprise small cell network (SCN) may include an enterprise small cell gateway (ESCG), which handles local access to resources such as the enterprise Intranet and telephony systems and other specialized functions. An enterprise small cell concentrator (ESCC) aggregates individual small cells together and can handle local mobility amongst cells, avoiding the load on the core network and improving performance. Thus an entire enterprise small cell network may present to the operator's core as if it were a single cell. Management functions may also be split between the operator management system and the local enterprise, reducing the operational burden on the operator and giving more flexibility to the local users.

10.6 Coverage, Capacity and Interference

A major technical challenge that femtocell designers initially faced was the need to manage potential interference. It takes up to two years to install conventional base stations, during which time radio engineers meticulously plan a station's position and radio characteristics to avoid interference. However, such an approach is not viable in the case of femtocells, deployed potentially in their millions at random. Automating a process conducted by radio engineers was no mean feat and simply would not have been

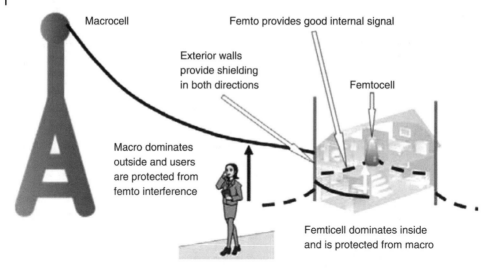

Figure 10.5 Managing interference between indoor small cells and macrocells.

possible a few years ago. Fortunately, the fact that the walls of buildings keep macrocell signals out and keep the femtocell signals in provides strong inherent interference mitigation for indoor femtocells (Figure 10.5).

Extensive studies (SCF003, SCF008 and SCF009) have shown that proper implementation of a few key techniques to reduce interference can take advantage of this attenuation in an intelligent manner. Such techniques include frequent monitoring of the cell's surrounding radio environment combined with adaptive power control. Indoor users gain faster data rates, as do outdoor users who now operate on less congested cells, while it costs less for operators to deliver higher overall network capacity. Large-scale, real-world deployments have demonstrated that these techniques work in practice and even allow new approaches, such as operating 3G or LTE small cell networks in the same spectrum as 2G macrocell networks, allowing a spectrum to be refarmed in a smoother and more rapid fashion.

Additionally, LTE incorporates a suite of intercell interference coordination (ICIC) features allowing radio resources to be controlled in an integrated manner with resources in the macrocell network. These have been progressively enhanced including ICIC (coordination of traffic channels in time and frequency), eICIC (coordination of traffic and control channels in time, frequency and power) and FeICIC (further enhancements including advanced receiver cancellation and support for carrier aggregation).

Once these techniques have been properly applied, reuse of spectrum between macro and small cells ceases to be a problem and becomes a positive opportunity for improving capacity, coverage and the user experience. For example, Figure 10.6 shows how the throughput experience of users is massively enhanced when femtocells are deployed co-channel with macrocells with appropriate interference management techniques. Throughput is enhanced by around 100 times for the users on the femtocells and around 10 times in total including the macrocell users, due to reduced loading on the macrocells.

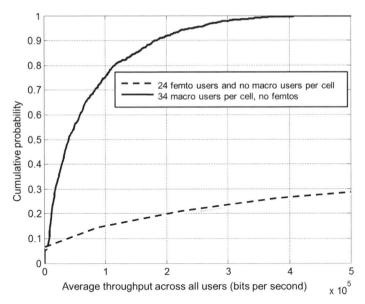

Figure 10.6 Average throughput of users with and without femtocells.

10.7 Business Case

The business case for small cell deployment differs according to the use case (home/ enterprise/urban/rural) and according to the operator's existing network and motivations for small cell deployment.

Residential femtocell deployments can enhance coverage for individual subscribers, reducing the cost of retaining or attracting customers in areas with poor coverage, which may include both rural areas but also densely populated areas where building penetration is poor. The cost of deployment is low on a per-customer basis and the costs of backhaul, power and sometimes the device itself may be borne by the customer. Thus femtocells can reduce churn, reduce the cost of attracting customers and act as a basis for cross-selling of additional mobile services and 'quad play' offers including broadband and TV services (SCF005 and SCF013).

Enterprises can benefit from enterprise-tailored voice capabilities on all mobile devices as well as a range of secure data services both delivered by the operator and locally over their own networks. Gaining good mobile coverage is important to enterprises: according to one study, 39% of UK IT managers complain of poor in-building mobile coverage and 35% of managers would switch operators to get better coverage (YouGov, 2013). The combination of different benefits results in a very attractive business case. A financial analysis of the business case for small cells across six enterprise environments demonstrates a highly positive business case for both enterprises and operators (SCF062). As an example, a medium enterprise case study demonstrates an increase of customer lifetime value (CLV) for the operator of 194% and for the enterprise of 73%.

10.8 Regulation

Broadly speaking, few new regulations are required to introduce small cells, since they fit under existing spectrum and telecoms regulations with little adaptation. Indeed small cells can help to meet the typical goals of regulators such as:

- **Spectrum efficiency.** Delivering very dense frequency reuse and consequent spectral efficiency in existing spectrum, small cells make good use of limited spectrum resources. They can also open up the use of higher frequencies where there may be vacant spectrum but where it is tricky to use traditional macrocells. They can equally be used at low frequencies given the use of appropriate interference management techniques.
- **Economic efficiency.** By reducing the cost of delivering services these benefits can ultimately be passed on to the national economy in terms of benefits generated for citizens/consumers and producers alike.
- **Enabling competition** by reducing the barriers to entry for new operators and enabling more operators to deliver services than would otherwise be possible in a given band. For example, in 2006 Ofcom in the UK auctioned a small block of 1800 MHz spectrum to 12 operators concurrently. Similarly, the US is making a large block of 3.5 GHz available for shared wireless broadband use via a tiered model including both licensed and unlicensed operation, enabled by small cells (FCC, 2015) and dynamic spectrum access systems. All of these give the opportunity for greater choice for consumers.
- **Broadening access** to services, for example in rural areas, avoiding any digital divide and offering the benefits of mobile broadband to a wider population of private users and businesses.

Given these positive impacts on regulatory goals, regulators should be motivated to adjust any regulations that may hinder the deployment of small cells. In 2008 the Radio Spectrum Committee of the European Union confirmed that, given that small cells operate as part of the operator's existing network, they can comply with existing technical licensing conditions and that the proliferation of such cells is supported in the context of more efficient use of spectrum. Also in 2008, the regulator in Japan simplified base station approval regulations to allow femtocells to be deployed by end-users.

One particular area of regulation is in ensuring that small cells can adequately report their location to support emergency calling and lawful intercept requirements. Several techniques are available for providing location capabilities to sufficient precision to allow meet such requirements to be met (SCF036).

Overall, while there have been some regulatory concerns surrounding small cells, most of these have been studied and found not to be an issue due to the femtocell remaining under operator control and the strict authentication and security mechanisms associated with small cells (SCF076).

10.9 Small Cells Compared With Other Indoor Wireless Technologies

10.9.1 Distributed Antenna Systems (DASs)

Femtocells do not overlap with a DAS for home environments, which are far too small for a DAS to be attractive or practical. In major public buildings, such as airports and shopping centres, where very high capacity and multioperator, multitechnology support is required, the DAS remains a natural choice.

There is a large number of buildings in between these two extremes, however, such as in offices and other workplaces, where the DAS has traditionally been too expensive but that are too large for a single femtocell deployment. In these environments picocells – enterprise small cells – are emerging as a natural choice and are seeing substantial growth.

By avoiding specialized cable installations in favour of conventional LAN cabling – potentially even sharing the existing office LAN – the installation costs can be reduced relative to the DAS. Further, the self-optimizing capabilities of small cells reduce the dependence on specialized RF planning, potentially allowing system integrators more used to Wi-Fi planning to create the design, although additional design guidelines must be followed.

In a cost comparison designed to challenge the opportunity for small cells, the total cost of deploying small cells in some real, relatively large, office environments was compared with the equivalent cost for a DAS in both single- and multioperator cases. In a 585 000 ft^2 building upgrading from an existing 3G DAS to LTE for four operators was 39% cheaper via small cells rather than via a traditional DAS approach. In the case of a new 4G-only system the savings increased to 48%.

Such savings are unlikely to mean that small cells will replace the DAS for the foreseeable future, but it is more likely that small cells will allow the large number of buildings where a DAS is not viable to gain dedicated coverage for the first time.

10.9.2 Wi-Fi

Wi-Fi is widely available and provides valuable services. While Wi-Fi was not originally intended for operator-style managed services, lacking security, authentication and mobility capabilities, several initiatives have sought to address these issues. While these initiatives have taken much longer to become widely established than was expected, Wi-Fi is progressively becoming a viable platform for operators, resulting in a new 'carrier Wi-Fi' ecosystem. In some cases, the applications for Wi-Fi create an alternative to licensed spectrum small cells. Overall, however, there are several reasons for the two technologies to continue to complement each other:

- In busy areas, cellular technologies can deliver a quality of service that is difficult to ensure with Wi-Fi, given its intrinsic 'listen before talk' nature (via carrier sense multiple access and collision avoidance – CSM/CA) and lack of centralized air interface management.
- Wi-Fi and LTE have different characteristics in terms of coverage and performance limitations under a high load as illustrated in SCF063. Small cells may provide a greater range in some cases, delivering an increased benefit per cell relative to the cost.
- Latency-critical services such as voice are also intrinsically well suited to cellular technologies for the same reason.
- While issues such as authentication and mobility can be addressed in Wi-Fi, there are several techniques that offer similar capabilities, leading to uneven support for these across phones, while small cells are uniformly supported by all phones.
- Given the increasing pressures on spectrum, only a combination of Wi-Fi and small cells allows all of the available spectrum resources to be used.

Most operators therefore see cellular small cells and Wi-Fi as complementary rather than competing technologies, which will often and increasingly be deployed together as mutually supportive elements of an integrated heterogeneous radio network (SCF089) and a variety of architectures exist for making the best use of this combination (SCWBA, 2016).

10.9.3 Repeaters and Relay Nodes

Repeaters were historically a common means of improving in-building coverage. They are essentially bidirectional amplifiers and therefore require some existing outdoor coverage and rely on professional installation to avoid feedback between the incoming 'donor' signal and the outgoing amplified signal. They are also unmanaged devices, which are difficult to retrieve and adjust when changes are made to the donor network. While they are a plausible tool for improving coverage, they use the macrocell capacity and so do not improve the overall system spectrum efficiency significantly. Finally, the cost of repeaters is now often greater than small cells, especially when installation and retrieval costs are taken into account, so many operators have replaced routine installation of repeaters with small cells.

However, in some circumstances no backhaul is available that is suitable for deployment of conventional small cells. In such cases, an emerging alternative is the use of 'relay nodes', which are standardized LTE devices that retransmit data from macrocells towards local users. They are more than simple repeaters since they actively decode and remodulate the data, allowing them to operate with weaker donor signals and without any risk of feedback. They are also an actively managed element of the network, allowing them to be retuned and optimized along with the rest of the network. They are in a sense therefore a form of small cell with off-air backhaul and a useful addition to the operator's options.

10.10 Market

While many positive market forecasts for small cells have been made over the years, it is certainly the case that deployments have been slower than expected. Several reasons can be identified for this:

- While the technology of small cells has been well-proven, the approach to deployment and management has been very new for operators, making it challenging to introduce them into existing organizational processes.
- The transition from 3G to LTE eased pressures on 3G capacity and caused operators to wait on mass deployments until their 4G macrocell network was established and LTE small cells were available at sufficiently low prices.
- Existing operators naturally and rationally seek to maximize their use of their existing investments in macrocells for as long as possible, so they tend to defer small cell deployments.
- In the case of urban small cells, the complexity of locating suitable sites and backhaul have made it more challenging than expected to achieve the necessary low costs for a viable business case.

- In the case of enterprise small cells, some operators have attempted to deploy them via traditional DAS system integrator routes, limiting the volumes due to the cost and slowness of that existing route.

Despite these challenges, small cells are now deployed in large numbers and are growing strongly. Overall 11.2 million small cells had been shipped as of March 2015, mainly femtocells (SCF050). Over 77 operators had deployed small cells, including 47 in the home and 71 in enterprise or public buildings. New deployment approaches and an increasing tendency for enterprise to self-fund deployments in their own buildings led to a doubling of enterprise shipments during 2015 and this is expected to be the area with strongest growth over the course of the next few years. Such growth will require a more complete 'industrialization' of deployment processes, putting design and installation into the hands of enterprise suppliers following simplified guidelines (SCF032) and using automated management approaches, leaving specialized cellular system integrators to concentrate on their traditional locations such as large-scale public buildings.

10.11 Future: New Architectures and Towards 5G

As small cells continue to be adopted more widely, the technology is developing to include a wider range of architectural options, allowing products to be selected to match the target environments more closely in performance and cost terms. For example, the trend towards virtualized and centralized RAN, following C-RAN, NFV and SDN principles is also being applied to small cells with appropriate architectural modifications (SCF159). Such architectures typically place greater stress on the quality and cost of the backhaul (often termed fronthaul in the context of C-RAN) (SCF158).

Another development is towards the deployment of small cells in the unlicensed/licence-exempt spectrum as an alternative to Wi-Fi via technologies such as license-assisted access LTE (LAA-LTE) and LTE-U (SCF097), although the ability for such technologies to co-exist with Wi-Fi in existing spectrum is controversial (Kwan *et al.*, 2015).

Finally, part of the expected distinction of 5G over LTE is the use of millimetre-wave spectrum bands where cell ranges are such that all cells are likely to class as small cells (SCF055).

Overall, it is clear that, over the course of around a decade, small cells have moved from a minor technology, viewed as a heretical approach by some operators, to an essential and in some cases dominant element in the evolution of mobile networks. They now have the potential to ensure a long-term sustainable approach to increasing network capacity and coverage to meet demand and to deliver these benefits at low cost and with high availability in even the most challenging locations.

References

FCC (2015) Federal Communications Commission (FCC), 3.5 GHz Citizens Broadband Radio Service, 14 April 2015, https://www.fcc.gov/rulemaking/12-354.

Kwan, R. *et al.* (2015) Fair co-existence of licensed assisted access LTE (LAA-LTE) and Wi-Fi in unlicensed spectrum, in *7th Computer Science and Electronic Engineering Conference (CEEC)*, Colchester, Essex, UK, September 2015, pp. 13–18.

Saunders, S.R., Carlaw, S., Guistina, A., Bhat, R.R., Rao, V.S. and Siegberg, R. (2009) *Femtocells: Opportunities and Challenges for Business and Technology*, 1st edition, John Wiley & Sons, Ltd, Chichester, UK. ISBN 978-047074816-9.

SCF003. Interference management in UMTS femtocells ('high-band'), Small Cell Forum document number SCF003. URL: http://scf.io/en/documents/003_Interference_management_in_UMTS_femtocells_high-band.php.

SCF005. Femtocell business case – signals research, Small Cell Forum document number SCF005. URL: http://scf.io/en/documents/005_Femtocell_business_case_Signals_Research.php.

SCF008. Interference management in UMTS femtocells: topic brief, Small Cell Forum document number SCF008. URL: http://scf.io/en/documents/008_Interference_management_in_UMTS_femtocells_topic_brief.php.

SCF009. Interference management in UMTS femtocells ('low-band'), Small Cell Forum document number SCF009. URL: http://scf.io/en/documents/009_Interference_management_in_UMTS_Femtocells_low-band.php.

SCF013. Business case for femtocells in the mobile broadband era – signals research, Small Cell Forum document number SCF013. URL: http://scf.io/en/documents/013_Business_case_for_femtocells_in_the_mobile_broadband_era_Signals_Research.php.

SCF025. HeNB (LTE femto) network architecture, Small Cell Forum document number SCF025. URL: http://scf.io/en/documents/025_HeNB_LTE_Femto_network_architecture.php.

SCF032. Enterprise femtocell deployment guidelines, Small Cell Forum document number SCF032 URL: http://scf.io/en/documents/032_Enterprise_femtocell_deployment_guidelines.php.

SCF036. Femtocell synchronisation and location: topic brief, Small Cell Forum document number SCF036. URL: http://scf.io/en/documents/036_Femtocell_synchronisation_and_location_topic_brief.php.

SCF050. Market status statistics Feb 2016 – mobile experts, Small Cell Forum document number SCF050. URL: http://scf.io/en/documents/050_Market_status_report_Feb_2016_Mobile_Experts.php.

SCF055. Small cells and 5G evolution: a topic brief, Small Cell Forum document number SCF055. URL: http://scf.io/en/documents/055_Small_cells_and_5G_evolution_a_topic_brief.php.

SCF062. Business case for enterprise small cells, Small Cell Forum document number SCF062. URL: http://scf.io/en/documents/062_Business_case_for_enterprise_small_cells.php.

SCF063. Small cell and Wi-Fi coverage study, Small Cell Forum document number SCF063. URL: http://scf.io/en/documents/063_Small_cell_and_W-iFi_coverage_study.php.

SCF067. E-SCN network architectures, Small Cell Forum document number SCF067. URL: http://scf.io/en/documents/067_Enterprise_small_cell_network_architectures.php.

SCF076. Regulatory aspects of small cells, Small Cell Forum document number SCF076. URL: http://scf.io/en/documents/076_Regulatory_aspects_of_small_cells.php.

SCF089. Next generation hotspot-based integrated small cell Wi-Fi, Small Cell Forum document number SCF089. URL: http://scf.io/en/documents/089_Next_generation_hotspot-based_integrated_small_cell_Wi-Fi.php.

SCF097. Small cells and license exempt spectrum: Carrier Wi-Fi, Wi-Fi Calling, LAA and LWA, Small Cell Forum document number SCF097 URL: http://scf.io/en/documents/097_Small_cells_and_license_exempt_spectrum_Carrier_Wi-Fi_Wi-Fi_Calling_LAA_and_LWA.php.

SCF158. Business case elements for small cell virtualization, Small Cell Forum document number SCF158. URL: http://scf.io/en/documents/158_Business_case_elements_for_small_cell_virtualization.php.

SCF159. Small cell virtualization functional splits and use cases, Small Cell Forum document number SCF159. URL: http://scf.io/en/documents/159_Small_cell_virtualization_functional_splits_and_use_cases.php.

SCWBA (2016) Industry perspectives, trusted WLAN architectures and deployment considerations for integrated Small-Cell Wi-Fi (ISW) networks, Small Cell Forum and Wireless Broadband Alliance, February 2016.

YouGov (2013) Bad mobile coverage for business, YouGov Survey of UK Businesses, February 2013. URL: http://bit.ly/YLLQBz.

11

In-Building Case Studies

Dr Vladan Jevremovic

Director of Research and Development, iBwave Solutions, Inc.

Once all the fundamentals of indoor wireless communications have been studied in the previous chapters, it is time to start practising the design of various in-building radio systems using practical design scenarios. This chapter aims at integrating theory with practical considerations that should be made for designing and installing wireless networks inside buildings, presented as case studies for various types of environments.

11.1 Public Venue

As mobile phones become more affordable, subscriber usage patterns continue to evolve. The second generation (2G) of mobile telephony primary focus was making voice calls on the move; third generation (3G) focus was on accessing emails and sending short text messages while being away from a desk. The fourth generation (4G) focuses on accessing the Internet on the move in the same fashion as it is accessed from a desk (fixed and steady location).

While voice users may make a call either on the go or sitting down in a restaurant or even at home, heavy Internet users are most likely to:

- Engage in Web surfing or data downloading while being stationary with nothing else to do.
- Access the Internet while on a commute to work using public transportation.
- Sit in airport terminals browsing for information.
- Take a break in a food court at shopping malls whilst accessing social networks.
- Check other scores or send videos/photos to friends at halftime during a sporting event in a stadium.

Everybody seems to be busy with their smart phones, watching videos or checking status updates. Thus, the use of mobile devices, some of which are broadband, demands service from operators that could fulfil users' needs.

Indoor Wireless Communications: From Theory to Implementation, First Edition. Alejandro Aragón-Zavala.
© 2017 John Wiley & Sons Ltd. Published 2017 by John Wiley & Sons Ltd.

11.1.1   Scenario

Wireless carriers are aware that user experience is shaped by the ability to access the network in public venues and are investing in public venue networks to address coverage and capacity. Most funding is allocated to venues with the highest density of subscribers, such as:

- Airports
- Stadiums and arenas
- Underground public transportation (subways)
- Shopping malls.

While each of those four types has its own design and implementation requirements, there are plenty of commonalities. As these are public venues, the venue network has to provide a signal for multiple wireless service providers (WSPs) that operate in the area. In many cases the venue manager may require that public safety ('first responders') and building operations trunked radio signals are also carried on the network. The 802.11 networks, commonly known as Wi-Fi, became so popular that venue managers insist on including them as well. These participants in venue networks are commonly referred to as *network tenants*.

11.1.2   Solution

In order to provide service to network tenants, a neutral host network capable of supporting a wide range of wireless technologies and spectrum bands needs to be built at the venue. The network also needs to be capable of delivering a high power signal to serving antennas at venues where distance between subscribers and antennas is large, such as football stadiums.

The best choice for such a network is a Distributed Antenna System (DAS), where network infrastructure – transmission cables, power amplifiers and antennas – are shared amongst network tenants. The tenants' base station (BTS) sectors are co-located within the venue and that location is commonly called a base station hotel, as explained in Chapter 8. At the base station hotel, the RF signal is delivered from a BTS sector to the DAS head-end. The head-end combines RF signals and transports them via an intermediate DAS network to remotely located antennas scattered throughout the venue. There are three types of DAS, which differ by the type of intermediate network that connects the head-end and remote antennas: passive, active and hybrid, which were fully described and explained in Chapter 9, Section 9.2.

11.1.3   Common Design Requirements

Whilst each public venue has its own design and implementation requirements, there are plenty of commonalities that can be found amongst them, which are presented in this section.

11.1.3.1   Multicarrier (Neutral Host)
The three types of networks that are most commonly included in public venues are:

- Commercial mobile networks
- Private mobile networks
- Wi-Fi networks.

To make the network less expensive and easier to deploy and maintain, these networks share hardware components whenever possible. The shared network is commonly known as the *neutral host network*. Commercial and private mobile networks usually have more than one tenant in the neutral host network.

Commercial mobile networks are operated by wireless services providers (WSPs), and WSPs that operate in the same frequency band often share power amplifiers (PAs) in remote units (RUs). As output power is shared equally amongst all active RF channels in the band, adding more channels decreases the output power per channel, thus decreasing coverage for all WSPs in the band. To prevent this it is important to include point of interface (POI) equipment between WSP source and the neutral host network, as the POI limits the total amount of power given to each WSP. This effectively ensures that the WSP that adds active channels is the only WSP that is affected by the change.

First responders and venue operations and maintenance operate as private networks. This network uses two-way radio technology and its design coverage scope is typically more extensive, as it must cover areas that are not accessible to the general public. Examples of such areas are delivery docks, electrical and equipment rooms, and areas that need to be accessed during periodical maintenance of the facility.

With the rise of consumer data usage, Wi-Fi networks became very popular as an inexpensive alternative to cellular data coverage. The technology is now commonly included in in-building networks and are often used to offload cellular data traffic. Wi-Fi is a common name for IEEE 802.11 technologies that were developed to provide a wireless local area network service. Wi-Fi is mostly deployed and operated by the venue, to improve customer experience and to advertise. Even though Wi-Fi is technically a tenant in the public venue network, due to transmission constraints it only shares remote antennas with the other tenants.

11.1.3.2 Multiband

Neutral host networks transmit both licensed and unlicensed bands. The actual number and type, frequency-division duplex (FDD) or time-division duplex (TDD) of licensed and unlicensed bands vary by region, as summarized in Table 11.1.

Table 11.1 Typical spectrum bands found in public venue networks.

North America	Europe	Asia
VHF FDD (150 MHz)	VHF FDD (150 MHz)	VHF FDD (150 MHz)
UHF FDD (450 MHz)	UHF FDD (450 MHz)	UHF FDD (450 MHz)
Public safety FDD (800 MHz)	Cellular FDD (900 MHz)	Public safety FDD (800 MHz)
4G FDD (700 MHz)	PCS FDD (1.8 GHz)	Cellular FDD (850 MHz)
Cellular FDD (850 MHz)	AWS FDD (1.9/2.1 GHz)	UMTS FDD (1.7/1.8 GHz)
AWS FDD (1.7/2.1 GHz)	4G TDD (2.6 GHz)	PCS FDD (1.9 GHz)
PCS FDD (1.9 GHz)	2.4 GHz TDD (Wi-Fi)	UMTS FDD (1.9/2.1 GHz)
2.4 GHz TDD (Wi-Fi)	5.7 GHz TDD (Wi-Fi)	4G TDD (2.3 GHz)
5.7 GHz TDD (Wi-Fi)		4G FDD (2.5 GHz)
		2.4 GHz TDD (Wi-Fi)
		5.7 GHz TDD (Wi-Fi)

Current worldwide licensed bands range from 700 MHz to 2.6 GHz. While different regions of the world may use the same name for a spectrum band, frequency of operation may not be the same. Even more significant is the fact that some spectra, like 4G, are FDD in North America and TDD in Europe and Asia. It is clear that while the advent of LTE has helped to streamline mobile technology across the world, it has done nothing to streamline frequency of operation or the type of licensed spectrum set aside for mobile networks.

The VHF (150 MHz) and UHF (450 MHz) bands are used mostly for two-way radio communications amongst venue operations and maintenance personnel, and occasionally for public safety. The SMR (800 MHz) band is used exclusively for public safety. Deploying the VHF band in public venue neutral host networks is rare and may increase complexity since it is difficult to find quality DAS equipment that includes the 150 MHz band.

The consequence of having to cover a wide range of frequency bands is that an active DAS may have different values in different frequency bands for gain flatness, maximum intermodulation distortion (IMD) and the input third-order intercept point (input IP3).

On the other hand, the most common unlicensed bands are two ISM bands: 2.4 GHz and 5.7 GHz. These bands are shared with other devices such as microwave ovens and point-to-point links and are used for various Wi-Fi technologies (802.11a, b/g, n, ac), as was explained in Chapter 2. The increasing congestion in these bands has made frequency coordination/frequency planning necessary when planning Wi-Fi networks.

11.1.3.3 Multitechnology

Public venue networks carry multiple wireless technologies. Since power amplifier linearity requirements differ for different technologies, so does maximum output power. The type and number of technologies vary with region, but the most common across the world are GSM, UMTS, LTE and Wi-Fi. In North America, CDMA2000 and EvDo are common, while PHS and AXGP can be found in Asia. Occasionally, WiMAX (802.16e) can be found throughout the world. The most popular trunked radio systems used by first responders and venue operating personnel are: TETRA, Tait, Motorola iDEN and Ericsson EDACS. A summary of such technologies found in North America, Europe and Asia is given in Table 11.2.

Table 11.2 Common technologies implemented in public venues.

North America	Europe	Asia
Trunked radio system	Trunked radio system	Trunked radio system
GSM	GSM	GSM
UMTS (WCDMA)	UMTS (WCDMA)	UMTS (WCDMA)
HSPA	HSPA	HSPA
cdma2000	LTE	cdma2000
EvDo	802.11	EvDo
LTE		LTE
802.11		WiMAX
		PHS
		AXGP

11.1.4 Common Best Practices

Whilst each type of public venue has its own specific best practices, they share some common best practices as expanded upon in the following paragraphs.

11.1.4.1 Passive Intermodulation (PIM)

Passive intermodulation (PIM) is a phenomenon that occurs in passive devices (cables, splitters, antennas, etc.) where two or more high power signals mix. As signal amplitude increases, intermodulation effects become more noticeable. If the spurious signal falls in the uplink frequency range, it may increase noise level, degrade signal quality and reduce uplink capacity. Public venue networks are especially vulnerable because many signals propagate through cables and antennas, and also because some public venues, like stadiums and arenas, use high-power amplifiers.

PIM sources can be external or internal. An external PIM source can be created if an antenna is located near rusty bolts or rusty mounts, such as air conditioning ducts (Anritsu, 2014). Internal PIM sources are at the conductor. To locate a PIM source, a recommended practice is to tap antennas and connectors lightly during PIM testing to see if a PIM spike occurs. Performing periodic PIM inspections and keeping antennas and equipment clean is essential for good performance of neutral host networks.

A bad connector is one with an improper attachment of connector to coaxial cable or a connector that is corroded. When a bad connector is identified, it needs to be disconnected, taken apart and inspected for physical damage or contamination. When reassembling connectors, care should be taken not to twist them. Small scratches caused by twisting can generate both VSWR and PIM. All tightening of connectors should be done using a torque wrench. Inadequate torque will leave gaps, which may cause PIM; excessive torque may damage the central connector.

Coaxial cables cause PIM if damaged or poorly terminated. If coaxial cables are cut at installation, care should be taken to clean debris from the cable because debris inside a connector may create PIM. To properly terminate cables, a connector clamping tool should be used to set the centre pin depth correctly. It is recommended to use 7/16 DIN connectors for termination because they are made specifically to counteract PIM and therefore are preferred over N-type connectors.

11.1.4.2 Downlink Design

Thermal noise in LTE is referenced to the physical resource block (PRB), which has a channel width of 180 kHz. Thermal noise referenced to PRB is equivalent to −121 dBm. If the PIM signal is kept at least 6 dB lower than thermal noise (−127 dBm), then the combined PIM signal and thermal noise is approximately −120 dBm. The difference between thermal noise and thermal noise combined with PIM is 1 dB, which means that the presence of PIM increases the noise level by 1 dB. This noise increase is deemed acceptable, and therefore the goal is to keep PIM at −127 dBm or lower in LTE networks.

In UMTS passive networks the BTS sector transmits at full power, 20 W (43 dBm) per channel, in order to overcome passive network losses. In those networks, the required PIM rating of combiners used at the headend when two 43 dBm carrier signals are applied needs to be 155 dBc (Rogers Canada Webinar, 2013) or 155 dBc (decibels below the input carrier signal). An acceptable UMTS PIM level is then $43 - 155 = -112$ dBm.

Since the acceptable LTE PIM level (−127 dBm) is lower than the acceptable UMTS PIM level (−112 dBm), the LTE transmit power per PRB also has to be lower than the UMTS transmit power per channel. For LTE networks, combiners with 162 dBc rating at 2×35 dBm input power are used at the headend (Rogers Canada Webinar, 2013). These are able to meet LTE PIM requirements as $35 - 162 = -127$ dBm. Note that the maximum LTE transmit power per PRB (35 dBm) is 8 dB lower than the maximum UMTS transmit power per channel (43 dBm). Assuming the same antenna EIRP in both UMTS and LTE passive DAS networks, this means that in LTE networks the maximum passive network loss is reduced by 8 dB. If the passive network loss is reduced by 8 dB, the passive LTE DAS will need more sectors than the passive UMTS DAS to cover the same area.

Unlike in passive DAS, in active and hybrid DAS the BTS sector signal is re-amplified in the DAS network before being sent to the remote antennas. For that reason, the BTS sector signal can be comfortably reduced to lower levels to satisfy LTE PIM requirements at the headend without impacting the remote antenna coverage radius.

11.1.4.3 Uplink Design

For LTE networks, low latency and high data rates are the key to customer satisfaction. To achieve high data rates, the SINR has to be high. Figure 11.1 shows the relationship between the uplink (UL) data rate per physical resource block (PRB) and signal-to-noise ratio (SNR).

As the PRB is the primary building block in LTE, UL data rates are increased by giving a subscriber more PRBs to transmit. Although a large number of aggregated PRBs with low individual data rates may achieve the desired composite uplink data rate, the goal is to minimize the number of PRBs needed by keeping the signal-to-interference-plus-noise ratio (SINR) high. In the next two examples, we examine uplink signal-to-noise ratio (SNR) values for comparable neutral passive and active DAS networks. We assume that the DAS has 16 antennas and is powered by four WSPs, that each WSP has one sector that has an output power of 35 dBm and that the LTE network operates in a channel that is 10 MHz wide (50 resource blocks).

The passive DAS architecture is shown in Figure 11.2. It has one 4×4 hybrid combiner and four 4×1 splitters that have a 6.5vdB insertion loss each, so the combined

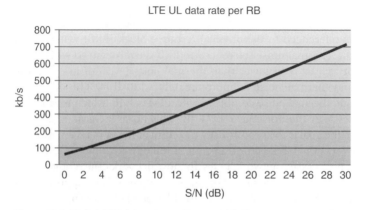

LTE UL data rate per RB

Figure 11.1 Uplink LTE data rate per resource block.

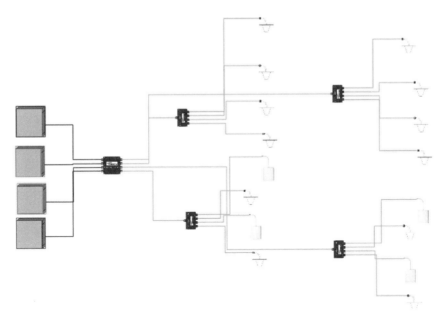

Figure 11.2 Passive DAS architecture.

combiner and splitter loss is 13 dB. Passive DAS loss is different for each antenna and, to calculate the link budget, one DAS antenna needs to be selected to calculate the loss. The total coaxial cable length from the BTS to the selected antenna is 87 m and, with a cable loss of 11.5 dB per 100 m, the coaxial cable loss is 10 dB. In the link budget shown in Table 11.3, the passive DAS loss (combiner + splitter + coaxial cable) for the selected antenna is 23 dB.

In the uplink, the noise per channel at the remote antenna is at the thermal noise level. It does not change from antenna to RF source because the DAS is passive and cannot generate noise by itself. Therefore, the noise level at the DAS antenna input and at the BTS sector input are the same. However, the UE signal does change because the passive DAS attenuates the signal as it passes from the DAS antenna to the BTS sector. The amount of signal attenuation is equal to the difference between the DAS antenna gain (3 dBi) and the passive loss (23 dB), which is 20 dB. This signal attenuation also reduces the uplink SNR by 20 dB, as the uplink SNR drops from 28.1 dB at the input of the DAS antenna to 8 dB at the BTS sector input. From Figure 11.1, we see that SNR = 8 dB gives a data rate per RB of 200 kHz.

For the hybrid DAS, we use a high-power Andrew ION™-B active DAS. Four remote units are used, each of which has a single output port that connects to a four-way splitter. Each splitter output connects to an antenna using coaxial cable. The hybrid DAS is shown in Figure 11.3.

The composite transmit power of the remote units is specified for each spectrum band and is shared amongst all WSPs that transmit in that band. The power per RF channel depends on the number of RF channels in the band. Composite RF power per remote unit (30 dBm) is divided equally amongst six RF channels, giving a transmit power per

Table 11.3 Example of passive DAS link budget calculation.

	Parameter	Value	Calculation
A	BTS power	35 dBm	
B	Passive losses	23 dB	
C	DL signal at antenna input	12 dBm	$A - B$
D	DAS antenna gain	3 dBi	
E	DL EIRP	15 dBm	$C + D$
F	DL RSRP threshold	−85 dBm	
G	Antenna to UE propagation losses	100 dB	$E - F$
H	Thermal noise at 10 MHz channel	−104 dBm	
I	UE Tx composite power	24 dBm	
J	UL signal at DAS antenna	−76 dBm	$I - G$
K	SNR at DAS antenna	28.1 dB	$J - H$
L	UL signal at BTS input	−96 dBm	$J + D - B$
M	Noise at BTS input	−104 dBm	H
	SNR at BTS input	8 dB	$L - M$

channel of 22.8 dBm. The remote unit uplink gain is 15 dB and noise is 7 dB. The uplink filter and combiner losses at the headend are 11 dB. Since coaxial cable runs differ from antenna to antenna, one antenna needs to be chosen to perform the link budget calculations. In the link budget example shown in Table 11.4, a coaxial cable loss of 3.7 dB is considered, which corresponds to a cable length of 32 m.

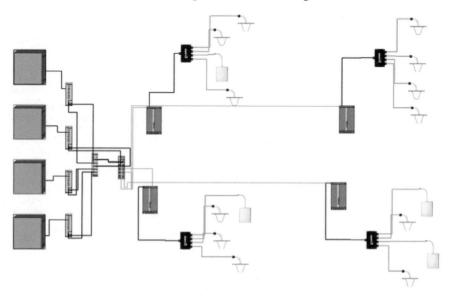

Figure 11.3 Hybrid DAS architecture.

Table 11.4 Hybrid DAS link budget.

	DL link budget	Value	Calculation
A	BTS power per channel	35 dBm	
B	RU composite power	30 dBm	
C	Number of channels/band	6	
D	RU power per channel	22.2 dBm	$B - 10 \log C$
E	Splitter losses	6.5 dB	
F	Cable + jumper losses	3.7 dB	
G	Antenna gain	3 dB	
H	Downlink EIRP	15 dBm	$D - E - F + G$
I	RSRP threshold	−85 dBm	
J	Antenna to UE propagation loss	100 dB	$H - I$

(a) Downlink

	UL link budget	Value [unit]	Noise level [dBm]	Calculation	Signal level [dBm]	Calculation	SNR [dB]
A	UE Tx composite power	24 dBm					
B	UE to antenna propagation loss	100 dB	−104	$-174 + 10 \log(10\,\text{MHz})$	−76	$A - B$	28
C	DAS antenna gain	3 dBi					
D	Cable + jumper loss	3.7 dB					
E	Splitter	6.5 dB					
F	At RU input		−104		−83	$A - B + C - D - E$	20.8
G	RU gain	15 dB					

(continued)

Table 11.4 (*Continued*)

	UL link budget	Value [unit]	Noise level [dBm]	Calculation	Signal level [dBm]	Calculation	SNR [dB]
H	RU NF	7 dB					
I	No. of RUs/fibre hub	4					
J	Composite NF at fibre hub	13 dB		$G + H + 10 \log(I)$			
K	At fibre hub output		−76	$F + J$	−68	$F + G$	7.8
L	Jumper cable loss	0.5 dB					
M	Splitter loss	7.1 dB					
N	At splitter output		−83.6	$K - L - M$	−76	$K - L - M$	
O	Jumper cable loss	0.5 dB					
P	Filter loss	4 dB					
Q	Jumper cable loss	0.5 dB					
R	BTS input		−89	$N - P - Q - R$	−81	$N - O - P - Q$	7.8

(b) Uplink

Unlike a passive DAS, a hybrid DAS generates uplink noise through uplink amplifiers. In this example, four remote amplifiers (remote units) generate a composite NF of 13 dB at a fibreh. This composite NF and the amplifier gain increase the noise per channel from the thermal level (-104 dBm) to -76 dBm. However, splitters, filters and cable jumpers insert a 13 dB loss in the uplink, thereby reducing the noise level to -89 dBm at the base station input. While UL noise increases 15 dB, the UL signal traversing the same path gains only 2 dB, so the SNR drops 13 dB from 20.8 dB at the RU input to 7.8 dB at the base station input.

If we compare the hybrid DAS UL SNR with the passive DAS UL SNR we may conclude that the two have a similar performance, as their UL SNR is almost the same. However, in this passive DAS example, the distance from the antenna to the BTS is 87 m and extending this distance further would reduce downlink antenna coverage. In the hybrid DAS, it is the distance between the antenna and the RU that limits downlink antenna coverage. The RU may be placed up to a few km away from the optical/RF converter, which is usually collocated with the base station at the headend. Therefore, the hybrid DAS can reach areas that are further from the base station than can the passive DAS, which gives hybrid DAS more deployment flexibility with the same uplink performance. The active DAS UL SNR calculation, omitted for the sake of brevity, is very similar to that for the hybrid DAS, the only major difference being that the active DAS does not use coaxial cables, only jumper cables.

11.1.5 Summary

Although public venues vary in size and RF morphology, neutral host networks that provide coverage at those venues share some common design requirements and best practices. Within the neutral host network, PIM generation is an important concern as poor installation or poor choice of antenna location can generate PIM that otherwise cannot be detected through RF design. LTE networks are most sensitive because they require very low PIM levels to operate properly. BTS sectors that feed a passive DAS need to transmit at lower than usual power levels to satisfy LTE PIM requirements, which increases the number of antennas and sectors in a passive DAS. While passive and active DASs may have comparable uplink data rates, passive DAS antennas need to be closer to the headend to maintain downlink coverage. The passive DAS needs double resources (coaxial cables, splitters) for LTE MIMO. All these factors imply that the passive DAS is not a good choice if an LTE network is to be included in a public venue neutral host network.

11.2 Stadium

As discussed in Chapter 2, a stadium is a venue that consists of a field/track surrounded by a bowl-shaped seating area. The largest stadium in the world is in Pyongyang, North Korea, with a capacity of 150 000 spectators. The second biggest is in Kolkata, India, with 120 000 spectators. There are only a handful of stadiums with more than 100 000 seats.

As of 2015, 934 stadiums worldwide have 30 000 or more seats: 228 in North America; 129 in Central and South America; 243 in Europe; 98 in the Middle East and Africa and 236 in the Asia-Pacific region. The USA alone has 217 stadiums with 30 000 seats or more and about two-thirds of these are used primarily for American Football. See Figure 11.4 for a worldwide high-capacity stadium breakdown.

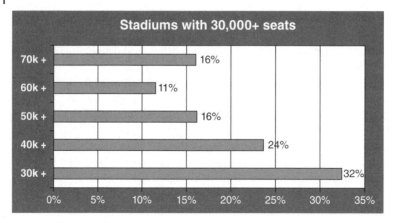

Figure 11.4 Worldwide high-capacity stadium breakdown.

Although the size and configuration of stadiums vary widely, there are design and deployment considerations common to all stadiums, which are discussed in this section.

11.2.1 Scenario

The case study venue has a capacity of 60 000 seats with five different seating levels, as shown in Figure 11.5. General public retail shops and concession stands are located on the first level, between the stadium entrance and the entry points to the seating bowl. On the same level, but not accessible to the general public, are conference rooms. One level below is ground level where the press rooms and team dressing rooms are located.

Figure 11.5 Stadium seating plan in 2-D.

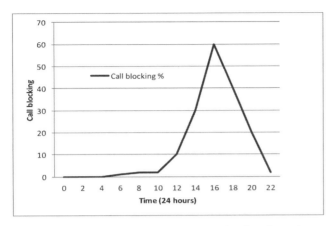

Figure 11.6 Typical call blocking statistics on the day of a stadium event.

The existing RF coverage coming from surrounding macro sites is fair to good in the bowl, but poor below it, in back offices, in conference rooms and in concession areas. The macro sites that cover the stadium report high call blocking during venue events, as illustrated in Figure 11.6. For this example, note that a blocking peak of 60% is reported in the busy-hour peak; i.e. during the event.

The call blocking problem is not restricted only to the stadium building, but also occurs in surrounding areas where spectators congregate before and after events. These transit areas include car parks, train stations, pedestrian footpaths (above or below ground), etc. Examples of such transit areas are shown in Figure 11.7.

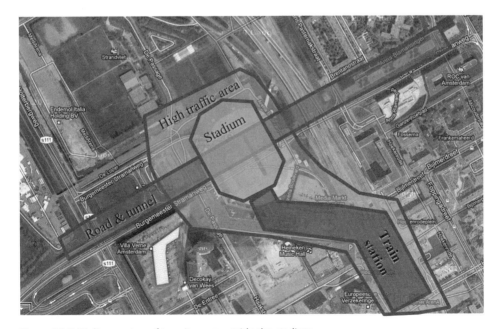

Figure 11.7 Bird's-eye view of transit areas outside the stadium.

The stadium network needs to provide coverage and capacity both to the stadium and to nearby transit areas. The network should be able to support traffic when the stadium is at full capacity during an event and also the traffic near the venue before and after the event.

11.2.2 Solution

Although there is some residual macro RF coverage in the seating area, the main problem is the lack of capacity from the surrounding macro sites. The stadium therefore needs its own RF network to satisfy demand during events. All commercial wireless service providers (WSPs) want to be included in the stadium network. The network must also include public safety and the venue operations trunked radios. An IEEE 802.11 (Wi-Fi) network needs to be included as well.

In order to provide service to multiple commercial and non-commercial carriers, a neutral-host mobile network needs to be built at the stadium. The best choice for such a network is a distributed antenna system (DAS), one in which fibre, power amplifiers, coaxial cables and antennas are shared amongst the network tenants. The DAS is also capable of providing a high power signal to the serving antennas, critical at large venues where the distances between subscribers and antennas are large.

Although transit areas outside the stadium have good coverage, they suffer from inadequate capacity before and after events. The DAS therefore has to be extended to those areas as well. The DAS signal has to be slightly stronger than the residual macro signal to guarantee that subscribers connect to the stadium network rather than to the surrounding macro network.

11.2.3 Design Requirements

The design requirements for this 60000-seat venue are as follows.

11.2.3.1 RF Coverage

- The stadium network signal must be dominant throughout the venue, even if there are areas where the existing macro coverage gives a 'five-bar' reading on a phone.
- To achieve this dominant signal requirement, the RF design should provide a signal that is everywhere 5 to 7 dB stronger than the residual macro signal. This requirement applies to the nearby transit areas as well (see Figure 11.7, above).

11.2.3.2 Capacity

- The stadium network should be designed to address all service types, from voice to streaming video.
- The pattern of voice and data calls may be different at the stadium than at typical macro networks: voice calls may be shorter due to stadium crowd noise, there may be far more file uploads to social media than usual and some venues may limit streaming video or, as was the case at the 2014 Super Bowl, ban it completely.

11.2.3.3 Handoff Management

- Establish clear handoff areas between the macro network and the stadium network in the transit areas outside the stadium.

- Once a subscriber hands off from the macro network, he should remain with the stadium network throughout the duration of his visit to the venue.
- Handoff traffic is handled by control channels. Extensive handoff traffic may use up control channel capacity and result in call blocking, even if sufficient physical resources are available to carry voice and data traffic.

11.2.3.4 Interference Management

- Non-serving sectors are a source of interfering signal. Interference can be internal, from non-serving DAS sectors, and external, from the macro network.
- Minimizing interference improves both network capacity and maximum achievable data rate (MADR).

11.2.4 Site Survey

When designing a network, certain rules need to be followed to achieve an optimum design, as specified in Chapter 7. One of these is related to the gathering of information for the site survey.

During the initial site visit, information about the physical structure, architecture and the different morphologies within the venue is gathered. A lot of information is captured in the form of photos, videos, measurements from data collection tools, voice memos and text annotation. Potential locations for antennas, cable runs and install equipment are also scouted.

During the survey the existing macro radio coverage at the venue should be recorded for all wireless carriers that are to be included in the network and at all frequency bands of interest. This is an important part of the survey because the actual stadium network has to overcome the residual macro coverage by a comfortable margin. Otherwise, spectators' user equipment (UE) may be registered with the macro network while at the venue, a highly undesirable situation because an important requirement is to offload the macro networks. Figure 11.8 shows a field engineer collecting RF and other venue data.

Figure 11.8 Site survey engineer collecting RF and other venue data.

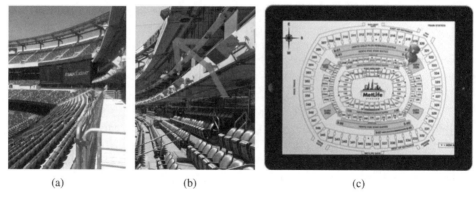

(a)   (b)   (c)

Figure 11.9 (a) General area of an antenna location; (b) an arrow marks the location; (c) a pushpin indicates the location on a site layout.

During the site survey, in addition to RF data, the engineer needs to identify spots where antennas may be mounted, to identify a room large enough to house the headend for the in-building network and to identify potential cable paths between the headend and the antennas.

Recording the information directly on to a floor plan saves time and facilitates information exchange with other departments and stakeholders. Plenty of information is captured and making sense of it is of crucial importance to reducing deployment timelines and costs. It is also important to identify several locations at which to mount candidate antennas so that different alternatives to control radio signals and provide capacity can be considered during the detailed RF coverage design phase later on.

Figure 11.9 shows an example of data collected by an engineer during a site survey. Figure 11.9(a) is a photograph of a general area of a potential antenna location in the seating area. Figure 11.9(b) zooms in on the location, marked with an arrow. The location needs to be identified on a layout plan as well so that the RF design engineers can know where they can put DAS antennas. This is illustrated by Figure 11.9(c), which shows the venue layout displayed on a tablet; a pushpin indicates the antenna location.

11.2.5 Detailed 3-D Modelling

Stadiums are multilevel structures that contain a wide variety of RF propagation environments. The most significant is the seating bowl, which is modelled as an *inclined surface* so as to take into account the difference in elevation between rows of seats. UE in this area will have a clear LOS to the serving antennas, positioned above the seats. An example of a 3-D model of a seating bowl is shown in Figure 11.10.

Capacity hotspots outside the bowl area need to be included in the design. Corporate and news media boxes, retail shops and concession stands are examples of capacity hotspots. Capacity hotspots that are not accessible to the general public such as conference rooms, press rooms, and team locker rooms also need to be included. Capacity hotspot examples are shown in Figure 11.11.

These areas are generally well isolated from the seating bowl so they need their own antennas for coverage. RF propagation characteristics in stadium hotspots vary considerably. Inside the bowl, a signal from the DAS is clearly in LOS with UE. Underneath the

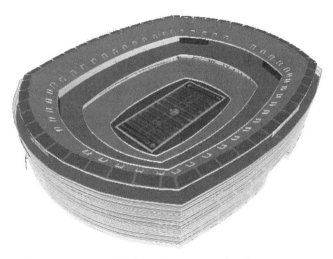

Figure 11.10 3-D model of a stadium seating bowl.

bowl, in the retail area, LOS is prevalent but there are numerous reflected signals due to the bowl's concrete walls. In the back-of-the-house areas, where the conference and locker rooms are located, wall density is much higher so there is significant signal diffraction and non-line-of-site (NLOS) propagation.

As propagation characteristics at hotspots differ from those in the seating area, it is necessary to properly model the hotspots in 3-D as well. Figure 11.12 shows the retail area immediately below the seating area.

Making a 3-D model of a venue can take anywhere from five hours for smaller venues to twenty hours for larger, more complex venues. Availability of the drawings in electronic form (CAD), rather than in the form of paper drawings, also affects the time to complete the 3-D model. Often a system is designed for a building under construction; hence there is the need to work using a 3-D model as no access is available

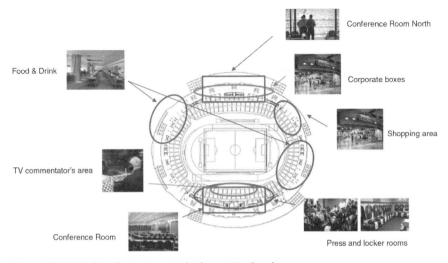

Figure 11.11 Stadium hotspots outside the seating bowl.

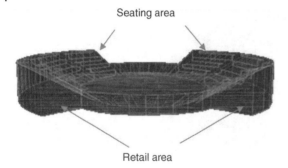

Seating area

Retail area

Figure 11.12 3-D 'wire-mesh' model of the stadium.

to the venue. Proper modelling of the venue is important not only for propagation analysis but also for good bill of materials estimates, such as for lengths of coax or fibre.

11.2.6 Sectorization

Sectorization has a dual purpose in the design of a radio network. First and foremost, it increases network capacity because each sector has its own channel cards capable of carrying voice and data traffic. Sectors are assigned a specific area to cover and serve a specific number of subscribers. The actual area to be covered and the actual number of subscribers that a sector should serve are a part of capacity sizing, where parameters such as the number of channels per sector, data rates, and the type and duration of calls and data connections are taken into account.

The second purpose of sectorization is to minimize the number of signals present in the area by limiting sector coverage. Limiting the coverage also limits interference from non-serving sectors, which improves capacity, the signal-to-interference-plus-noise ratio (SINR) and the maximum achievable data rate (MADR). In LOS areas like the seating bowl, sector overlap minimization is achieved by using highly directional antennas.

There are a few common sectorization types. Horizontal (ring) or vertical (wedge) sectorization is used most commonly where the number of sectors is not very high. Examples of such sectorization schemes are shown in Figure 11.13. The advantage of

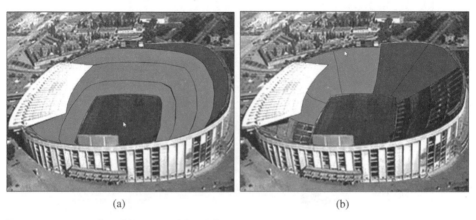

(a)   (b)

Figure 11.13 Examples of (a) ring and (b) wedge stadium sectorization.

Figure 11.14 Mixed sectorization example with 24 sectors.

these methods is that horizontal or vertical movements by spectators do not result in extensive UE handover.

If, however, a significant number of sectors is required, then hybrid ring-and-wedge sectorization is the best option. An example of mixed sectorization with 24 sectors is shown in Figure 11.14.

11.2.7 Macro Coverage Management

Early on in macro network deployments, WSPs recognized the revenue potential of venues with high subscriber density, such as stadiums. At first they tried to provide good stadium coverage by pointing sectors of nearby macro sites towards the venue. However, as networks became more data-centric, the data traffic became more congested and capacity at the venue became the primary concern. Nowadays many stadiums have good coverage in the seating bowl, but need a dedicated stadium network to direct the traffic away from the neighbouring cell sites that would otherwise be overloaded during stadium events.

When designing stadium network coverage, a common practice is to design so that the DAS signal will be stronger than the residual macro signal by at least 5 to 7 dB. If the residual macro signal is already strong, a DAS may need many antennas to achieve this goal, which may make the cost of deploying the DAS prohibitive. The most effective way to reduce the residual macro coverage is to downtilt antennas at nearby sectors that point towards the venue, as shown in Figure 11.15.

The upper line in Figure 11.15 shows the RF signal path with the initial antenna downtilt. The signal is diffracted from the stadium roof, which causes little attenuation, thus providing a strong signal in the bowl. When the sector is downtilted further, the signal path shown in the lower line penetrates the concrete wall. This is often preferable because the concrete wall significantly attenuates the signal before it reaches the bowl.

Figure 11.15 Sector antennas downtilt.

11.2.8 Passive Intermodulation Management

Neutral-host networks with high transmit power are susceptible to passive inter-modulation (PIM) noise generation. If the PIM noise level is sufficiently high, it can reduce coverage, slow the network, cause dropped calls and reduce battery life. LTE networks are especially susceptible to PIM noise because SINR is referenced to the resource block, which is 180 kHz wide. Thermal noise referenced to 180 kHz is −121 dBm and, for an LTE network to function properly, maximum PIM noise must be −127 dBm or lower. To ensure that PIM noise is kept in check, the following steps should be undertaken during the design stage:

- Combiners near the power amplifier should have a PIM specification of −162 dBc *at* 2 × 35 dBm to achieve the required PIM noise.
- Braided coaxial cables and N-type connectors are a known source of PIM noise and should not be used.
- Silver-plated 7/16 DIN connectors should be used as they have a low PIM rating (165 dBc).
- Consider only antenna locations that are away from metal objects, as metal near antennas generates PIM.
- Do not use equipment for which the manufacturer does not specify a PIM rating.
- Perform on-site PIM testing prior to antenna installation, using an antenna on a pole.

11.2.9 Design for Stadium Capacity

In order to properly size stadium networks it is necessary to determine the number of sectors required to support each carrier's capacity requirements. The number of sectors per carrier depends on the number of seats, the carrier's subscriber penetration rate (i.e. the percentage of that carrier's subscribers as a proportion of the general population) and the carrier's mobile traffic profile. For this case study, the stadium has 60 000 seats and the stadium network needs to carry three WSPs, public safety, stadium operations network and Wi-Fi. The characteristics of the three WSPs are as follows:

WSP A:
- Cellular band (850 MHz), 2 UMTS channels
- AWS band (2100 MHz), 2 UMTS channels

- 700 MHz band (700 MHz), 10 MHz LTE-FDD channel
- 40% subscriber penetration rate.

WSP B:
- PCS band (1900 MHz), 2 UMTS channels
- 2.5 GHz band, 10 MHz LTE-TDD channel
- 10% subscriber penetration rate.

WSP C:
- AWS band (1900 MHz), 2 UMTS channels
- PCS band, 5 MHz LTE-FDD channel
- 20% subscriber penetration rate.

Let us define HSPA and LTE traffic distribution per user at the stadium as shown in Table 11.5, which lists, for each service type, the duration of the network connection during the busy-hour expressed in millierlangs (mE) per subscriber, the fixed data rate in kbps and the probability, expressed as a percentage, that a subscriber is using, or will attempt to use, that service type during the busy-hour. It is important to note that a subscriber is not limited to one service type attempt per busy-hour; rather he may use, or attempt to use, any or all of the service types listed.

The probability of video streaming or video conferencing is low because it is assumed that they are very rarely used at the venue; most of the traffic at the venue is Internet browsing and data downloading, with some email. It is further assumed that half the subscribers use HSPA and the other half use LTE, reasonable assumptions for practical stadium scenarios. Thus:

- WSP A: 60 000 × 40% = 24 000 customers
- WSP B: 60 000 × 10% = 6000 customers
- WSP C: 60 000 × 20% = 12 000 customers

It is assumed that voice traffic is carried over the WCDMA (R99) protocol, while 3G and 4G data are carried over HSPA and LTE protocols, respectively. When defining the subscriber profile for R99, we take into account that data traffic would switch to R99 data only if both HSPA and LTE are unavailable. Therefore, the probability of an R99 data call is very low, which is reflected in the R99 traffic distribution shown in Table 11.6.

For HSPA and LTE, SINR coverage in the seating area is calculated and broken down into SINR intervals based on the modulation scheme that can be achieved in each SINR interval. The example in Table 11.7 shows that, in a region where LTE PDSCH SINR is

Table 11.5 Data traffic distribution at the stadium during the busy-hour by service type: duration (millierlangs per user), data rate (kbps) and probability of an attempt (%).

Service type	mE/user	kbps	Probability
Emails	5	100	0.50%
Browsing	15	200	1.50%
Video conferencing	1	600	0.10%
Data download	15	1000	1.50%
Video streaming	2	2000	0.20%

Table 11.6 R99 traffic distribution at the stadium during the busy-hour by service type: duration (millierlangs per user), data rate (kbps) and probability of an attempt (%).

Service type	mE/user	kbps	Probability
Voice	33	12.2	3.30%
Emails	3	64	0.30%
Browsing	3	128	0.30%
Data download	3	384	0.30%

20 dB or more, 64-QAM modulation with the coding rate $R = 0.93$ is possible and gives a spectral efficiency of 5.5 bit/s/Hz. With SINR between 15 and 20 dB, spectral efficiency is 3.9 bit/s/Hz; with SINR between 9 and 15 dB, the efficiency is 2.4 bit/s/Hz, etc.

By knowing the relationship between the signal modulation scheme, spectral efficiency and SINR, the number of resources needed to support each service type listed in Table 11.5 may be calculated. These 'resources' are different for different technologies: LTE resources are physical sesource blocks (PRBs), UMTS resources are HSPA orthogonal codes, etc. As spectral efficiency varies with SINR, so does the number of resources needed to support a certain service type in each SINR zone. For example, if SINR is high, a single PRB may be sufficient to support email but if SINR is low, more than one PRB may be required.

11.2.9.1 Data Capacity Sizing

For LTE capacity dimensioning, a downlink LTE SINR coverage map must be calculated. As the sector overlap affects the SINR, an assumption must be made about the number of sectors in the network. Table 11.8 shows how to estimate the number of sectors based on the user profile, estimated the MADR coverage, and estimated the call blocking rate.

The MADR coverage breakdown assumption is driven by high density sectorization, which is quite common at stadiums. The highest data rates are only possible near antennas and thus the 58.3 Mb/s data rate gets only 20% coverage. The lowest data rates (8.8 Mb/s) are where sectors overlap, and estimating 30% of the stadium to be in sector overlap is reasonable. The other two MADR rates are equally split, with 25% coverage each.

Table 11.7 LTE example showing the relationship between the modulation scheme, MCS efficiency (bit/s/Hz) and SINR (dB).

Modulation	MCS efficiency	SINR
QPSK	1.18	3
16 QAM	2.40	9
64 QAM	3.90	15
64 QAM	5.55	20

Table 11.8 Estimated number of sectors at the stadium.

Item	Symbol	Value	Unit	Comments
Number of spectators	a	60,000	users	
Market share	b	40	%	For one operator
% of LTE subscribers	c	75	%	75% of 40% of 60,000
LTE subscribers	d	18,000		data only; voice on UMTS
MIMO data rates	e			
64 QAM	e1	58.3	Mbps	Taken from MADR MIMO
% coverage	e2	20%		coverage map
64 QAM	e3	37.9	Mbps	
% coverage	e4	25%		
16 QAM	e5	21.2	Mbps	
% coverage	e6	25%		
QPSK	e7	8.8	Mbps	
% coverage	e8	30%		
Average data rate per sector	f	29.1	Mbps	e1* e2+e3*e4+e5 *e6+e7*e8
Email users usage	g			
connection duration	g1	50	mErlangs/ user	Taken from user profile
data rate	g2	100	kbps	
call blocking	g3	3%		Desired email blocking
Web browsing usage	h			
connection duration	h1	100	mErlangs/ user	Taken from user profile
data rate	h2	100	kbps	
call blocking	h3	4%		Desired web browsing blocking
Video conferencing	i			
connection duration	i1	5	mErlangs/ user	Taken from user profile
data rate	i2	600	kbps	
call blocking	i3	8%		Desired video conferencing blocking
Data download	j			
connection duration	j1	150	mErlangs/ user	Taken from user profile
data rate	j2	1000	kbps	
call blocking	j3	10%		Desired data download blocking
Video streaming	k			
connection duration	k1	1	mErlangs/ user	Taken from user profile

(continued)

Table 11.8 (*Continued*)

Item	Symbol	Value	Unit	Comments
data rate	k2	2000	kbps	
call blocking	k3	20%		Desired video streaming blocking
Delay between consecutive data Tx	l	10	ms	A subscriber receives data once per frame
Transmission duration	m	1	ms	A subscriber data is in one subframe
duty cycle	n	40%		this is also called cell load
Required data throughput	o			
email	o1	21.8	Mb/s	d*g1/1000*(1-g3)*g2/1000*m/l/n
browsing	o2	43.2	Mb/s	d*h1/1000*(1-h3)*h2/1000*m/l/n
Video conferencing	o3	12.4	Mb/s	d*i1/1000*(1-i3)*i2/1000*m/l/n
Data download	o4	607.5	Mb/s	d*j1/1000*(1-j3)*j2/1000*m/l/n
Video streaming	o5	7.2	Mb/s	d*k1/1000*(1-k3)*k2/1000*m/l/n
Total data throughput in the stadium	t	692.1	Mb/s	o1+o2+o3+o4+o5
Sectors required for total throughput	u	23.8		t/f
Number of sectors rounded	w	24.0		roundup (u)

A general rule of thumb for the call blocking rate is that the higher data rate services have a higher blocking rate and that the call blocking ratio of two service types is the same order of magnitude as their data rate ratio. For example, if we assume that the average call blocking for video streaming (2 Mb/s) is 20%, then call blocking for the data download (1 Mb/s) should be about 10%.

With these two sets of assumptions we calculate 24 LTE sectors at the stadium. Propagation analysis with 24 sectors has produced an SINR coverage map of the stadium bowl that can be split into four SINR ranges, as shown in Figure 11.16. Each SINR range has a specific modulation scheme with a specific spectral efficiency value, as seen in Table 11.8. The dependence of spectral efficiency on SINR is important because spectral efficiency ultimately determines the maximum achievable data rate (MADR) for each SINR range.

As we see from Figure 11.16, the LTE SINR Range 1 ($3 \leq$ SINR ≤ 9) covers 30% of the area, the SINR Range 2 ($9 \leq$ SINR ≤ 15) covers 25% of the area, the SINR Range 3 ($15 \leq$ SINR ≤ 20) covers 25% of the area and the SINR Range 4 (SINR > 20) covers 20% of the area. Assuming a uniform distribution of spectators, the percentage of LTE users within a particular SINR range is the same as the SINR coverage percentage for that range. For the sake of brevity, repeating this exercise with HSPA SINR is omitted; calculating the HSPA SINR distribution yields 50% 3G users in the SINR HSPA Range 1, 30% in Range 2, 15% in Range 3 and 5% in Range 4.

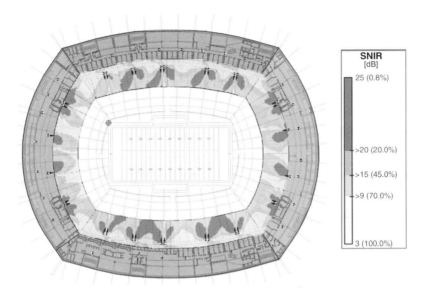

Figure 11.16 LTE PDSCH SINR coverage used for the data sizing example.

Let us assume that there are 1000 subscribers in a sector (which is the case for WSP A) and that they are split equally between HSPA and LTE. We then calculate HSPA and LTE busy-hour traffic (in erlangs) based on the number of subscribers in each SINR range and busy-hour traffic per subscriber, as given in Table 11.5. The results are shown in Table 11.9.

The number of resources needed to support the service types across the ranges is calculated next. This is readily determined if the relationship between SINR versus spectral efficiency is known, which can be taken either from research papers or directly from vendors. Table 11.10 shows, for this example, the distribution of resources by the SINR range for HSPA and LTE networks for WSP A.

Based on Tables 11.9 and 11.10, and given the total number of HSPA and LTE resources in a sector, the blocking probability for each service type may be calculated. The blocking rate is defined as the percentage of attempted network connections that are denied due to insufficient network resources. The blocking rate formula for multiple services used for this calculation is taken from ITU (2013). The resultant blocking rates for HSPA and LTE technologies, by service type and by SINR range, are as shown in Table 11.11.

Table 11.11 is the key for sizing the network because it shows the blocking rate for each service type throughout the seating area (SINR ranges 1 to 4). From Table 11.11 we see that 27.3% of the attempts to stream video using LTE from the area where LTE SINR exceeds 20 dB (Range 4) are blocked due to insufficient LTE resources. In contrast, only 0.2% of attempts to stream video from Range 4 areas are blocked due to insufficient HSPA resources. While the blocking rates in areas where SINR is very good are clearly much higher for the LTE network, it should be noted that WSP A has four UMTS channels, which is equivalent to 20 MHz of spectrum. On the other hand, the LTE channel is only 10 MHz wide.

Table 11.9 HSPA (above) and LTE (below) users and busy-hour traffic (erlangs) by service type and by SINR range.

Metrics	Range 1	Range 2	Range 3	Range 4
SNIR	5	11	22	24
Percentage coverage	50.0%	30.0%	15.0%	5.0%
Users	250	150	75	25
Emails	1.25	0.75	0.38	0.13
Browsing	3.75	2.25	1.13	0.38
Video conferencing	0.25	0.15	0.08	0.03
Data download	3.75	2.25	1.13	0.38
Video streaming	0.50	0.30	0.15	0.05

Metrics	Range 1	Range 2	Range 3	Range 4
SNIR	3.1	8.7	14.3	19.9
Percentage coverage	30.0%	25.0%	25.0%	20.0%
Users	150	125	125	100
Emails	0.75	0.63	0.63	0.50
Browsing	2.25	1.88	1.88	1.50
Video conferencing	0.15	0.13	0.13	0.10
Data download	2.25	1.88	1.88	1.50
Video streaming	0.30	0.25	0.25	0.20

Table 11.10 HSPA (above) and LTE (below) resources by service type and by SINR range.

Metrics	Range 1	Range 2	Range 3	Range 4
HSPA codes for emails	3	1	1	1
HSPA codes for browsing	6	3	1	1
HSPA codes for video	19	8	1	1
HSPA codes for download	31	13	2	1
HSPA codes for streaming	63	25	4	2

Metrics	Range 1	Range 2	Range 3	Range 4
PRB for emails	6	3	2	2
PRB for browsing	11	5	4	3
PRB for video	31	15	10	7
PRB for download	51	25	16	11
PRB for streaming	101	50	31	22

Table 11.11 HSPA (above) and LTE (below) blocking rate by service type and by SINR range.

Service type	Range 1	Range 2	Range 3	Range 4
Emails	0.3%	0.1%	0.1%	0.1%
Browsing	0.7%	0.3%	0.1%	0.1%
Video conferencing	2.5%	0.9%	0.1%	0.1%
Data download	4.6%	1.6%	0.2%	0.1%
Video streaming	13.7%	3.5%	0.4%	0.2%

Service type	Range 1	Range 2	Range 3	Range 4
Emails	8.0%	4.1%	2.7%	2.7%
Browsing	14.4%	6.7%	5.4%	4.1%
Video conferencing	36.7%	19.2%	13.1%	8.0%
Data download	54.6%	30.5%	20.4%	14.4%
Video streaming	82.9%	53.8%	36.7%	27.3%

If the calculated blocking rates shown in Table 11.11 are not acceptable, the number of sectors should be increased in order to reduce the number of subscribers per sector, the SINR map recalculated and the capacity calculations repeated. This is an iterative process that is continued until acceptable blocking rates are found.

Carried busy-hour traffic is calculated based on offered traffic (Table 11.9) and blocking rate (Table 11.11) for each service type. Results are shown in Table 11.12.

Table 11.12 Carried HSPA (above) and LTE (below) busy-hour traffic by service type and by SINR range.

Service type	Range 1	Range 2	Range 3	Range 4
Emails	1.24	0.75	0.37	0.12
Browsing	3.67	2.23	1.12	0.37
Video conferencing	0.23	0.15	0.07	0.02
Data download	3.26	2.14	1.12	0.37
Video streaming	0.35	0.27	0.15	0.05

Service type	Range 1	Range 2	Range 3	Range 4
Emails	0.74	0.62	0.62	0.50
Browsing	2.20	1.86	1.86	1.49
Video conferencing	0.14	0.12	0.12	0.10
Data download	1.98	1.77	1.81	1.47
Video streaming	0.21	0.22	0.23	0.19

To determine offered busy-hour traffic, the entries in Table 11.9 are summed for LTE and HSPA. Offered traffic is 19 erlangs for both HSPA and LTE. Similarly, to determine carried busy-hour traffic, the entries in Table 11.12 are summed for LTE and HSPA. Carried HSPA traffic is 18.05 erlangs and carried LTE traffic 18.26 erlangs. The composite call blocking rate (BR) is calculated as

$$BR = 1 - \frac{\text{Carried traffic}}{\text{Offered traffic}} \qquad (11.1)$$

This yields blocking rates of 5% for HSPA and 3.9% for LTE. While both technologies have similar statistics, it should be noted that UMTS has four channels (20 MHz of spectrum) while the LTE channel is only 10 MHz wide.

Duty cycle is defined as the ratio of carried traffic to theoretical maximum traffic when all resources are used for the busy-hour. It is 6% for HSPA and 4.1% for LTE. HSPA data usage is 4.42 gigabytes in a busy-hour, while LTE data usage is 4.51 gigabytes. Since WSP A has 24 000 subscribers, the traffic and data usage numbers need to be multiplied by 24 to get the total WSP A traffic for the whole stadium.

Similar calculations can be done for WSP B and WSP C. WSP C has half the number of subscribers, but also half as many UMTS channels and half the LTE bandwidth, and therefore needs 12 sectors. WSP B has half the subscribers that WSP C has but, under the assumption that the 10 MHz LTE TDD channel is configured symmetrically in uplink and downlink, has the same LTE capacity. Under that assumption, WSP B needs half the sectors, but with a lower blocking rate, lower data traffic in erlangs and lower data usage in gigabytes than WSP C. The final sector breakdown, based on data traffic sizing only, is as follows:

- WSP A: 24 sectors
- WSP B: 6 sectors
- WSP C: 12 sectors

11.2.9.2 Voice Capacity Sizing

Voice capacity is sized through the WCDMA portion of the UMTS signal. As was done for data capacity calculations, E_b/N_0 coverage needs first to be determined, then separate the coverage into four different E_b/N_0 ranges and finally identify service types that can be used in each range. The E_b/N_0 map for one stadium level is shown in Figure 11.17.

Assuming a uniform subscriber distribution, the percentage of subscribers connecting to the service in a particular E_b/N_0 range is the same as the percentage of coverage for that range. The resulting user distribution in the E_b/N_0 range and R99 traffic in each range (in erlangs) is shown in Table 11.13.

Only OVSF codes with a spreading factor up to SF128 are used for the service types shown in Table 11.13. The required number of OVSF codes per service type and E_b/N_0 range is as shown in Table 11.14.

As was the case with HSPA and LTE technologies, call blocking rates are calculated as in ITU (2013) and are as shown in Table 11.15.

For a sector with 1000 subscribers, the R99 voice blocking rate is 1.6% throughout the bowl seating area. Most macro UMTS networks use a busy-hour call blocking rate target

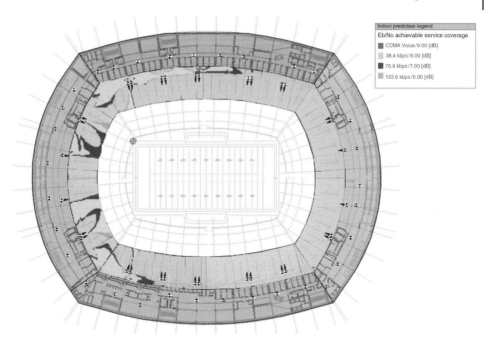

Figure 11.17 E_b/N_0 coverage map used for the voice sizing example.

Table 11.13 R99 busy-hour data traffic (in erlangs), by service type, in each E_b/N_0 range.

Metrics	Range 1	Range 2	Range 3	Range 4
Eb/No	9	8	7	6
Distribution	1.0%	1.0%	9.0%	90.0%
Users	10	10	90	900
Voice	0.3	0.3	3.0	29.7
Emails	-	0.0	0.3	2.7
Browsing	-	-	0.3	2.7
Data download	-	-	-	2.7

Table 11.14 OVSF codes, by service type, in each E_b/N_0 range.

Service type	Range 1	Range 2	Range 3	Range 4
Voice	1	1	1	1
Emails	-	4	4	4
Browsing	-	-	8	8
Data download	-	-	-	16

Table 11.15 R99 blocking rates, by service type, in each E_b/N_0 range.

Service type	Range 1	Range 2	Range 3	Range 4
Voice	1.6%	1.6%	1.6%	1.6%
Emails	-	6.5%	6.5%	6.5%
Browsing	-	-	13.4%	13.4%
Data download	-	-	-	28.0%

between 1% and 2%. Other R99 service types have higher blocking rates, but this is not of much concern because they are supported with better rates in 3G and 4G networks. The conclusion is that the call blocking rate for R99 traffic is acceptable and therefore the 24 sector configuration is sufficient to support voice traffic for 3G and 4G subscribers for WSP A.

11.2.10 RF Coverage Design

For the RF signal to be dominant at the venue, it has to be slightly stronger than the residual signal coming from surrounding macrocell sites. As most stadiums are open-air, the residual macro signal itself is usually fairly strong. However, the large number of sectors required for WSPs implies that highly directional high-gain antennas need to be deployed, which means that high received power in the seating area can easily be achieved. Figure 11.18, which shows LTE reference signal received power (RSRP) coverage at the bowl, makes it clear that an RSRP of −75 dBm or better is easily achieved over 90% of the bowl.

The modulation scheme used in LTE networks is directly related to PDSCH SINR, as high SINR makes possible high-order modulation such as 64-QAM. High-order modulation has high spectral efficiency, which allows for a high maximum achievable data rate (MADR) in the network. However, a large number of sectors also implies numerous sector overlaps, which may cause interference and lower SINR. An example of a 24-sector LTE PDSCH SINR plot is shown in Figure 11.19.

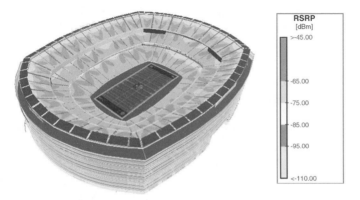

Figure 11.18 LTE RSRP coverage at the stadium.

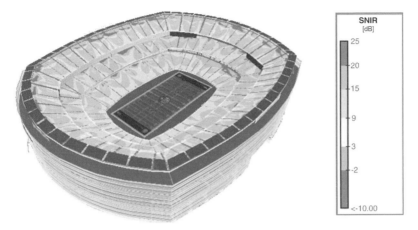

Figure 11.19 LTE PDSCH SINR coverage at the stadium.

Based on the SINR coverage distribution, the downlink MADR distribution across the stadium is then calculated, as shown in Figure 11.20.

11.2.11 Summary

A neutral-host DAS solution is cost-effective for stadium networks in which multiple commercial and non-commercial networks must share infrastructure. Stadium networks are characterized by a very high density of users who need many sectors to satisfy their data needs. The high sectoriszation requirement is addressed by using highly directional DAS antennas, which provide good spatial signal isolation. This also helps to control sector overlap and minimizes intersector interference.

The RF propagation environment differs vastly within the stadium, from pure LOS in the seating area, to LOS with a lot of reflections in retail areas underneath the bowl, to NLOS in locker rooms and conference rooms. To properly model the coverage, 3-D modelling of the venue is essential. Most stadiums have open-air seating areas and

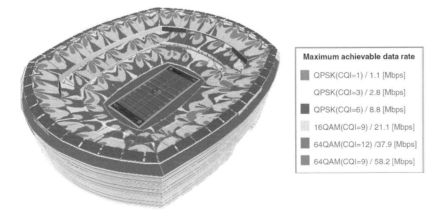

Figure 11.20 Downlink LTE MADR coverage at the stadium.

therefore many have significant residual macro coverage in those areas. As the stadium network signal must be dominant everywhere inside the venue, it is essential to perform an RF survey to determine the residual signal strength prior to designing the DAS. Since spectators tend to spend time outside the venue before and after events, the design area should be extended to parking lots, side streets and nearby bus and train stations.

Finally, neutral-host DAS networks with high-power amplifiers are susceptible to PIM generation and PIM can severely impact the stadium network. LTE is particularly sensitive to PIM because it has low thermal noise power (121 dBm). Care must be taken at the design stage to avoid PIM generation by using equipment with a PIM rating of −162 dBc at 2 × 35 dBm. Also, antennas must not be placed near metallic structures because this tends to generate PIM as well.

11.3 Shopping Centre

Contemporary shopping malls offer more than just shopping opportunities, as many include individual restaurants and bars, a food court, movie theatre, gym, skate ring, etc. Malls nowadays serve as anchors of social life in suburban areas and visitors consider wireless connectivity an important factor of their shopping mall experience. Mall management recognizes this and therefore considers an in-building network that provides superior mall wireless coverage and quality as a value-added commodity.

11.3.1 Scenario

In this case study, our shopping mall is a two-storey enclosed structure with concrete outer walls and a glass sunroof. The mall is 200 metres long end to end and up to 60 metres wide. It has three big anchor retail stores, which also have separate entrances from the open-air parking lot, and many small retail shops that can only be accessed from inside the mall. The upper level floor has an open view of the lower level pedestrian traffic, similar to what is represented in Figure 2.8, Chapter 2. A 3-D presentation of the mall is shown in Figure 11.21.

While the parking lot has good signal reception coming from the nearby macrocells, the mall management has received many complaints about signal coverage and quality within the mall. To that end, the mall management wants to have an indoor wireless

Figure 11.21 3-D representation of a shopping mall.

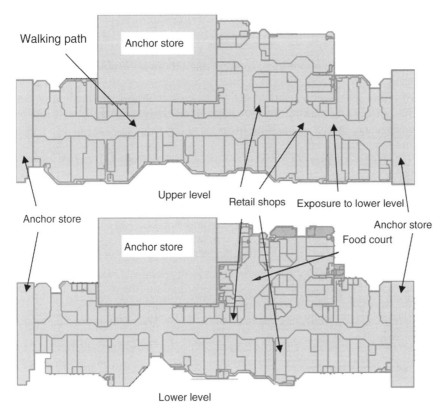

Walking path

Anchor store

Upper level

Retail shops   Exposure to lower level

Anchor store

Anchor store

Anchor store
Food court

Lower level

Figure 11.22 Lower and upper level mall floorplans.

network built that would improve customer experience inside the mall for three major wireless service providers (WSPs). The network also needs to carry a first responders (E911) network. Wi-Fi installation is not needed as the mall management has already installed a Wi-Fi network. The management does not want antennas installed inside any shops. The antennas need to be as small and inconspicuous as possible, so as not to interfere with interior mall aesthetics. The lower and upper level mall floor plans are shown in Figure 11.22.

11.3.2   Design Requirements

Specific design requirements for the venue are as follows.

11.3.2.1   RF Coverage

- RF coverage should be provided for the following technologies:
  - UMTS
  - LTE
  - Trunked radio (Tait technology).
- The following frequency bands should be included in the network:
  - SMR band (800 MHz)

- Cellular band (850 MHz)
- PCS band (1900 MHz)
- AWS band (2100 MHz).
- The coverage should include common areas and retail shops accessible to mall visitors on both levels.
- For trunked radio, the coverage should extend to back hallways and loading docks.
- The mall has an outside parking lot that has sufficient coverage and need not be targeted for coverage by the in-building network.
- Target signal strength is:
  - UMTS CPICH = −85 dBm
  - LTE RSRP = −95 dBm
  - Trunked radio Rx = −95 dBm.
- As the mall has a glass sunroof, a residual macro signal may be present in the middle of the mall. To ensure in-building signal dominance in the presence of a macro signal, the in-building signal should be 5–7 dB stronger than the residual macro signal.
- Handoff between in-building and macro networks should be limited to mall entrance areas.
- For trunked radio, the coverage needs to be 100% throughout the mall including back hallways and stores; for cellular technologies the target coverage is 95%.
- The in-building signal should not exceed −100 dBm at the distance of 30 metres outside the mall.

11.3.2.2 Antenna Placement Restrictions

- Antennas may not be placed inside any stores, including anchor stores.
- WSPs and first responders must share antennas.
- The mall management requires that any in-building antennas mounted in the common areas should be as small and inconspicuous as possible so as to not to disturb the mall interior aesthetics.
- This requirement does not apply to back hallways, access docks, freight elevator, etc.

11.3.3 Solution

As the in-building system has to include multiple technologies and bands, the in-building network should use a neutral host distributed antenna system (DAS). Due to the proximity of antennas to customers, and to comply with ICNIRP guidelines for limiting exposure to time-varying EM fields (ICNIRP, 1998), low power DAS remote units should be used. In this context, low transmit power means up to 24 dBm composite transmit power per amplifier.

11.3.4 Antenna Choice and Placement

An important design requirement is that antennas must be small and not immediately obvious to mall visitors. A good choice is the selection of antennas manufactured by Andrew. The omnidirectional Andrew Cellmax O-25 has 0.85 dBd of gain, a V-plane beamwidth of 40° and an H-plane beamwidth of 360°. The directional Andrew Cellmax D-25 has 4.85 dBd of gain, a V-plane beamwidth of 60° and an H-plane beamwidth of

Figure 11.23 Omnidirectional Andrew Cellmax antenna in front of a retail store.

70°. Vertical beamwidths are chosen to illuminate higher and lower floors without causing interference, whereas the horizontal beamwidth for the directional antennas permits restriction of the coverage horizontally in areas such as corridors, as well as shaping such coverage for specific sectors.

Another important requirement is to maximize coverage inside stores while placing antennas outside stores. Omnidirectional antennas should be mounted flush against the ceiling in front of individual retail stores in common access areas. An example of an Andrew Cellmax omnidirectional antenna installation is shown in Figure 11.23.

Occasionally, mall management may require indoor antennas to be completely invisible. This requirement can be satisfied by mounting antennas behind the ceiling, using a behind-ceiling mounting kit. An example of an antenna that can be mounted behind the ceiling is Galtronics PEAR M4773 omnidirectional antenna, shown in Figure 11.24.

The second requirement is that, while anchor stores are excluded from coverage requirements, there should be an indoor signal present in the vicinity of the store entrance, to facilitate handover with the macro network. To facilitate that requirement, a directional antenna should be mounted in the common mall area, opposite the anchor store and pointing toward the store entrance. A good choice is the Andrew Cell-Max™ D-25 directional antenna with 4.85 dBd gain, a V-plane beamwidth of 60 degrees and an H-plane beamwidth of 70 degrees. An example of an Andrew Cell-Max™ directional antenna installation is shown in Figure 11.25.

11.3.5 RF Coverage Design

The suggested DAS antenna plan layout is shown in Figure 11.26. Antenna locations on both levels were chosen to maximize coverage inside the shops, while keeping in mind antenna installation restrictions outlined by the mall management. This explains why

Antenna

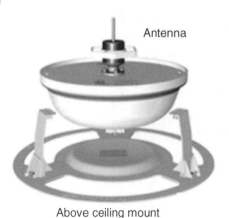

Figure 11.24 Galtronics PEAR M4773 with an above-ceiling mounting kit.

Above ceiling mount

antennas are placed in almost exact locations on both levels. Looking at UMTS CPICH and LTE RSRP coverage, it can be verified that required signal levels were attained with this design.The predicted UMTS CPICH coverage is also shown in Figure 11.26.

If the cumulative percentage distribution is analysed in Figure 11.26, the UMTS CPICH signal is greater than −85 dBm at 94.8% of the lower level area and at 96.2% of the upper level area. It can be concluded that over 95% of the total area has a UMTS CPICH signal greater than −85 dBm, which satisfies the UMTS design target at the AWS frequency band.

On the other hand, the LTE RSRP coverage is shown in Figure 11.27. Similarly, the LTE RSRP is greater than −95 dBm over 94.9% of the lower level area and over 96.6% of the upper level area. It can be concluded that over 95% of the total area has an LTE RSRP

Figure 11.25 Directional Andrew Cellmax antenna pointing towards an anchor store.

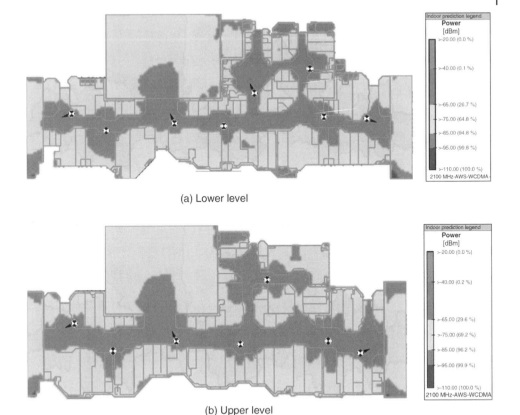

(a) Lower level

(b) Upper level

Figure 11.26 DAS antenna locations and AWS UMTS CPICH coverage in a shopping centre.

signal greater than −85 dBm, which also satisfies the design target at the PCS frequency band.

11.3.6 Capacity Dimensioning

In order to properly dimension the in-building network, the number of sectors required to support each carrier's capacity needs to be determined. The number of sectors per carrier depends on the number of visitors during the busiest time of the day, carrier's subscriber penetration rate and carrier's mobile traffic profile.

First, the number of visitors at the mall during the busiest time of day is estimated. This information may be obtained from a public online source, such as the Travel and Leisure newsletter, which lists the annual number of visitors for most visited shopping malls in the USA (TLM, 2014). Alternatively, this information may be obtained directly from the mall management. Let us assume that 5000 visitors are at the mall during the busiest hour. Let us also assume that the three WSPs that are included in the in-building network have the following characteristics:

WSP A:
  – Cellular band (850 MHz), 2 UMTS channels
  – AWS band (2100 MHz), 2 UMTS channels

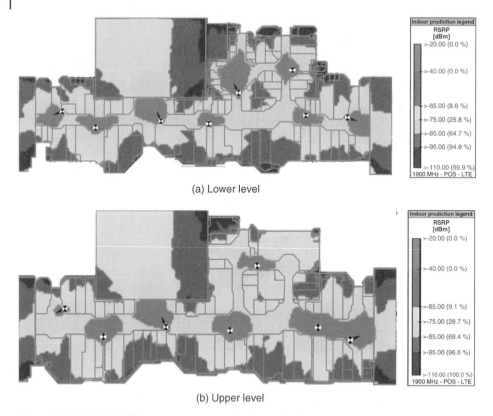

(a) Lower level

(b) Upper level

Figure 11.27 PCS LTE RSRP coverage at upper and lower levels.

  – PCS band (1900 MHz), 10 MHz LTE-FDD channel
  – 40% subscriber penetration rat
WSP B:
  – Cellular band (850 MHz), 2 UMTS channels
  – AWS band (2100 MHz), 2 UMTS channels
  – PCS band (1900 MHz), 10 MHz LTE-FDD channel
  – 35% subscriber penetration rate.
WSP C:
  – AWS band (1900 MHz), 2 UMTS channels
  – PCS band, 10 MHz LTE-FDD channel
  – 25% subscriber penetration rate.

The *subscriber penetration rate* is the percentage of the WSP subscribers among the general population. Let us assume that 3G and 4G traffic distribution at the mall per service type per visitor is as shown in Table 11.16. For each service type, the duration of the network connection during the busy-hour is expressed in mE (millierlangs) per subscriber. The probability that a subscriber will attempt to use that service type during the busy-hour is expressed in a percentage. Finally, the fixed data rate in kb/s is defined for each service type.

Table 11.16 Data traffic distribution at the mall by service type, call duration (in mE), data rate and call probability (in %) during the busy-hour.

Service type	mE/user	kbps	Probability
Emails	5	100	0.50%
Browsing	15	200	1.50%
Video conferencing	5	600	0.50%
Data download	15	1000	1.50%
Video streaming	5	2000	0.50%

It is assumed that voice traffic is carried over the WCDMA (R99) protocol, while 3G and 4G sata are carried over HSPA and LTE protocols. Further, it is also assumed that data subscribers are equally divided between HSPA and LTE networks. Based on the subscriber penetration, the number of customers at the venue for each WSP is calculated as follows:

- WSP A: $5000 \times 0.4 = 2000$ customers
- WSP B: $5000 \times 0.35 = 1750$ customers
- WSP C: $5000 \times 0.25 = 1250$ customers.

Next, the HSPA and LTE SINR coverage throughout the mall is calculated and broken down into intervals based on the modulation scheme that can be achieved in each interval. An example is shown in Table 11.17, where in the region where LTE PDSCH SINR > 20 dB, 64 QAM modulations with coding rate $R = 0.93$ is possible, which gives a spectral efficiency of 5.5 b/s/Hz. With an SINR between 15 and 20 dB, the spectral efficiency is 3.9 b/s/Hz; with an SINR between 9 and 15 dB, the efficiency is 2.4 b/s/Hz; etc.

The LTE SINR coverage at both levels is shown in Figure 11.28, with the same SINR intervals as in Table 11.17.

By knowing the relationship between signal modulation, spectral efficiency and SINR, the number of resources needed to support each service type listed in Table 11.17 may be calculated. The 'resources' have different names for different technologies: in LTE a resource is a physical resource block (PRB); in UMTS a resource is the orthogonal code. As spectral efficiency varies with SINR, so does the number of resources needed to support a certain service type in each SINR zone. For example, if the SINR is high, only

Table 11.17 Relationship between modulation, MCS efficiency and SINR.

Modulation	MCS efficiency	SINR
QPSK	1.18	3
16 QAM	2.40	9
64 QAM	3.90	15
64 QAM	5.55	20

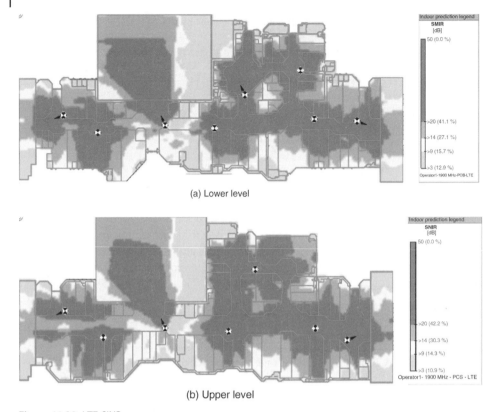

(a) Lower level

(b) Upper level

Figure 11.28 LTE SINR coverage.

one PRB may be needed to support email, but if the SINR is low, two or more PRBs are needed to provide the service.

A detailed data capacity dimensioning example is presented in Section 11.2.9 for a stadium. For the shopping centre in consideration, only critical steps are outlined. A uniform user distribution throughout the mall is assumed; that is the percentage of users within a particular SINR range is the same as the SINR coverage percentage in that same range. For WSP A, 500 HSPA and 500 LTE subscribers per sector are assumed. Let us also assume the same LTE SINR coverage shown in Figure 11.28. Based on this information and on the busy-hour traffic per subscriber depicted in Table 11.16, the total busy-hour traffic in erlangs for LTE technology is calculated and presented in Table 11.18.

For each SINR range, the required number of primary resource blocks needed to carry the service needs to be calculated. An example of such calculations can be found in the stadium case study in Section 11.2. Once we have that information, we proceed to calculate the encountered blocking rate as stated in ITU (2013), per service type and SINR range, which is shown in Table 11.19.

Based on the information in Tables 11.18 and 11.19, carried traffic per service type and SINR range can be calculated, as depicted in Table 11.20.

The composite blocking rate for multiple services is defined as 1-carried/offered traffic. As the total carried traffic from Table 11.20 is 21.83 erlangs and the total offered traffic

Table 11.18 Offered LTE busy-hour traffic (erlangs) per SINR range.

Metrics	Range 1	Range 2	Range 3	Range 4
SNIR	3	9	14	20
Percentage coverage	14.0%	14.0%	30.0%	42.0%
Users	70	70	150	210
Emails	0.35	0.35	0.75	1.05
Browsing	1.05	1.05	2.25	3.15
Video conferencing	0.35	0.35	0.75	1.05
Data download	1.05	1.05	2.25	3.15
Video streaming	0.35	0.35	0.75	1.05

Table 11.19 Carried LTE busy-hour traffic (percentile) per SINR range.

Service type	Range 1	Range 2	Range 3	Range 4
Emails	1.1%	0.5%	0.4%	0.4%
Browsing	2.0%	0.9%	0.7%	0.5%
Video conferencing	6.3%	2.8%	1.8%	1.3%
Data download	11.5%	5.0%	3.0%	2.0%
Video streaming	28.3%	11.2%	6.3%	4.3%

Table 11.20 Carried LTE busy-hour traffic (erlangs) per SINR range.

Service type	Range 1	Range 2	Range 3	Range 4	Total
Emails	0.35	0.35	0.75	1.05	0.10
Browsing	1.03	1.04	2.23	3.13	0.62
Video conferencing	0.33	0.34	0.74	1.04	0.61
Data download	0.93	1.00	2.18	3.09	3.02
Video streaming	0.25	0.31	0.70	1.00	1.90

from Table 11.18 is 22.5 erlangs, the composite blocking rate is 3%; the duty cycle is 4.9% and the data usage is 6.26 GB. If these figures are deemed acceptable, we may conclude that two LTE sectors, each supporting up to 500 subscribers, are sufficient to carry LTE traffic at the mall for WSP A. The same calculation is done for HSPA technology with a similar conclusion: two sectors are sufficient to carry HSPA traffic at the mall for WSP A. With that, we completed our capacity calculation for WSP A and may nowproceed with the sectorization. We can also carry out similar calculations for WSP B and C, to verify the number of sectors needed to support their traffic requirements.

11.3.7 Sectorization

The sectorization plan for the two-sector DAS within the shopping centre is shown in Figure 11.29. This is the so-called *wedge* sectorization where areas of different floors stacked on top of each other belong to the same sector. By utilizing this plan, handoff between adjacent in-building sectors is isolated to the same area on both levels.

The sectorization plan divides coverage almost equally between the sectors, while keeping the handoff area outside the food court and large anchor store, having a large number of visitors.

11.3.8 Data Rate Coverage

Finally, the LTE maximum achievable data rate (MADR) coverage plots are shown in Figure 11.30. It is assumed that the LTE channel is 10 MHz, the macro signal is

(a) Lower level

(b) Upper level

Figure 11.29 Two-sector sectorization plan.

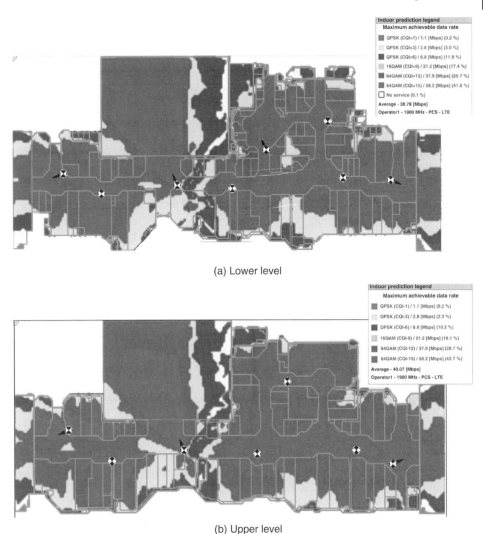

(a) Lower level

(b) Upper level

Figure 11.30 LTE MADR coverage.

at −90 dBm just outside the mall and that a 2×2 MIMO was deployed. As expected, the MADR is the highest (dark) near DAS antennas, where UE is in LOS with the antennas. On the other hand, the MADR is the lowest (brighter) in between sectors, since the intersector interference is the highest at the edge of the sector.

11.3.9 Summary

A large two-story shopping centre has a poor existing macro signal coverage and there is a need for a dedicated in-building wireless system. As an example, three WSPs are included in the in-building network, as well as a first responders (public safety) network. The three WSPs support UMTS and LTE technologies at the cellular, PCS and AWS band, which dictates the use of a neutral host DAS as the solution. Since the mall has two levels, 3-D modelling of the venue is essential.

Aesthetic aspects dictate the DAS design at the shopping mall more than at any other public venue. A typical mall management request is that antennas may not be installed inside any stores. To overcome this restriction, directional antennas were used to provide better signal penetration inside anchor and retail stores. Antennas also must be visually appealing and must 'blend in' with the surrounding environment. Good choices to fulfil this requirement are the Andrew Cellmax O-25 and D-25 antennas. As the DAS serves multiple in-building base station sectors, a sectorization plan must be developed for each operator. The plan should minimize DAS sector overlap, as the interference is the highest at sector boundaries. In this example, the wedge sectorization plan was adopted as it minimizes the sector overlap on each level.

11.4 Business Campus

Office buildings are very diverse venues. These buildings vary in layout, size and number of floors, as discussed in Chapter 2. Office buildings are often grouped together in a *business campus*. A typical business campus has a cluster of multilevel office buildings located off public streets. Very often a business campus has a single tenant that uses the campus as its corporate headquarters. An example business campus with a dozen multilevel office buildings is shown in Figure 11.31.

Business campuses that have multiple tenants are less common. In a multitenant campus, the in-building network is most often built piecewise for each tenant or building instead of all at once across the whole campus. This is not a preferred way to deploy a network because it is time-consuming and therefore more costly.

11.4.1 Scenario

In our specific case study, the business campus consists of two buildings. The larger building has an 80 metre by 20 metre rectangular cross-section and ten floors. The

Figure 11.31 A major business campus with a dozen office buildings.

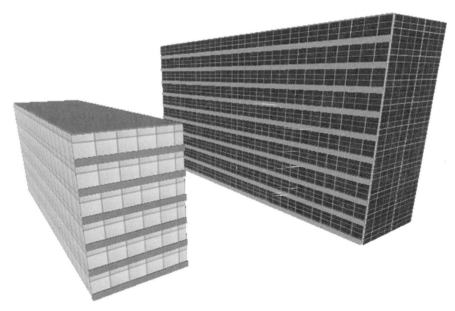

Figure 11.32 Three-dimensional representation of the case study business campus.

smaller building has a 60 metre by 17 metre rectangular cross-section and six floors. Each floor is 3 metres high. The buildings are 30 metres apart and directly across from each other. A three-dimensional representation of the venue is shown in Figure 11.32.

The two buildings have a single tenant who requires good voice and data coverage throughout both buildings. As the company has adopted a 'bring your own device to work' policy, the employees use their own smart phones for business, for which they are reimbursed, monthly, by the company. In order to provide good coverage for everybody, all major wireless service providers (WSP) must be included in the network.

11.4.2 Design Requirements

Specific design requirements for the venue are as follows.

11.4.2.1 RF Coverage

- RF coverage should be provided for the following technologies:
  - UMTS
  - LTE
  - Trunked radio (public safety).
- The coverage should include stairs and four lifts in two elevator shafts.
- The in-building wireless (IBW) target signal level should be 5 to 7 dB greater than the residual macro signal for all WSPs. This is to ensure that the IBW predominates throughout both buildings.
- It is preferable to measure residual macro coverage on all floors. However, if it is not feasible to perform a detailed RF survey on all floors, the coverage may be approximated by adding 1 to 2 dB per floor (Martijn and Herben, 2003). This approximation

applies only to floors that are lower than the height of neighbouring macro sites. For floors above the height of neighbouring macro sites, residual macro coverage does not significantly increase with floor height.

- Minimum signal strength is UMTS CPICH = −85 dBm, LTE RSRP = −95 dBm and Trunked Radio Rx = −95 dBm, and may be adjusted depending on the residual macro coverage.

11.4.2.2 Handoff Management

- Establish a clear handoff area between the macro network and the in-building network in the areas where most of the traffic occurs (entrance lobby, parking area entrance, etc.).
- Once a user hands off from the macro network, he should remain on the in-building network throughout the duration of his stay in the building.
- The number of handoffs from one in-building sector to another should be minimized. If sectors are designed horizontally, the only areas where a handoff can occur are in a stairwell area or inside an elevator. Handoffs in stairwell areas are allowed because the users are moving at pedestrian speed, thus allowing plenty of time for a handoff to complete. Allowing handoffs to occur in a moving elevator is undesirable because elevators generally move too fast for handoffs to complete. Figure 11.33 shows a cross-section of the taller building showing the two elevator shafts.

Each elevator shaft holds two elevator cars (marked with an 'X'). Only one elevator car per shaft is visible in Figure 11.33 because the lifts are side by side in the elevator shaft. The slim area between the shafts is the lift bank area. The signal at lift banks is not strong

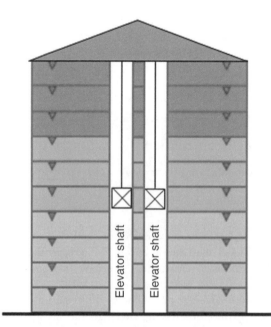

Figure 11.33 Cross-section of the taller building with two elevator shafts.

enough to penetrate into the shafts, and that lack of coverage is represented by the white background in the shafts.

11.4.2.3 Interference Management

- To minimize interference between the macro and in-building network, many WSPs mandate that the IBW signal level must drop to a specified value at a specified distance outside the building. This is called *signal leakage*. An example of signal leakage outside a building is shown in Figure 11.34.
- Many office buildings have large panoramic windows that have low (less than 5 dB) penetration loss, making it a challenge to meet the above requirement.
- In-building signal containment is especially important in a campus environment because buildings may be close to each other. Overlapping coverage from one building to another may cause excessive interference, lower capacity and reduced data rate.
- For this particular project, UMTS CPICH at a distance of 30 metres from the exterior of each building must be −90 dBm or less, LTE RSRP must be −100 dBm or less and Trunked Radio Rx must be −100 dBm or less.

In the following sections the design of an in-building wireless network to meet the common requirements is discussed, while the most common design mistakes are highlighted.

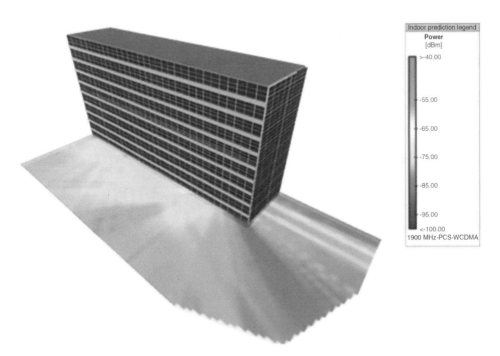

Figure 11.34 Example of a signal leakage from antennas on the ground floor.

11.4.3 Solution

As neutral host small cells are not commercially viable yet, the best solution is a distributed antenna system (DAS). In the shorter building, the DAS is fed by one trunked radio sector, one UMTS sector and two LTE sectors. The taller building has one trunked radio sector, two UMTS sectors and three LTE sectors. Details of capacity calculations are omitted for the sake of brevity.

Public safety (PS) has additional technological, regulatory and jurisdictional requirements, summarized in Spindler (2015). PS and WSPs may be deployed in a converged DAS or two separate DAS may be built, one for PS and one for all WSPs. The decision whether to deploy the converged or the discrete DAS architecture should be based on EIRP, spectrum bands and technologies that are being deployed.

11.4.4 Interference Control

One of the most popular choices for in-building DAS antennas is *omnidirectional*, as a ceiling mounted installation is normally available in office corporates, as depicted in Figure 11.35. The Andrew Cell-Max™ O-25 is a good option for this. This antenna has a gain of 0.85 dBd and a V-plane beamwidth of 40°. As with all omnidirectional antennas, the H-plane beamwidth is 360° and the front-to-back ratio is 0 dB.

The signal level is greater than −85 dBm over 97.6% of the floor plan. This coverage may seem acceptable at first. However, the RF coverage requirement also includes the signal leakage outside the buildings. Let us examine the RF leakage when omnidirectional antennas are used (Figure 11.36). The ground-floor antenna locations are indicated by the arrows.

The RF signal is not well contained inside the building; the UMTS signal level just outside the smaller building is around −75 dBm. As the two buildings are 30 metres apart, it is clear that the requirement for the leaked signal to be limited to −90 dBm is not satisfied.

The solution to minimize this unacceptable leakage is to use *directional* antennas. A very popular choice is the V-polarized Andrew Cell-Max™ D-25 antenna, which has a

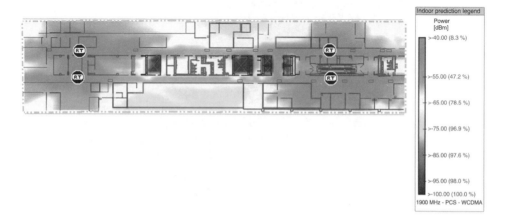

Figure 11.35 In-building RF coverage using omnidirectional antennas.

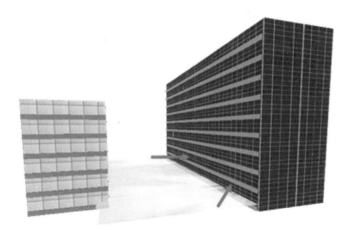

Figure 11.36 In-building RF leakage from two ground-floor antennas.

maximum boresight gain of 4.85 dBd, an H-plane beamwidth of 70 dB, a V-plane beamwidth of 60 dB and a front-to-back ratio of 20 dB. This antenna can be mounted vertically against a wall or column. Figure 11.37 shows the resultant RF coverage, with an arrow pointing in the direction of the main lobe of each antenna.

Using directional antennas has reduced the RF coverage to 91.4%. However, coverage at the edges is also reduced, which helps in suppressing RF leakage. Most of the energy is directed towards the middle, the direction in which the antennas are aimed. A good front-to-back ratio ensures that energy coming from sidelobes is low, which further reduces leakage.

Figure 11.38 shows a three-dimensional view of coverage between the buildings with the use now of directional antennas. Note that leakage levels have been significantly reduced.

The leaked RF coverage with directional antennas is 10 to 15 dB lower than with the omnidirectional antennas. This is a significant reduction and shows that directional antennas can help with interference control.

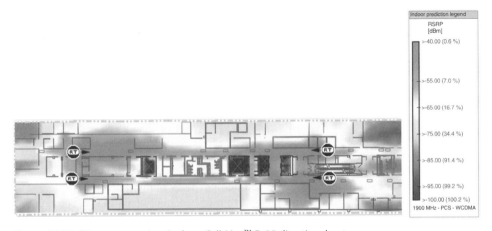

Figure 11.37 RF coverage using Andrew Cell-Max™ D-25 directional antennas.

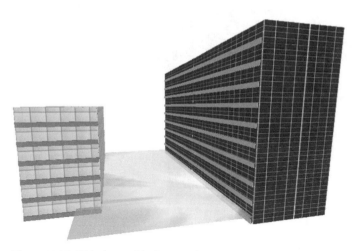
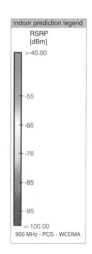

Figure 11.38 RF leakage with directional antennas.

11.4.5 Lift Coverage

Another important aspect of the design is the provision of coverage inside the lifts. As can be seen from the in-building RF coverage in Figure 11.35 and Figure 11.37, there is no RF coverage in the elevator shafts due to the high penetration loss of the elevator doors. A much simplified vertical cross-section of the building showing a three-sector in-building RF coverage with two elevator shafts with no coverage is shown in Figure 11.33.

The simplest solution to provide the required coverage may seem to be to simply add omnidirectional antennas on each floor at the elevator banks. Doing so would extend each sector's coverage into the shaft, as shown in Figure 11.39, in which omnidirectional antennas are represented by dots. However, applying this solution would increase the risk of dropped calls in moving elevators. A handoff takes 2 to 3 seconds to complete, and user equipment (UE) that goes from sector 1 (lower floors) straight to sector 3 (upper floors) would not have enough time to execute the handoff from one sector to another; calls would be dropped.

An alternative solution is to place RF equipment inside the lifts and/or the elevator shafts. A clear advantage of this approach is that the coverage does not rely on signal penetration through the metal doors. A disadvantage is that some municipalities in North America do not allow any electronic equipment inside elevator shafts due to the fire hazard. There are a few types of coverage solution with electronics inside the shaft. One, as shown in Figure 11.40, is to place an indoor antenna inside the lift, connected to a DAS by means of a cable. Here the handoff can occur only when the lift is stopped and passengers are entering or leaving it, greatly reducing the possibility of a dropped call. However, the antenna needs to be connected to a DAS remote unit via a cable, which makes this type of solution impractical in buildings with many floors.

Another solution, shown in Figure 11.41, is to install a directional DAS antenna on top of the elevator shaft, pointing down the shaft. A donor directional antenna is located on the top of each lift, pointed upward at the DAS antenna. The donor antenna is connected

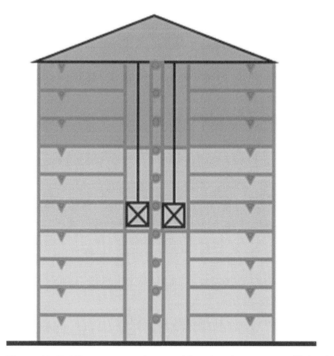

Figure 11.39 Lift coverage using omnidirectional antennas at lift banks.

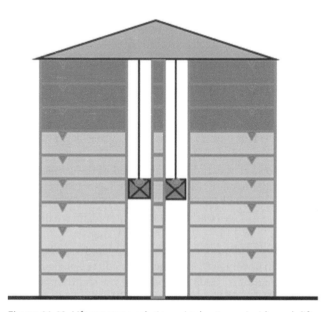

Figure 11.40 Lift coverage solution: wired antenna inside each lift.

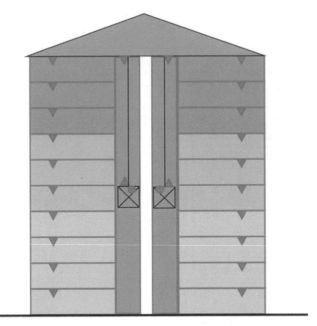

Figure 11.41 Lift coverage solution: passive repeater at each lift.

via a jumper cable to an omnidirectional serving antenna located inside the lift. The antenna pair used at each lift is purely used as a passive repeater system.

If the DAS antenna has a CPICH EIRP of 20 dBm, and if the target signal inside each elevator is −85 dBm one metre away from the serving antenna, the maximum path loss is 20 + 85 = 105 dB. A typical antenna gain is 7 dBi for directional antennas and 3 dBi for omnidirectional, ones while a 1 dB loss is typical for a 1 metre RF jumper cable. It is of interest to calculate the maximum distance between the antenna on top of the shaft and the lift. Some simple free-space path loss calculations at 2.1 GHz, also taking into account the 8 dB gain due to the waveguide effect inside the shaft (Spindler, 21015), show this maximum distance to be approximately 115 metres.

If the shaft is longer than 115 metres, another DAS antenna may be mounted at the bottom of the shaft, pointing upwards. This antenna sends a signal that is captured by another directional donor antenna located at the bottom of the elevator, pointing down the shaft. This donor antenna at the bottom of each lift is connected to the same serving antenna inside the lift. To ensure a proper handoff, RF coverage overlap between the two DAS antennas in the shaft must be accounted for. Typically, 10 − 15% of the antenna range is given to overlap, so with 15% coverage overlap the maximum distance between the top and bottom DAS antennas is 2 × 115 − 2 × 15 = 200 metres.

A third possible solution is to install radiating cable along the length of the shaft. This approach provides more uniform coverage than using a DAS directional antenna (Stamopoulos, Aragón-Zavala and Saunders, 2003). Depending on shaft length and cable attenuation per metre, the signal may be strong enough to penetrate the lift, making the passive repeater unnecessary. This solution is illustrated in Figure 11.42, where the radiating cable is shown as a solid black line between the lift and the lift bank.

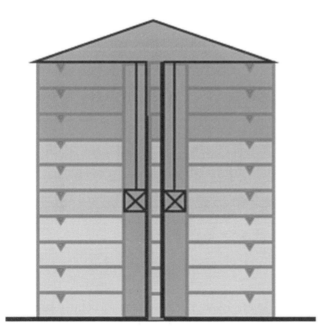

Figure 11.42 Lift coverage solution: radiating cable in shaft.

11.4.6 Detailed RF Coverage Design

To properly represent RF coverage at the venue, the coverage leakage from both buildings into the area between the buildings must be included in the analysis of the ground-floor level coverage. Examples of ground-floor level LTE RSRP coverage are shown in Figure 11.43 and Figure 11.44.

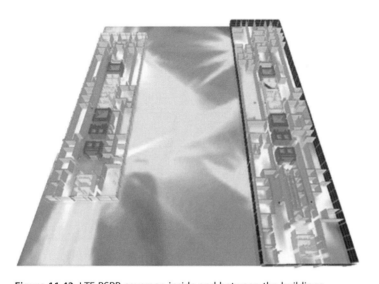
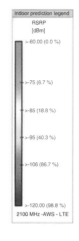

Figure 11.43 LTE RSRP coverage inside and between the buildings.

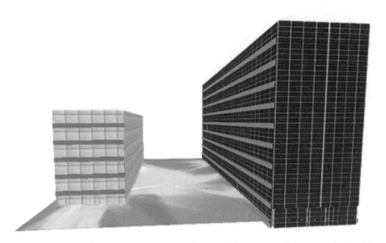

Figure 11.44 Full 3-D representation of LTE RSRP coverage between the buildings.

For UMTS, a pilot signal (CPICH) coverage and E_c/I_0 coverage are of interest. As the signal from the same sector is multicast on the same floor, E_c/I_0 is very uniform as there is no intersector interference. The signal from the other building does interfere because it belongs to a different sector. The result is shown in Figure 11.45.

SINR and maximum achievable data rate (MADR) coverage are also of interest. As is the case with UMTS E_c/I_0, antenna simulcast removes interference between antennas on the same floor, which causes high SINR. However, the signal from the other building does interfere and causes poor SINR between the buildings, as shown in Figure 11.46.

Good SINR coverage also means that MADR is very high inside buildings. MADR for a 10 MHz LTE SISO is shown in Figure 11.47.

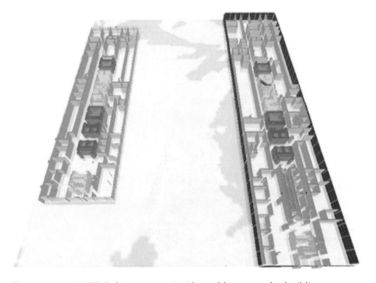
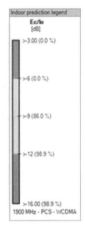

Figure 11.45 UMTS E_c/I_0 coverage inside and between the buildings.

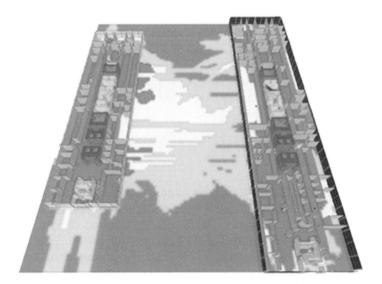

Figure 11.46 LTE SINR coverage inside and between the buildings.

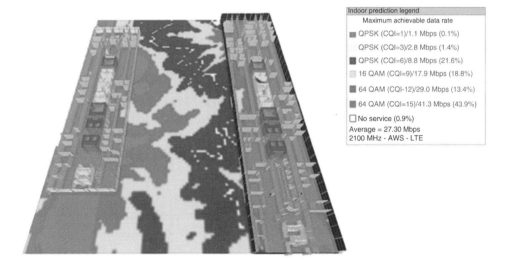

Figure 11.47 LTE MADR coverage inside and between the buildings.

11.4.7 Summary

In-building coverage is required in a business campus with a single tenant who occupies two buildings. The tenant has adopted a 'bring your own device to work' policy, so all major WSPs need to be included in the system. First responders (public safety) also need to be included. Based on the requirements that multiple technologies and multiple frequency bands need to be included, a hybrid neutral-host DAS is chosen. To model the coverage properly, three-dimensional modelling of the venue is essential. Due to the close proximity of the buildings, the overlapping in-building coverage from one building to the other must be taken into account. During the design process, care must be taken to contain the RF signal within the buildings as much as possible. This is most efficiently done by choosing directional antennas rather than omnidirectional ones. Coverage often needs to be provided in stairwells and lifts. Lifts are especially challenging in municipalities that prohibit electronic equipment inside elevator shafts due to the fire hazard. Where such equipment is permitted, the most effective solution is radiating cable or a combination of a DAS antenna in each shaft and a passive repeater in each elevator.

11.5 Underground (Subway)

Rapid rail transit is a common term for the type of local passenger train service found in many large cities throughout the world. As of 2013, 178 cities worldwide had rapid rail transit, with 7000 train stations and 8000 km of rail. The world's busiest systems are Tokyo with 3.2 billion passengers annually, followed by Seoul and Beijing with 2.5 billion each. The largest system is in New York, which has 468 rail transit stations, closely followed by Seoul with 429, with Paris a distant third with 300 (iBwave, 2013).

While rapid rail transit differs from city to city, there are some common characteristics. A rapid rail transit system always covers the core urban area of the city (town centre), the main commercial and/or tourist area, and must also reach major surrounding areas where commuters live (suburbs). As most traffic occurs when commuters travel to and from work, the busiest times (rush-hours) are in the morning hours and in late afternoon and early evening. In a typical system, most stations in the core urban area are located below ground and are interconnected with tunnels while, in suburban areas, some stations may be below ground and others at ground level. Distances between urban stations are generally shorter than distances between suburban stations.

A train station in a core urban area typically serves more than one rail line. Some stations operate two rail lines from a single platform. Other stations accommodate more rail lines by digging deeper underground, and have multiple platforms, one stacked above the other. These platforms are interconnected by stairs, escalators and lifts. A simplified view of a station with multiple platforms is shown in Figure 11.48.

There are many examples worldwide of rapid transit stations sharing facilities with regional passenger train services; among the best known are Union station in Washington DC, Amsterdam Centraal railway station and London Victoria station. These stations also have a large number of retail shops and cafés located in the general access area between the station entrance and the entrance to the train platforms. This must be taken into account when sizing the wireless networks, as the shopping and dining opportunities cause commuters to spend more time at the station than they would otherwise.

11.5.1 Scenario

While macro coverage near the station entrance and inside the entry hall is acceptable, the coverage beyond ticket booths at escalators, hallways, train platforms and inside tunnels is often poor to non-existent. The two systems are located on separate platforms and each platform ends with a wide tunnel with two sets of tracks that carry traffic in both directions. The new wireless network at the station has to support major wireless

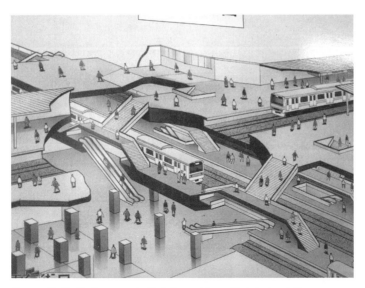

Figure 11.48 Underground rail station with multiple train platforms.

service providers (WSP) in the city, as well as the city Emergency Service network and Wi-Fi. The network also has to cover inside tunnels linking stations; however, Wi-Fi does not need to be provided in tunnels. The Emergency Service network has to be available inside trains at all times, even when two trains are stopped next to each other inside a tunnel. Target coverage areas at the venue are illustrated in Figure 11.49.

11.5.2   Design Requirements

11.5.2.1   RF Coverage
RF coverage at the station must provide the following wireless technologies:

- GMS
- UMTS
- LTE
- Trunked Radio (Public Safety)
- Wi-Fi (802.11n).

The following frequency bands must be included in the network:

- 700 MHz
- SMR band (800 MHz)
- Cellular band (850 MHz)
- PCS band (1900 MHz)
- AWS band (2100 MHz)
- 2.4 GHz.

The coverage at the station must include the entry hall with retail shops, train platforms, ticket booths, stairs, escalators and hallways. A 3-D view is shown in Figure 11.50. The coverage must also extend into tunnels for all technologies except Wi-Fi and for all bands except the 2.4 GHz band. An important requirement is that

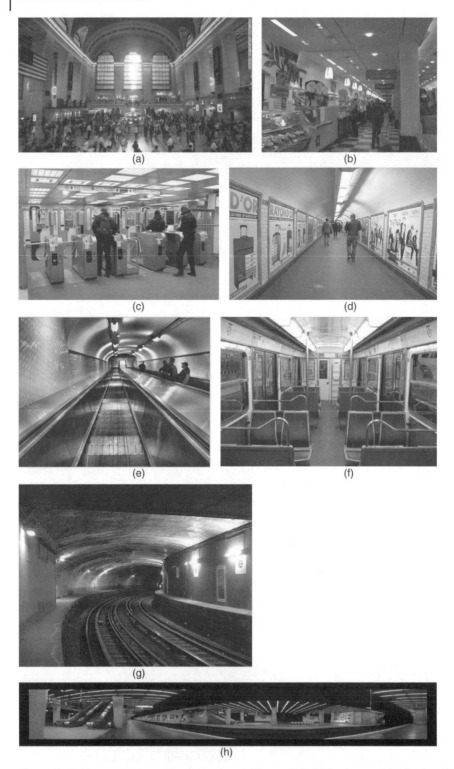

Figure 11.49 Target coverage areas at the venue: (a) entry hall; (b) food court; (c) ticket booths; (d) hallways; (e) escalators; (f) train car interiors; (g) tunnels; (h) train platforms.

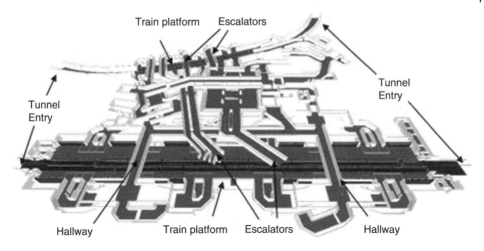

Figure 11.50 3-D view of the underground station.

Emergency Service coverage inside trains must be present even when trains are stopped in a tunnel next to each other.

Target RF signal strength values are as follows:

- GSM BCCH −85 dBm
- UMTS CPICH −85 dBm
- LTE RSRP −95 dBm
- Trunked Radio Rx −95 dBm
- Wi-Fi Rx −75 dBm.

For trunked radio, coverage is required at 100% of the station and tunnels. For cellular technologies, coverage is required at over 95% of the station and tunnels. For Wi-Fi, coverage is required at over 95% of the station only. These design values are typical for wireless networks.

11.5.2.2 Capacity

During the rush-hour there are 120 train stops at the station. On average, a train spends 30 seconds at the station loading and unloading passengers. A typical train has 10 cars, and each car can carry up to 150 passengers. At the entry hall are a dozen retail stores and several small restaurants. During the rush-hour there are 1000 commuters, shopkeepers and restaurant staff in the entry hall area. In this example, four WSPs require coverage at the venue; their technology and customer penetration breakdown is as follows:

- WSP A: 30% penetration rate
  - GSM 5%
  - UMTS 35%
  - LTE 60%.
- WSP B: 25% penetration rate
  - GSM 5%
  - UMTS 40%
  - LTE 55%.

- WSP C: 25% penetration rate
  - GSM 5%
  - UMTS 45%
  - LTE 50%.
- WSP D: 20% penetration rate
  - GSM 10%
  - UMTS 60%
  - LTE 30%.

A WSP network is sized to support rush-hour traffic with specified call blocking rates. For a voice-only network such as GSM, the target voice call blocking rate is between 1% and 5%, with 2% most commonly used. For voice and data networks such as UMTS and LTE, the call blocking rate varies with the data rate; an application that requires a high data rate, such as data download or video streaming, will have a higher call blocking rate. In such cases, a composite call blocking rate is calculated that takes into account call blocking rates for individual applications. A typical composite call blocking rate is 5% or more.

11.5.2.3   Handoff Management

The goal of handoff management is to establish a clear handoff area between two sectors. Two types of handoff exist: handoffs between macro networks and the in-building network and those between in-building network sectors. As most of the station is underground, there is very little residual macro network signal overlap, but is mostly near the station entrance where the retail shops and restaurants are located. Depending on the number of subscribers at the station, a WSP may require multiple sectors to support subscriber traffic. In such cases, the handoff area between the sectors should be confined to transit areas such as walkways, escalators or stairways. Another potential handoff area is inside tunnels where handoff must be completed while the train is moving at full speed.

11.5.3   Solution

As the in-building system has to include multiple WSPs and the Emergency Service (ES) network, an optimum solution is the use of a neutral host distributed antenna system (DAS). ES and WSPs may be deployed in a converged DAS or two separate DAS may be built, one for ES and one for all WSPs. The decision whether to deploy the converged or the discrete DAS architecture should be based on EIRP, spectrum bands and technologies that are being deployed. At the station, DAS amplifiers (remote units) are connected via coax cables to point-source antennas. In tunnels, coverage is provided by radiating cables, sometimes called leaky feeders, which are connected to remote units. Wi-Fi access points are deployed throughout the station but do not share antennas and cabling with the DAS. Because Wi-Fi does not support full mobility, there is no requirement for Wi-Fi coverage in tunnels. As a general principle, Wi-Fi coverage can be provided inside train cars only if Wi-Fi access points use a cellular network as wireless backhaul. In that case, the cellular network shares capacity with the Wi-Fi network and the maximum achievable data rate of the Wi-Fi inside cars is limited by the maximum achievable data rate for the network in tunnels.

11.5.4 RF Coverage Design

Coverage at the station is best achieved with multiple point-source antennas. At a long platform with a high or open ceiling, it is best to use directional antennas placed at opposite ends. In locations with low ceilings, such as walkways, entry halls and stairs like those shown in Figure 11.49, omnidirectional antennas are best. Antenna equivalent isotropically radiated power (EIRP) should be kept under 30 dBm so as to limit passive intermodulation (PIM).

As outlined earlier, the station has two platforms. The Regional Rail platform (Platform 1), equipped with two directional antennas, is shown in Figure 11.51. An illustration of Platform 1 with two trains stopped next to each other is shown in Figure 11.52. The Rapid Rail Transit platform (Platform 2) with one train at the platform is shown in Figure 11.53.

If line of sight (LOS) occurs between tunnel exits, directional point-source antennas mounted at opposite ends may provide sufficient coverage. If not, radiating cables are recommended. In a tunnel with two tracks, radiating cables carrying public safety signal should be mounted along opposite walls to provide coverage even when trains are stopped next to each other. Figure 11.54 illustrates this case, with radiating cables mounted on opposite walls.

Let us now examine radiating cable length limits. Eupen RMC 78-HLFR cable has 69 dB coupling loss at 1900 MHz and 6.4 dB loss per 100 metres. Assuming that a high power (10 W) remote unit (RU) is used to feed the cable, −85 dBm received power inside

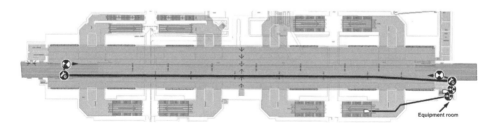

Figure 11.51 Platform 1 with directional antennas at opposite ends.

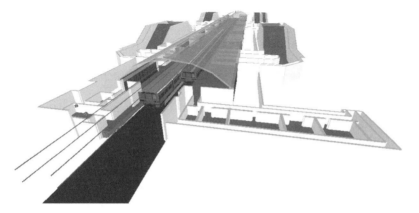

Figure 11.52 Platform 1 with two trains present.

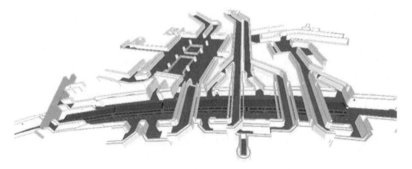

Figure 11.53 Platform 2 with one train present.

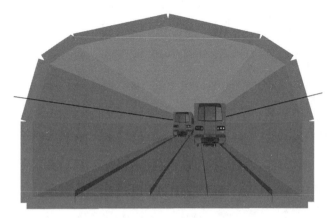

Figure 11.54 Radiating cables mounted along opposite walls in a tunnel.

the train and approximately 5 dB penetration loss for 1 cm thick windows (De Backer *et al.*, 1996), the maximum path loss is $40 + 85 - 5 = 120$ dB. Taking into account the 69 dB coupling loss, the maximum permissible loss due to cable length is $120 - 69 = 51$ dB. At 6.4 dB loss per 100 metres, this maximum cable loss of 51 dB is reached after only 800 metres.

In tunnels longer than 800 metres, cascading RUs may be used to extend coverage. However, the use of cascading RUs increases uplink noise per channel, which degrades uplink data rate and capacity. To maintain uplink performance, carrier-to-noise ratio (C/N) for the sector with cascaded RUs must be calculated and compared against the target uplink C/N.

Search window is a UMTS parameter that defines the maximum path distance that can be resolved at the receiver. According to the 3GPP RRC Protocol Specification (3GPP, 1999), UMTS search window has one of three values: 40 chips (± 20 chips), 256 chips (± 128 chips) and 2560 chips (± 1280 chips). Taking into account UMTS chip duration (0.81 μs), the maximum resolvable path difference if the search window is set at 40 chips is 1560 metres. If the search window is set at 256 chips then the maximum path difference is 10 kilometres and for 2560 chips it is 100 kilometres. If multipath is greater than the maximum path difference, it is treated as interference. To combat interference caused by excessive signal delay, the search window size must be adjusted. For example, a UMTS sector covering a 2-kilometre tunnel should have a search window time set at 256 chips, not 40.

LTE technology counters multipath by introducing a 'guard period', a downlink transmission gap between consecutive OFDM symbols. This transmission gap is called the *cycle prefix* (CP) and, according to Sesia, Toufik and Baker (2009), the guard period is 5.2 μs for normal CP and 16.7 μs for extended CP if the subchannel spacing is 15 kHz. A CP duration of 5.2 μs is equivalent to a maximum path difference of 1560 metres, while a 16.7 μs CP duration is equivalent to a 5 kilometre path difference. Therefore a sector covering a tunnel that is less than 1.56 km long may use normal CP, while extended CP should be used for tunnels between 1.56 km and 5 km long. On tunnels longer than 5 km, more than one LTE sector should be used.

If multiple LTE sectors are used in a tunnel, a sufficient coverage overlap must be maintained to allow for multiple handover tries. When calculating coverage overlap, one should take into account the train speed at the handover area, and that the signal strength from both sectors must be above the target level during consecutive handoff attempts (Tolstrup, 2011). As a handoff attempt takes 2–3 seconds, and at least three handoff attempts should be possible in the overlapping area, a train should spend at least 10 seconds in the handoff zone. For example, if the train speed is 36 km/h, the handoff zone should be 100 metres long and the last 100 metres of the radiating cable connected to one tunnel sector should overlap with the first 100 metres of the radiating cable connected to the next.

11.5.5 Capacity

Capacity sizing determines the number of sectors required to support peak traffic at the venue. As discussed earlier, on average there are two trains per minute and each train spends 30 seconds at the station. This implies that there are at least two trains at the station during the rush-hour. However, these are average numbers and to calculate maximum peak capacity one must take into account that occasionally as many as four trains may be at the station simultaneously. Each train has up to 10 cars and each car has up to 150 passengers, so four trains may carry up to 6000 rush-hour commuters. When we take into account the presence of 1000 more people in the entry hall shopping and food area, the maximum number of people at the station during the rush-hour is 7000.

During the rush-hour, according to the customer penetration breakdown, WSP A has 2100 subscribers at the station, WSP B and C have 1750 subscribers each and WSP D has 1400. If we assume two trains in the tunnel during the rush-hour, there are another 3000 commuters there; that is WSP A has 900 subscribers in the tunnel, WSP B and C have 750 each and WSP D has 600.

The next step is to break down the number of subscribers per technology, both inside tunnels and at the station, taking into account the technology percentage distribution shown in Table 11.21.

11.5.5.1 Data

The first step is to define subscriber profiles for each technology and each WSP. For each service type, the duration of the network connection during the busy-hour is expressed in millierlangs (mE) per subscriber. The probability that a subscriber will attempt to use

Table 11.21 Number of subscribers by technology and location.

	GSM		UMTS		LTE	
	Station	**Tunnel**	**Station**	**Tunnel**	**Station**	**Tunnel**
WSP A	105	45	735	315	1260	540
WSP B	88	38	700	300	962	413
WSP C	88	38	787	337	875	375
WSP D	70	30	840	360	420	180

each service type during the busy-hour is expressed as a percentage. It is important to note that a subscriber is not limited to one service attempt per busy-hour; rather, he or she may attempt to use all of the service types listed. A fixed data rate (in kbps) must be defined for each service type. An example of data traffic distribution at a subway station is shown in Table 11.22.

It is assumed that voice traffic is carried over the WCDMA (R99) protocol, while 3G and 4G data are carried over the HSPA and LTE protocols, respectively. LTE SINR coverage in the venue is calculated and broken down into intervals based on the modulation scheme that can be achieved in each interval. The example in Table 11.23 shows that, in the region where LTE PDSCH SINR $\geq$ 20 dB, 64-QAM modulation with coding rate $R = 0.93$ is possible, which gives a spectral efficiency of 5.5 bit/s/Hz. With SINR between 15 and 20 dB, spectral efficiency is 3.9 bit/s/Hz, with SINR between 9 and 15 dB, the efficiency is 2.4 bit/s/Hz, etc. These MCS efficiency values assume SISO configuration.

By knowing the relationship between the signal modulation scheme, spectral efficiency and SINR, the number of resources needed to support each service type listed in Table 11.22 can be calculated. These 'resources' are different for different technologies: LTE resources are physical resource blocks (PRBs); UMTS resources are HSPA orthogonal codes; etc. As spectral efficiency varies with SINR, so does the number of resources needed to support a certain service type in each SINR zone. For example, if SINR is high, a single PRB may be sufficient to support email but, if SINR is low, more than one PRB may be required.

Table 11.22 Data traffic distribution at the venue during the busy-hour by service type: call duration (millierlangs per user), data rate (Kbps) and call probability (%).

Service type	mE/user	kbps	Probability
Email	5	100	0.50%
Browsing	15	200	1.50%
Video conferencing	10	600	1.00%
Data download	15	1000	1.50%
Video streaming	10	2000	1.00%

Table 11.23 LTE example showing the relationship between the modulation scheme, MCS efficiency (bit/s/Hz) and SINR (dB).

Modulation	MCS efficiency	SINR
QPSK	1.18	3
16 QAM	2.40	9
64 QAM	3.90	15
64 QAM	5.55	20

Next, we produce the downlink LTE SINR coverage map. As sector overlap affects SINR, an assumption must be made about the number of sectors in the network. The station has two train platform levels and two additional levels. It is reasonable to assume that one sector is needed for each train platform. We will also assume that each sector also covers one additional level. The resultant LTE SINR coverage at the station is shown in Figure 11.55.

LTE SINR Range 1 ($3 \leq$ SINR ≤ 9) covers 10% of the area, SINR Range 2 ($9 \leq$ SINR ≤ 15) covers 20% of the area, SINR Range 3 ($15 \leq$ SINR ≤ 20) covers 20% of the area and SINR Range 4 (SINR ≥ 20) covers 50% of the area. Assuming a uniform distribution of spectators (a reasonable assumption), the percentage of LTE users within each SINR range is the same as the SINR coverage percentage for that range.

Note that LTE SINR Range 1 ($3 <$ SINR < 9) covers 10% of the area, SINR Range 2 ($9 <$ SINR < 15) covers 20% of the area, SINR Range 3 ($15 <$ SINR < 20) covers 20% of the area and SINR Range 4 (SINR < 20) covers 50% of the area. Assuming a uniform distribution of commuters (a reasonable assumption), the percentage of LTE users within each SINR range is the same as the SINR coverage percentage for that range.

Let us assume that LTE has a 10 MHz channel and that there are 500 subscribers per LTE sector for a total of 1000 LTE subscribers at the station. The LTE busy-hour traffic

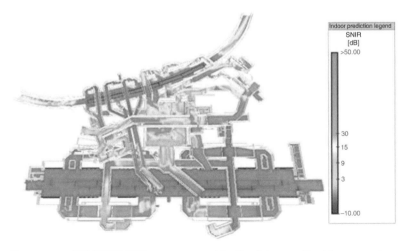

Figure 11.55 LTE PDSCH SINR coverage used for the data capacity sizing example.

Table 11.24 Offered LTE busy-hour traffic (erlangs) by service type and by SINR range.

Metrics	Range 1	Range 2	Range 3	Range 4
SINR	3	9	15	20
Percentage coverage	10.0%	20.0%	20.0%	50.0%
Users	100	200	200	500
Emails	0.50	1.00	1.00	2.50
Browsing	1.50	3.00	3.00	7.50
Video conferencing	1.00	2.00	2.00	5.00
Data download	1.50	3.00	3.00	7.50
Video streaming	1.00	2.00	2.00	5.00

(in erlangs) is calculated based on the number of subscribers per SINR range and on busy-hour traffic per subscriber, as given in Table 11.22. The results are shown in Table 11.24.

The number of resources needed to support the service types across the ranges is calculated next. This is readily determined if the relationship between SINR and spectral efficiency is known, which can be taken either from research papers (De Backer *et al.*, 1996) or directly from vendors. Table 11.25 shows, for this example, the distribution of LTE PRBs for each SINR range.

Based on Tables 11.24 and 11.25, and given the total number of PRBs in the two 10 MHz channel sectors, the blocking probability for each service type can be calculated. The blocking rate is defined as the percentage of attempted network connections that are denied due to insufficient network resources. The blocking rate formula for multiple services used for this calculation is taken from the ITU-R recommendation (3GPP, 1999). The encountered blocking rate for each service type and SINR range is shown in Table 11.26.

Table 11.26 is the key for sizing the network as it shows the blocking rate for all service types throughout the station (Ranges 1 to 4). For example, we see that 7.5% of the attempts to stream video using LTE from the area where LTE SINR > 20 dB (Range 4) are blocked due to insufficient LTE resources.

Table 11.25 Number of LTE PRBs by service type and by SINR range.

Service type	Range 1	Range 2	Range 3	Range 4
Emails	6	3	2	2
Web browsing	11	5	4	3
Video conferencing	31	15	10	7
Data download	51	25	16	11
Video streaming	101	50	31	22

Table 11.26 Encountered LTE blocking rate (%) during the busy-hour by service type and by SINR range.

Service type	Range 1	Range 2	Range 3	Range 4
Emails	2.0%	1.0%	0.7%	0.7%
Browsing	3.7%	1.7%	1.3%	1.0%
Video conferencing	10.7%	5.1%	3.4%	2.4%
Data download	17.9%	8.6%	5.4%	3.7%
Video streaming	36.4%	17.6%	10.7%	7.5%

If the calculated blocking rates shown in Table 11.26 are deemed unacceptable, then the number of sectors must be increased, the SINR map recalculated and the capacity calculations repeated. This is an iterative process that is continued until acceptable blocking rates are found.

Carried busy-hour traffic is calculated next based on offered traffic (Table 11.24) and encountered blocking rate (Table 11.26) for each service type. Results are shown in Table 11.27.

To determine total offered busy-hour traffic at the venue, the entries in Table 11.24 are summed. To determine carried busy-hour traffic, the entries in Table 11.27 are summed. Offered LTE traffic is 55 erlangs while carried LTE traffic is 52.07 erlangs. The composite call blocking rate (CCBR) is calculated as in Equation (()), which, for this example, yields 5.3%.

The *duty cycle* is defined as the ratio of carried traffic to theoretical maximum traffic when all resources are used for the full busy-hour and is calculated to be 5.8%. Data usage is carried traffic during the busy-hour and is calculated to be 17.03 gigabytes (GB).

If the capacity statistics for two LTE sectors carrying traffic for 1000 subscribers are deemed acceptable, then WSP B and WSP C may use this same sectorization plan as they have close to 1000 LTE subscribers. For WSP A (1200 subscribers), the SINR coverage map and capacity statistics need to be recalculated for three LTE sectors. As WSP D has only 500 subscribers, its sectorization plan calls for one sector only.

The same exercise can be repeated for HSPA. Assuming the same two-sector configuration, the HSPA SINR coverage map is as shown in Figure 11.56.

Table 11.27 Carried LTE busy-hour traffic (erlangs) by service type and by SINR range.

Service type	Range 1	Range 2	Range 3	Range 4	Total
Emails	0.49	0.99	0.99	2.48	0.21
Browsing	1.44	2.95	2.96	7.42	1.24
Video conferencing	0.89	1.90	1.93	4.88	2.42
Data download	1.23	2.74	2.84	7.22	5.88
Video streaming	0.64	1.65	1.79	4.62	7.29

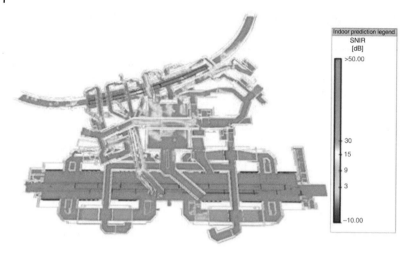

Figure 11.56 HSPA SINR coverage used for the data capacity sizing example.

HSPA SINR Range 1 ($5 \leq SINR \leq 11$) covers 6% of the area, SINR Range 2 ($11 \leq SINR \leq 22$) covers 19% of the area, SINR Range 3 ($22 \leq SINR \leq 24$) covers 17% of the area and SINR Range 4 ($SINR > 24$) covers 58% of the area. Let us assume one 5 MHz RF channel and 400 subscribers per sector for a total of 800 HSPA subscribers. We then calculate busy-hour traffic in erlangs based on the number of subscribers per SINR range and on busy-hour traffic per subscriber as given in Table 11.22. The results are shown in Table 11.28.

The number of HSPA codes needed to support the service types across the zones are calculated in a similar manner to that used for LTE resource blocks (RBs). The distribution of HSPA codes per SINR range is shown in Table 11.29.

Based on Table 11.28 and Table 11.29, and given the total number of HSPA codes in the two sectors, the blocking probabilities for each service type are calculated next. Encountered blocking rate values for each service type and SINR range are as shown in Table 11.30.

Table 11.28 Offered HSPA busy-hour traffic (erlangs) by service type and by SINR range.

Metrics	Range 1	Range 2	Range 3	Range 4
SINR	5	11	22	24
Percentage coverage	6.0%	19.0%	17.0%	58.0%
Users	48	152	136	464
Emails	0.24	0.76	0.68	2.32
Web browsing	0.72	2.28	2.04	6.96
Video conferencing	0.48	1.52	1.36	4.64
Data download	0.72	2.28	2.04	6.96
Video streaming	0.48	1.52	1.36	4.64

Table 11.29 Number of HSPA codes by service type and by SINR range.

Service type	Range 1	Range 2	Range 3	Range 4
Emails	3	1	1	1
Web browsing	6	3	1	1
Video conferencing	19	8	1	1
Data download	31	13	2	1
Video streaming	63	25	4	2

Carried busy-hour traffic is calculated based on offered traffic (Table 11.28) and encountered blocking rate (Table 11.30) for each service type. Results are shown in Table 11.31.

Offered HSPA traffic (Table 11.28) is 44 erlangs and carried traffic (Table 11.31) is 40 erlangs. The composite blocking rate is 9%, duty cycle is 26.7% and data usage is 12.9 gigabytes (GB). Although these results show that the HSPA has a higher blocking rate for fewer subscribers than the LTE, it should be noted that an LTE RF channel is 10 MHz wide, twice the spectrum of an HSPA channel (5 MHz).

Table 11.30 Encountered HSPA blocking rate by service type and by SINR range.

Service type	Range 1	Range 2	Range 3	Range 4
Emails	6.7%	2.3%	2.3%	2.3%
Browsing	13.3%	6.7%	2.3%	2.3%
Video conferencing	39.3%	17.5%	2.3%	2.3%
Data download	59.4%	27.8%	4.5%	2.3%
Video streaming	91.6%	49.9%	8.9%	4.5%

Table 11.31 Carried HSPA busy-hour traffic (erlangs) by service type and by SINR range.

Service type	Range 1	Range 2	Range 3	Range 4	Total
Emails	0.22	0.74	0.66	2.27	0.16
Browsing	0.62	2.13	1.99	6.80	0.97
Video conferencing	0.29	1.25	1.33	4.54	1.86
Data download	0.29	1.65	1.95	6.80	4.48
Video streaming	0.04	0.76	1.24	4.43	5.43

11.5.5.2 Voice

Voice capacity sizing is based on the assumption that voice traffic will be carried over the WCDMA (R99) part of the UMTS signal. As an assumption was made that VoIP is not implemented in the LTE network, voice capacity has to account for both 3G and 4G subscribers. The number of voice subscribers is shown in Table 11.32.

When defining the subscriber profile for R99, we take into account the fact that data traffic will switch to R99 data only if both HSPA and LTE are unavailable. Therefore, the probability of an R99 data call is very low, as reflected in the R99 traffic distribution shown in Table 11.33.

As we did for the data capacity calculations, we first determine E_b/N_0 coverage (E_b/N_0 is the ratio of energy per bit to noise power spectral density), separate the coverage into four different E_b/N_0 ranges and identify the service types that can be used in each range. The resultant E_b/N_0 coverage map is shown in Figure 11.57, assuming a two-sector configuration.

The coverage area in which a subscriber may connect to video telephony (dark grey in Figure 11.57) is also the area in which a connection to switched data and voice is possible. Assuming a uniform subscriber distribution, the percentage of subscribers connecting to the service in a particular E_b/N_0 range is the same as the percentage of coverage for that range. If we assume 1000 R99 subscribers per sector for a total of 2000 subscribers in the venue, the resulting user distribution and R99 traffic per E_b/N_0 range (in erlangs) are as shown in Table 11.34.

Table 11.32 Number of 3G and 4G voice subscribers.

	Voice	
	Station	**Tunnel**
WSP A	1995	855
WSP B	1662	713
WSP C	1662	712
WSP D	1260	540

Table 11.33 R99 traffic distribution at the venue during the busy-hour by service type: call duration (millierlangs per user), data rate (kbps) and call probability (%).

Service type	mE/user	kbps	Probability
Voice	33	12.2	3.30%
Emails	3	64	0.30%
Browsing	3	128	0.30%
Data download	3	384	0.30%

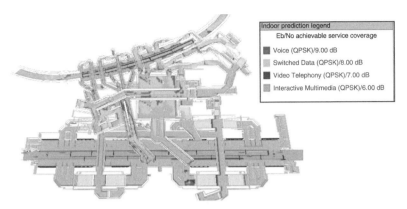

Figure 11.57 E_b/N_0 coverage map used for the voice sizing example.

Only OVSF codes with a spreading factor up to SF128 are used for the service types shown in Table 11.34. The required numbers of OVSF codes for each service type and E_b/N_0 range are as shown in Table 11.35.

As was the case with HSPA and LTE technologies, call blocking rates are calculated as in 3GPP (1999). Results are as shown in Table 11.36.

The R99 blocking rate is defined as the percentage of attempted network connections that are denied, during the busy-hour, due to insufficient network resources. For two

Table 11.34 R99 busy-hour traffic (erlangs) by service type and E_b/N_0 range.

Metrics	Range 1	Range 2	Range 3	Range 4
E_b/N_0	9	8	7	6
Distribution	0.1%	0.5%	4.5%	95.0%
Users	2	10	90	1900
Voice	0.1	0.3	3.0	62.7
Emails	-	0.0	0.3	5.7
Browsing	-	-	0.3	5.7
Data download	-	-	-	5.7

Table 11.35 Number of OVSF codes by service type and E_b/N_0 range.

Service type	Range 1	Range 2	Range 3	Range 4
Voice	1	1	1	1
Emails	-	4	4	4
Browsing	-	-	8	8
Data download	-	-	-	16

Table 11.36 R99 encountered blocking rates (%) by E_b/N_0 range.

Service type	Range 1	Range 2	Range 3	Range 4
Voice	1.1%	1.1%	1.1%	1.1%
Emails	-	4.4%	4.4%	4.4%
Browsing	-	-	8.9%	8.9%
Data download	-	-	-	18.6%

sectors with 2000 subscribers, the R99 voice blocking rate is just over 1% throughout the venue. Most macro UMTS networks use a busy-hour call blocking rate target between 1% and 2%. Other R99 service types have higher blocking rates, but this is not of much concern because they are supported with better rates in HSPA and LTE networks. The conclusion is that the call blocking rate for R99 traffic is acceptable and therefore the two-sector configuration is sufficient to support voice traffic for both 3G and 4G subscribers.

Although Wi-Fi coverage need not be provided in tunnels because Wi-Fi does not support high speed and handoff, Wi-Fi can be provided inside the train by placing an access point inside each car. These access points may use the existing cellular signal in the tunnel and at the station as wireless backhaul. This is known as *train Wi-Fi* and is separate from the Wi-Fi provided at the station. The *train Wi-Fi* shares capacity with the WSP that provides wireless backhaul and its maximum achievable data rate is limited by the backhaul data rate.

11.5.6 Environmental Challenges

Train stations and tunnels are some of the most challenging environments in which to deploy an in-building wireless system. At these types of venues, more than at any other, the environment dictates RF design and network maintenance. Some of the proven best practices are as follows:

- Equipment locations should be chosen for ease of access, not for ease of installation. If a choice is to be made between having one DAS remote unit in the middle of a tunnel to feed a bidirectional antenna, or having one RU and one point-source antenna at each tunnel entrance, the latter should be chosen as two RUs at the entrances are less expensive to maintain than one RU inside a tunnel.
- When choosing an equipment room location, ease of access should be the most important criterion; a small room that has 24/7 access is preferable to a bigger room accessible only when the station is closed to the public.
- When choosing RU locations, consider whether vandals can easily spot and damage or steal the equipment.
- Each RU should have battery backup power, as power outages at stations and in tunnels are common. When sizing battery backup power, it should be taken into account that these power outages are frequent but short.
- If radiating cable is used inside a tunnel, it must be securely attached to the walls; otherwise, wall vibrations caused by passing trains may loosen the cable and degrade its performance.

- Tunnels are occasionally exposed to water leaks, so all connections should be waterproofed as rust can cause passive intermodulation (PIM). Passive components with high ingress protection (IP) liquid rating are preferred, as they are more water-resistant.
- Another source of PIM is dust, of which there is plenty inside tunnels.
- Some stations may experience AC instabilities, which may reduce the lifetime of the equipment. In such cases, power converters should be used to convert from AC to DC and back to AC, thereby filtering out the instabilities.
- Point-source antennas at the station should be placed away from metallic objects, as these are a known source of PIM. Train stations have such objects hanging from ceilings and attached to walls. The most significant metallic objects, however, are the trains themselves as they have metal walls. Therefore antennas that cover platforms should not be mounted immediately above the train tracks.

11.5.7 Radio Coverage Maps

GSM coverage for WSP A in a tunnel, both at train platforms and at the station, is shown in Figure 11.58 to Figure 11.61. Note that signal strength inside trains is at −85 dBm or greater, even when trains are next to each other – the worst-case scenario.

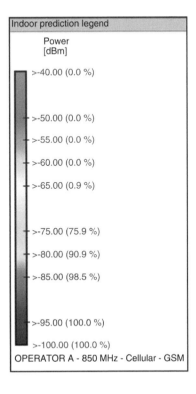

Figure 11.58 GSM coverage in a tunnel and inside trains.

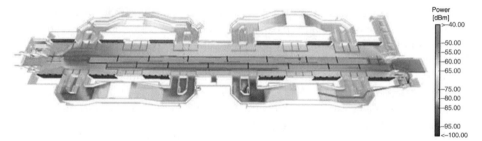

Figure 11.59 GSM coverage at platform 1 with two trains at the platform.

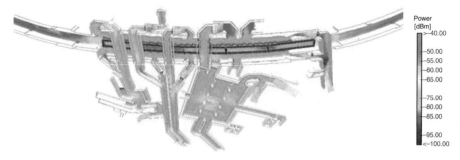

Figure 11.60 GSM coverage at platform 2 with a train at the platform.

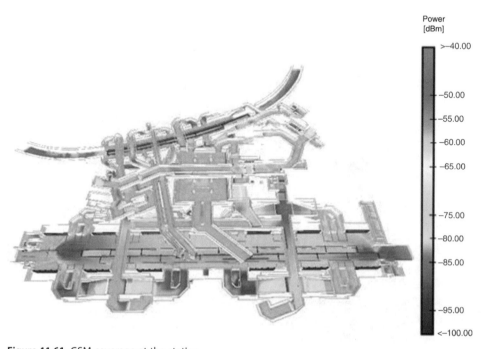

Figure 11.61 GSM coverage at the station.

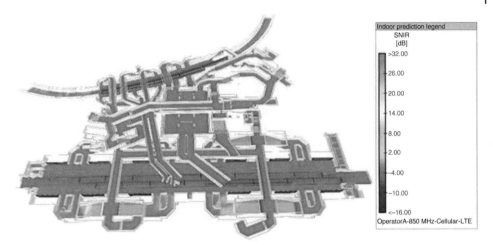

Figure 11.62 LTE SINR coverage.

For data-centric networks, signal-to-interference-plus-noise ratio (SINR) coverage is key because good SINR coverage facilitates high data rates. Figure 11.62 shows LTE SINR at the station.

LTE coverage at the station is provided with 14 antennas. At opposite sides of train platforms, directional antennas are used (Amphenol Jaybeam model 7478000 with 11.5 dBi gain); omnidirectional antennas are used elsewhere (Kathrein model 741 572 with 2.15 dBi gain). Macro network interference is limited to the station entry hall (not shown). SINR is very high at the platforms and lower where in-building sectors overlap. Consequently, the maximum achievable data rate (MADR) is also very high at the platforms and lower elsewhere, as shown in Figure 11.63.

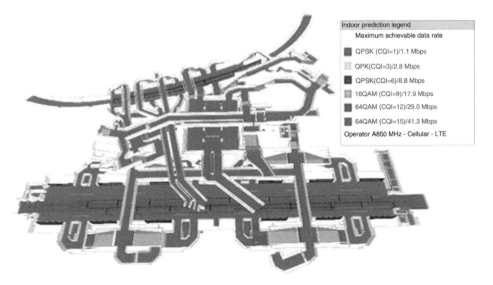

Figure 11.63 LTE MADR coverage.

11.5.8 Summary

A rail station providing service to both regional trains and local rapid transit trains has poor existing macro coverage and is in need of a dedicated in-building wireless system. As an example, at this station four WSPs must be included in the in-building network, as well as Wi-Fi and a First Responders (public safety) network. Wi-Fi coverage at the station does not extend into the tunnels, while all other wireless services require coverage in connecting tunnels as well. All four WSPs support 2G, 3G and 4G technologies across different frequency bands, which dictates the use of a neutral-host DAS as the solution. Since the station is a multilevel structure, 3-D modelling of the venue is essential. Inside tunnels, the network has to provide coverage inside trains even when trains are stopped next to each other. For most tunnels, this requirement dictates the use of radiating cables, also known as leaky feeders.

There are several environmental factors specific to such venues that make design and maintenance a challenge. Rail stations have many metallic objects and, because proximity to metal objects causes PIM, it is often challenging to find good locations for antennas. In tunnels, damp conditions may cause connectors to rust and vibrations caused by passing trains may loosen radiating cables; both conditions generate PIM and therefore degrade network performance. Long tunnels may require cascaded amplifiers to maintain coverage; however, cascading amplifiers may increase uplink noise, which affects uplink capacity and data rate. Another way to maintain coverage in tunnels is to install multiple sectors; however, sufficient sector coverage overlap must be provided to allow for multiple handoff attempts at nominal train speeds.

As a rule, it is preferable to locate equipment so that it is easy to access rather than less expensive to deploy, as the equipment requires frequent maintenance. One must also be mindful of vandalism and make the equipment as inconspicuous as possible. The numerous requirements make this venue type one of the most expensive to design and operate.

References

3GPP (1999) 3GPP TS 25.331 V3.5.0 (2000-12), RRC Protocol Specification (Release 1999), p. 397, Section 10.3.7.60.

Anritsu (2014) PIM technical paper, Understanding PIM. URL: http://www.anritsu.com/en-US/Products-Solutions/Solution/Understanding-PIM.aspx.

De Backer, B., Borjeson, H., Olyslager, F. and De Zutter, D. (1996) The study of wave-propagation through a windowed wall at 1.8 GHz, in *IEEE 46th Vehicular Technology Conference on Mobile Technology for the Human Race*, Vol. 1.

iBwave (2013) iBwave webinar, designing seamless networks in subway systems.

ICNIRP (1998) International Commission on Non-Ionizing Radiation Protection, Guidelines for limiting exposure to time-varying electric, magnetic, and electromagnetic fields (up to 300 GHz), *Health Physics*, **75** (4), 494–522.

ITU (2013) International Telecommunication Union, ITU-R Recommendation M.1768-1: Methodology for calculation of spectrum requirements for the terrestrial component of International Mobile Telecommunications, Geneva.

Martijn, E.F.T and Herben, M.H. (2003) Characterization of radio wave propagation into buildings at 1800 MHz, *IEEE Antennas and Wireless Propagation Letters*, **2**, 122–125.

Rogers Canada Webinar (2013) In-building system design and installation strategies for LTE. URL: http://www.ibwave.com/Resources/PastWebinars/WebinarJuly172013.aspx.

Sesia, S., Toufik, I., and Baker, M. (2009) *LTE – The UMTS Long Term Evolution: From Theory to Practice*, 2nd edition, John Wiley & Sons, Ltd, Chichester. ISBN 978–0470660256.

Spindler, J. (2015) *Public Safety and Cellular DAS: Converged or Discrete?* Antenna System and Technology, April 7, 2015. URL: http://www.antennasonline.com/main/blogs/public-safety-and-cellular-das-converged-or-discrete.

Stamopoulos, I., Aragón-Zavala, A. and Saunders, S.R. (2003) Performance comparison of distributed antenna and radiating cable systems for cellular indoor environments in the DCS band, in *International Conference on Antennas and Propagation, ICAP 2003*, April 2003.

TLM (2014) Travel and Leisure online magazine, America's most visited shopping malls. URL: http://www.travelandleisure.com/articles/americas-most-visited-shopping-malls.

Tolstrup, M. (2011) *Indoor Radio Planning: A Practical Guide for GSM, DCS, UMTS, HSPA and LTE*, 2nd edition, John Wiley & Sons, Ltd, Chichester. ISBN 0-470-71070-8, 2011.

Index

Indoor Wireless Communications: From Theory to Implementation, First Edition. Alejandro Aragón-Zavala.
© 2017 John Wiley & Sons Ltd. Published 2017 by John Wiley & Sons Ltd.